Organizational Risk Management

Organizational Risk Management

An Integrated Framework for Environmental, Health, Safety, and Sustainability Professionals, and their C-Suites

Charles F. Redinger
Institute for Advanced Risk Management
Harvard, MA, USA

Copyright © 2025 by John Wiley & Sons, Inc. All rights reserved, including rights for text and data mining and training of artificial technologies or similar technologies.

Published by John Wiley & Sons, Inc., Hoboken, New Jersey.
Published simultaneously in Canada.

No part of this publication may be reproduced, stored in a retrieval system, or transmitted in any form or by any means, electronic, mechanical, photocopying, recording, scanning, or otherwise, except as permitted under Section 107 or 108 of the 1976 United States Copyright Act, without either the prior written permission of the Publisher, or authorization through payment of the appropriate per-copy fee to the Copyright Clearance Center, Inc., 222 Rosewood Drive, Danvers, MA 01923, (978) 750-8400, fax (978) 750-4470, or on the web at www.copyright.com. Requests to the Publisher for permission should be addressed to the Permissions Department, John Wiley & Sons, Inc., 111 River Street, Hoboken, NJ 07030, (201) 748-6011, fax (201) 748-6008, or online at http://www.wiley.com/go/permission.

Trademarks: Wiley and the Wiley logo are trademarks or registered trademarks of John Wiley & Sons, Inc. and/or its affiliates in the United States and other countries and may not be used without written permission. All other trademarks are the property of their respective owners. John Wiley & Sons, Inc. is not associated with any product or vendor mentioned in this book.

Limit of Liability/Disclaimer of Warranty: While the publisher and author have used their best efforts in preparing this book, they make no representations or warranties with respect to the accuracy or completeness of the contents of this book and specifically disclaim any implied warranties of merchantability or fitness for a particular purpose. No warranty may be created or extended by sales representatives or written sales materials. The advice and strategies contained herein may not be suitable for your situation. You should consult with a professional where appropriate. Further, readers should be aware that websites listed in this work may have changed or disappeared between when this work was written and when it is read. Neither the publisher nor authors shall be liable for any loss of profit or any other commercial damages, including but not limited to special, incidental, consequential, or other damages.

For general information on our other products and services or for technical support, please contact our Customer Care Department within the United States at (800) 762-2974, outside the United States at (317) 572-3993 or fax (317) 572-4002.

Wiley also publishes its books in a variety of electronic formats. Some content that appears in print may not be available in electronic formats. For more information about Wiley products, visit our web site at www.wiley.com.

Library of Congress Cataloging-in-Publication Data Applied for:

Hardback ISBN: 9781119538745

Cover Design: Wiley
Cover Image: © Ellagrin/Shutterstock

Set in 9.5/12.5pt STIXTwoText by Straive, Pondicherry, India

SKY10091432_112024

Dedicated to my parents,
Rev. Joseph Dallas Redinger, MDiv, and Elaine Van Aken Redinger, MEd
Champions of justice, equity, and fairness.
Warriors for health and well-being.
Anchors for humanity.
On whose shoulders this body-of-work builds.

Contents

Foreword *xvii*
Preface *xix*
Acknowledgments *xxv*
Acronyms *xxix*

1 Introduction: Leverage to Create *1*
Imperatives *3*
Becoming Aware *5*
Postulates and Design Principles *5*
Who You Are *7*
What You Will Learn *7*
1.1 A New Era *8*
    1.1.1 Value and Purpose *8*
    1.1.2 Social–Human Dimension of Risk *9*
    1.1.3 Environmental, Social, and Governance *10*
    1.1.4 Diversity, Equity, and Inclusion *11*
    1.1.5 Health – Organizational and Human *11*
1.2 Leverage *12*
    1.2.1 Levine's Lever *12*
    1.2.2 A → B, Current Reality and Where You Want to Go *14*
    1.2.3 Thinking in Systems *15*
    1.2.4 Shifts *15*
        1.2.4.1 Expanding Perspective and Awareness *16*
        1.2.4.2 Language as Currency *17*
        1.2.4.3 Integrated Capitals Perspective *18*
    1.2.5 Frameworks and Structures *19*
    1.2.6 Metrics and Indicators *20*
1.3 Integration *20*
    1.3.1 Integrating What? *21*
    1.3.2 Integrated Thinking *22*

viii Contents

      1.3.3 Integrated Risk Management *22*
1.4 Culture of Health *24*
1.5 Leadership and Seat at the Table *25*
      1.5.1 Motive Force and Organizational Energy *26*
      1.5.2 The Table *26*
1.6 Finding Leverage *27*
      1.6.1 Your A → B *27*
      1.6.2 Developing a Playbook *28*
      1.6.3 Creating a Project *28*
1.7 Book Logic and Chapter Summaries *29*
      1.7.1 Chapter Flow and Book Logic *29*
      1.7.2 Chapter Summaries *30*
            1.7.2.1 Chapter 2 – Risk Logics *30*
            1.7.2.2 Chapter 3 – Frameworks *30*
            1.7.2.3 Chapter 4 – Conformity Assessment and Measurement *31*
            1.7.2.4 Chapter 5 – Awareness in Risk Management *31*
            1.7.2.5 Chapter 6 – Field Leadership *32*
            1.7.2.6 Chapter 7 – Decision-Making *32*
            1.7.2.7 Chapter 8 – Risk Matrix – An Integrated Framework *33*
            1.7.2.8 Chapter 9 – Matrix in Action *33*
            1.7.2.9 Chapter 10 – Escalate Impact *33*
Suggested Reading *34*

Part I Foundations **35**

2 Risk Logics **37**
Defining Risk Logics *39*
Core Logic *39*
Premises *41*
Uncertainty *41*
2.1 Contexts, Drivers, Orientations *42*
      2.1.1 Evolutions *43*
            2.1.1.1 EHSS Management: Compliance, Performance, Impact *44*
            2.1.1.2 Frameworks: Process, Program, System, Field *45*
            2.1.1.3 Sustainability: ESG, Materiality, Double Materiality, Value, Capitals *46*
            2.1.1.4 Object/Foci: Shareholder, People/Workers, Stakeholders *46*
            2.1.1.5 Organizational Risk Management: Four Generations – Insurance, Regulatory Compliance, Consensus Standards, and Value and Purpose *47*

      2.1.2   Corporate Governance – Purpose and Value Creation/ Protection   *47*
      2.1.3   ESG   *51*
      2.1.4   Social and Human Capital   *52*
      2.1.5   Culture of Health   *53*
  2.2   Defining Risk   *54*
      2.2.1   Risk to Whom?   *54*
      2.2.2   Definitions   *55*
      2.2.3   Risk-Reward and Opportunity   *56*
      2.2.4   Fourth-Generation Risk Management   *57*
  2.3   Core Concepts   *58*
      2.3.1   Analysis, Assessment, Communication, and Management   *58*
      2.3.2   Risk Profile   *59*
      2.3.3   Owner – Risk Owner, Generator, and Source   *59*
      2.3.4   Acceptability – Acceptable Risk and Risk Appetite   *60*
      2.3.5   Tolerance – Risk Tolerance   *60*
      2.3.6   Transfer – Risk Transfer   *62*
      2.3.7   Risk-Based Thinking   *62*
  2.4   Conformity Assessment   *63*
  2.5   Systems Perspective   *64*
      2.5.1   Systems Thinking   *65*
      2.5.2   System Dynamics Iceberg   *66*
      2.5.3   Deeper Levels   *68*
  2.6   Risk Field   *68*
      2.6.1   Field Background   *70*
      2.6.2   Characterizing an Organizational Field   *71*
      2.6.3   Operationalization – Risk Field → Risk Matrix   *74*
  Suggested Reading   *74*

3   Frameworks   *77*
  Power of Structure   *79*
  Expanding Perspective → Expanding Awareness   *79*
  Learning Context   *80*
  "Next Generation" Frameworks – Evolution and Integration   *80*
  3.1   Types of Frameworks   *80*
      3.1.1   Regulatory   *81*
      3.1.2   Consensus Standards   *81*
      3.1.3   Evolved Organizational and Professional Practices   *82*
      3.1.4   Tailoring   *82*
  3.2   National Academy of Sciences and EPA: Risk Decision-Making Anchors   *83*
  3.3   International Organization for Standardization (ISO)   *86*
      3.3.1   Background   *86*
      3.3.2   Risk Management Evolution at ISO   *86*

Contents

- 3.3.3 ISO 31000 Overview *87*
- 3.3.4 ISO 37000 – Risk Governance, Principle 6.9 *89*
- 3.4 ISO Management System Standards *90*
  - 3.4.1 High-Level Structure *91*
  - 3.4.2 ISO MSS Demonstrative – Occupational Health and Safety (ISO 45001:2018) *92*
    - 3.4.2.1 Scope (1) *94*
    - 3.4.2.2 Terms and Definitions (3) *96*
    - 3.4.2.3 Context of the Organization (4) *96*
    - 3.4.2.4 Leadership and Worker Participation (5) *97*
    - 3.4.2.5 Planning (6) *100*
    - 3.4.2.6 Support (7) *103*
    - 3.4.2.7 Operation (8) *105*
    - 3.4.2.8 Performance Evaluation (9) *109*
    - 3.4.2.9 Improvement (10) *111*
- 3.5 COSO Enterprise Risk Management Framework *113*
  - 3.5.1 Evolution *113*
  - 3.5.2 Overview – 2017 Version *114*
- 3.6 Environmental, Social, and Governance *117*
  - 3.6.1 Overview and Terminology *117*
    - 3.6.1.1 Sustainability *117*
    - 3.6.1.2 Environmental, Social, and Governance *117*
    - 3.6.1.3 Corporate Social Responsibility *118*
    - 3.6.1.4 Materiality *118*
    - 3.6.1.5 Materiality Beyond Financial Reporting – Double/Impact Materiality *120*
  - 3.6.2 Human Capital *121*
  - 3.6.3 Reporting and Performance Criteria *123*
  - 3.6.4 Global Reporting Initiative (GRI) *124*
  - 3.6.5 International Sustainability Standards Board *125*
  - 3.6.6 Value Reporting Foundation, IIRC, SASB *126*
- 3.7 Transcending Paradigms *128*
  - 3.7.1 NIOSH Total Worker Health *129*
  - 3.7.2 Culture Health for Business (COH4B) *130*
  - 3.7.3 Capitals Coalition *131*
- 3.A ISO 3100:2018 Principles *133*
- 3.B COSO ERM (2017) Principles *134*
- Suggested Reading *136*

4 Conformity Assessment and Measurement *137*

- 4.1 Frameworks and Guidelines *139*
  - 4.1.1 National Research Council *139*

   4.1.2 ISO Committee on Conformity Assessment (CASCO) *140*
   4.1.3 Inference Guidelines and Decision-Making Currency *140*
  4.2 Measurement *141*
  4.3 Auditing *143*
   4.3.1 Historical Background *143*
   4.3.2 Types of Audits *144*
    4.3.2.1 First Party – Internal Audits *144*
    4.3.2.2 Second- and Third-Party External Audits *145*
    4.3.2.3 Hybrid Approaches *145*
  Suggested Reading *146*

Part II Leverage *147*

5 Awareness in Risk Management *149*
  5.1 Origins and Development *152*
   5.1.1 Genesis and Fourth-Generation Risk Management *152*
   5.1.2 Early Years, 1999–2018 *152*
    5.1.2.1 Integrated Model *153*
    5.1.2.2 Second-Order Change *153*
    5.1.2.3 360 Perspective *155*
    5.1.2.4 Stakeholder Domains *155*
   5.1.3 Current Iteration Risk Field → Risk Matrix (ABRM v.2) *155*
  5.2 Defining Awareness *156*
   5.2.1 Standards and Frameworks *157*
   5.2.2 Paying Attention *159*
  5.3 Orientations, Perspectives, and Mental Models *160*
   5.3.1 Decision-Making Prequel – Bias and Heuristics *161*
   5.3.2 Being-Doing *162*
  5.4 Leverage and Seven Risk Awareness Elements *162*
   5.4.1 Awareness *163*
   5.4.2 Internal State *163*
   5.4.3 Risk *164*
   5.4.4 Purpose *164*
   5.4.5 Value Creation and Preservation *164*
   5.4.6 Decision-Making Processes *165*
   5.4.7 Generative Field *165*
  5.5 Language as Currency *165*
   5.5.1 Future-Based Language *167*
   5.5.2 Carriers of Meaning *168*
  5.6 Shifting Mindset and Paradigms *169*
   5.6.1 Revisiting A → B *170*

　　　　5.6.2　A Learning Context　*171*
　　　　　　5.6.2.1　Organizational Learning　*171*
　　　　　　5.6.2.2　Double Loop Learning　*172*
　　　　　　5.6.2.3　Transformational Learning　*173*
　　　　5.6.3　Capitals Coalition's Four Shifts Model　*175*
　　　　5.6.4　Anatomy and Physiology of Shifts　*176*
　　Suggested Reading　*176*

6　Field Leadership – Motive Force　*179*
　　6.1　Motive Force　*181*
　　　　6.1.1　Organizational Energy　*181*
　　　　　　6.1.1.1　People　*183*
　　　　　　6.1.1.2　Structures　*183*
　　　　　　6.1.1.3　Contexts/Drivers　*183*
　　　　6.1.2　Culture of Health　*183*
　　6.2　Field "Actors" – Individuals, Teams/Departments, Enterprise, Community　*184*
　　　　6.2.1　Interiority, Accountability　*185*
　　　　6.2.2　The Hats You Wear – Designer, Builder, Operator, Participant　*185*
　　6.3　Creating Value　*186*
　　　　6.3.1　Why This Is Important　*186*
　　　　6.3.2　ISO 31000:2018 and COSO's ERM Framework　*186*
　　　　6.3.3　ISO 37000:2021, Section 6.2 – Value Generation　*187*
　　　　6.3.4　Capitals　*188*
　　　　　　6.3.4.1　Defining Capitals　*188*
　　　　　　6.3.4.2　Capitals Coalition and Value　*190*
　　6.4　Leadership and Participation in Frameworks　*191*
　　　　6.4.1　ISO 37000:2021　*191*
　　　　6.4.2　COSO's Enterprise Risk Management　*193*
　　　　6.4.3　ISO 31000:2018　*194*
　　　　6.4.4　ISO MSS Examples – ISO 14001:2015 and ISO 45001:2018　*195*
　　6.5　Emerging Leadership Paradigms　*196*
　　　　6.5.1　System Leadership – Senge, Hamilton, Kania　*196*
　　　　6.5.2　Responsible Leadership – Accenture, World Economic Forum　*197*
　　Suggested Reading　*198*

7 Decision-Making – Expanding Perspective *201*

Awareness – Process, Paradox, and Tension *203*
Types of Decisions *204*
Organizational Learning *204*
Expanded Platform *204*

- 7.1 Background and Anchors *205*
  - 7.1.1 Decision Science *205*
  - 7.1.2 The Human *206*
    - 7.1.2.1 Two-system Brain *207*
    - 7.1.2.2 Perception *208*
    - 7.1.2.3 Brain Function *209*
- 7.2 Systems Perspective *211*
  - 7.2.1 Systems 101 *212*
  - 7.2.2 Inputs and Processes *214*
    - 7.2.2.1 Data and Measurement Consideration *214*
  - 7.2.3 Output, Outcome, and Impact *215*
  - 7.2.4 Feedback *216*
  - 7.2.5 Stocks and Flows *216*
  - 7.2.6 Impact-Dependency Pathways *218*
- 7.3 Frameworks *219*
  - 7.3.1 ISO 37000:2021, Governance of Organizations – Guidance *219*
  - 7.3.2 ISO 31000:2018, Risk Management – Guidelines *220*
  - 7.3.3 COSO Enterprise Risk Management – Integrating with Strategy and Performance *222*
  - 7.3.4 ISO Management System Standards *222*
- 7.4 Key Considerations *222*
  - 7.4.1 Carriers of Meaning *223*
  - 7.4.2 Context, Framing, and Narrative – Or Is it the Number? *223*
  - 7.4.3 Defining Risk *224*
  - 7.4.4 Decision-Making Currency *224*
    - 7.4.4.1 Inference Guidelines *225*
    - 7.4.4.2 Residual and Acceptable Risk *225*
    - 7.4.4.3 Materiality, Value, and Purpose *226*
  - 7.4.5 Rates, Cycles, and Time Horizon *226*
  - 7.4.6 Delays and Buffers *227*
- 7.5 Risk Decision-Making Kernel *227*

Suggested Reading *228*

Part III Integrating Eras *229*

8 Risk Matrix: An Integrated Framework *231*

    Generating Organizational Energy – The Engine *232*

    New Language and Dimensionality *233*

Tailoring *233*

8.1 Risk Field to Risk Matrix *234*

8.2 Matrix Structure *236*

    8.2.1 Nomenclature *236*

    8.2.2 Cells, Rows, and Columns *238*

8.3 Contexts/Drivers (y-Axis) *239*

    8.3.1 Regulatory/Technical *240*

    8.3.2 Organizational *240*

    8.3.3 Social–Human *240*

8.4 Actors/Motive Force (z-Axis) *240*

    8.4.1 Individual *242*

    8.4.2 Team/Department *242*

    8.4.3 Enterprise/Company *242*

    8.4.4 Community *242*

8.5 Risk Management Elements (x-Axis) *243*

    8.5.1 Foundational Five *245*

        8.5.1.1 Risk Assessment (E2) *245*

        8.5.1.2 Emergency Preparedness and Response [EPR] (E3) *246*

        8.5.1.3 Management of Change (E4) *247*

        8.5.1.4 Communication – Systems and Practices (E5) *248*

        8.5.1.5 Competency and Capabilities (E6) *252*

    8.5.2 Trim Tabs *254*

        8.5.2.1 Purpose and Scope (E1) *254*

        8.5.2.2 Social-Human Engagement (E7) *257*

        8.5.2.3 Leadership (E8) *266*

        8.5.2.4 Decision-Making (E9) *266*

    8.5.3 Operational Elements *268*

        8.5.3.1 Frameworks (E10) *268*

        8.5.3.2 Auditing and Metrics (E11) *271*

        8.5.3.3 Operation (E12) *275*

        8.5.3.4 Escalating Impact (E13) *276*

        8.5.3.5 Future-Ready Strategy (E14) *276*

Suggested Reading *277*

9 Matrix in Action 279
- 9.1 Risk Matrix Applications 282
- 9.2 Matrix Dynamics 283
  - 9.2.1 Complexity and Tight Coupling 283
  - 9.2.2 Z-axis – Actors/Motive Force 284
  - 9.2.3 Y-axis – Contexts/Drivers 285
  - 9.2.4 X-axis – Risk Management Elements 286
  - 9.2.5 Gravitational Pulls 287
  - 9.2.6 Topographies/Ecosystems 288
- 9.3 Integrate and Integration 288
  - 9.3.1 Integrating What? 289
    - 9.3.1.1 Integrated Thinking → Integrated Decision-Making 290
    - 9.3.1.2 Risk Field 290
  - 9.3.2 Templating 290
  - 9.3.3 Integrated Thinking 291
    - 9.3.3.1 Rotman School of Management – Integrative Thinking 291
    - 9.3.3.2 The International Integrated Reporting Council 292
    - 9.3.3.3 COSO Enterprise Risk Management (ERM) Framework 293
- 9.4 Scorecards and Dashboards – Portals for Integration 293
  - 9.4.1 Risk Management Elements Dimension Example – Decision-Making Dashboard/Slice 295
  - 9.4.2 Contexts/Drivers Dimension Example – Social–Human Element Dashboard/Slice 297
  - 9.4.3 Actors/Motive Force Dimension Example – Enterprise/Company Element Scorecard/Slice 299
- Suggested Reading 302

10 Escalate Impact 303
- A → B 306
- 10.1 Transcending Paradigms 306
  - 10.1.1 Evolutions 307
  - 10.1.2 New Clearings 307
  - 10.1.3 Shifts 308
- 10.2 Generative Fields 308
- 10.3 Leverage – Creating Generative Fields 312

   10.3.1 Value Generation and Health *313*
   10.3.2 Interiority *314*
   10.3.3 Generative Field Engine – Social–Human Engagement (E7) and Leadership (E8) *316*
   10.3.4 Pedagogy of Evaluation *319*
  10.4 Carriers of a Field *322*
   10.4.1 Portal to Future-Ready *323*
   10.4.2 The Table and Its Seats *324*
   10.4.3 Trim Tab *324*
Suggested Reading *325*

Glossary *327*
Index *337*

Foreword

There is a vast literature on the topic of organizational or enterprise risk management. Any type of threat that generates uncertainty within an organization – economic, political, regulatory, cybersecurity, climate, culture, and other threats – can create risk and the potential for loss. Many books, business school research papers, and international consensus standards have all been directed to the twin goals of predicting and mitigating the broad spectrum of potential risks to an organization. Risk control efforts can yield the added benefit of promoting innovation and positive organizational growth.

Given the plethora of available books on the organizational risk management topic, is there room for yet another book? Yes, if the book is written by an organizational risk prevention practitioner, and for occupational and environmental health and safety risk prevention practitioners. *Organizational Risk Management: An Integrated Framework for Environmental, Health, Safety, and Sustainability Professionals, and Their C-Suites* is a volume that takes the reader through the lifelong learning of Dr. Redinger from the time of his doctorate at the University of Michigan through many schools of risk management practice which he has participated over the decades.

While leaving out none of the rich intellectual history of the environmental, health, safety, and sustainability risk management global literature, the author takes the reader with him as he recounts important stops along his journey to becoming a renowned management systems expert. More than a personal journey, Dr. Redinger teaches the reader about each of the important contributions that scholars and practitioners of the risk management school of practice have made over the decades. The scope of the journey Dr. Redinger takes us on is encyclopedic, with stops along the way to learn about the latest advances in cognitive and organizational science, as well as institutional theory.

Organizational Risk Management will serve as a deeply resourced tool-box for occupational and environmental health and safety professionals, risk managers and sustainability officers to succeed in the challenging business environment we are in now. More than that, though, *Organizational Risk Management* expands traditional thinking around risk management to include steps to create a culture of health in the organization by paying close attention to the well-being of the people who work in the organization from the bottom to the top and all levels in-between.

Organizational Risk Management has much to offer a broad scope of readers. From risk management students and professors to practitioners who want to enhance their fund of knowledge in the risk sciences, there is much to offer any reader. I think the reader will enjoy the journey *Organizational Risk Management* provides as it integrates state-of-the-art risk science learning with the author's compelling professional life story.

John Howard, MD
Director, U.S. National Institute for Occupational Safety and Health

Preface

Modeling from my parents set a trajectory that I had no idea was being set. My earliest memories are of my father's work in civil rights as an Episcopal minister in South Central Los Angeles, and my mother's work as a special education teacher. I helped my dad on Sundays at his church, saw the impact of the Watts Riots, threw rice at weddings, rode in the front seat of hearses with him and watched him officiate funerals. I spent days in my mother's classroom. The term "developmentally challenged" wasn't in the lexicon, nor had "mainstreaming" become a practice in school districts. The only criterion for her students was being toilet trained. Racial integration was evolving – our school district in Pasadena was one of the first to be "integrated" with bussing.

When I began my doctoral work with Professor Steven Levine at the University of Michigan, he was in the process of metaphorically "throwing his analytical instrumentation out the window." He was a full professor who had made significant contributions to occupational exposure science. An insight he had early on in our time together was that "we can do better." He was always on the lookout for ways to impact workplace health, and his attention turned from the analytical to organizational performance-related research. With his guidance, our group produced groundbreaking work.

The torch Professor Levine passed on began many inquiries. Early in my postdoctoral travels I read Willis Harman's book *Global Mind Change: The Promise of the 21st Century*. This book introduced me to his insight that to impact complex problems – such as workplace health – look to business (companies) for leverage. With a public policy background from master's work at the University of Colorado's Graduate School of Public Affairs, I have an orientation toward public policy (e.g. regulatory agency intervention) solutions. Harman's influence brought me back to the core of Professor Levine's work, engaging with, and influencing companies.

Publishing is currency in academia. There was a constant drumbeat in our group, and in the Industrial Health Department, to publish. I was the lead author on a management systems chapter we wrote for the fifth edition of *Patty's Industrial Hygiene and Toxicology*. Wiley representative Bob Esposito approached me in late 2012 about pursuing this book. He was familiar with my risk management expertise from chapters I had written in *Patty's*. I politely declined as I had bandwidth concerns. In addition to my "day job" I had significant volunteer engagements with two organizations; an officer at the American Industrial Hygiene Association (AIHA), serving as the board secretary, and serving on the Center for Safety and Health's (CSHS) Board of Directors.

Bob circled back in 2017 and I said "yes." Even though my bandwidth was limited for a project of this magnitude, I sensed an imperative to present a fresh perspective on risk management for environmental, health, safety, and sustainability (EHSS) professionals and their organizations. Professor Levine's admonition, "we can do better," was always in the background and impacted the yes to Bob.

The complexities of organizational EHSS risk management are well known and have historically been addressed from a technical and regulatory compliance perspective. In the years leading up to 2020, there was growing sentiment for organizations to focus more on the social dimensions of risk. This became an imperative in 2020 when it was pushed to the top of organization risk profiles. The trajectory I sensed in 2017 obviously didn't anticipate the pandemic, and all that cascaded from it.

The doctoral work and Professor Levine's mentoring brought my attention to EHSS risk management. Since then, my focus has been on both the improvement of organizational performance, and the skills and competencies of the professionals who drive EHSS's contributions. The management system work we did at the University of Michigan was groundbreaking and had global impact; we were pioneers in that arena. After the doctoral work, I began applying what we had developed in organizations and standards-development activities. It soon became clear with the organizational work that the success of a management system was a function of (1) how an organization's culture was conditioned for implementation, and (2) the culture's risk orientation.[1] While I was familiar with Peter Senge's work on organizational learning with the publication of the *Fifth Discipline* in 1990, it was in the 2005 timeframe that I began augmenting my technical skills with a deep dive into organizational science. I took classes at MIT's Sloan School of Management and participated in numerous workshops with Peter and others at the Society of Organizational Learning (SoL). Since that time, I've done extensive work on developing methods and techniques to impact and shift an organization's risk culture, so that at minimum, there is (1) increased awareness of the links

---

1 This notion of risk orientation is a central theme in this book; I use the word "orientation" to represent a collective bundle of individual cognitive factors, that include perspective, perception, attitude, mental models, etc.; and, the collective expression of them in the culture.

between actions, outputs, outcomes, and impacts; (2) an understanding about the risk management decision-making process; and (3) increased ability to align actions with desired goals. These methods and techniques are presented throughout this book, along with an integrated framework within which to use them. Beyond these three "minimum" points on impacting and shifting a risk culture, I suggest in the book's final chapter – given risk management's central role in organizational governance – it can serve as a platform to achieve goals beyond ones traditionally associated with it.

One of the writing projects in the mix when Bob circled back in 2017 was a book chapter on EHSS risk management that evolved into being titled "Decision Making in Managing Risk." The chapter's evolution was impacted by the work of Daniel Kahneman, Amos Tversky, and Paul Slovic.[2] Through our research, it became apparent to the co-authors and me that organizational risk management involved much more than its historic technical and regulatory compliance roots. In the chapter, we introduced the term "risk realms" and framed EHSS risk management as a decision-making endeavor that happens at the intersection of these risk realms. While any number of "realms" could be identified, we simplified them into three groups of drivers: (1) those external to the organization (e.g. regulations, laws), (2) those internal to the organization (e.g. mission, values), and (3) those related to people (e.g. employees, community, consumers). The chapter was also impacted through participating in a SoL workshop in 2018 where I learned about Generative Social Fields (GSF) and was invited to participate in a "gathering" convened as part of efforts to evolve a GSF lexicon and GSF constructs.[3] Learning about the generative field construct has enriched the development of the integrated risk management framework presented here. I touch on GSF roots and their contributions throughout the book.

My appreciation for and understanding of risk management's social dimensions and their affect on organizational risk profiles grew through research I did for this book, my participation as a strategic advisor to two standards-development activities, and trajectories that unknowingly began in my youth. For this project, I've conducted interviews and discussions with many professionals who were in, or had been in, a wide range of organizations, at all levels, including executive management. Included were academics, standards-developers, and public

---

2 Pioneers in decision and cognitive science. Kahneman received a Nobel Prize for work he and Tversky did together; his book *Thinking Fast and Slow* provides a good overview of their work, as well as does Michael Lewis's book, *The Undoing Project*.
3 In sociology, field theory examines how individuals construct social fields, and how they are affected by such fields. Social fields are environments in which competition between individuals and between groups takes place, such as markets, academic disciplines, musical genres, etc. This concept provides a lens through which to view organizations.

policymakers. Through the interviews, ideas presented in the risk decision-making chapter were validated and evolved. They began in late 2019, continued into early 2021, and have been ongoing into 2024. Participants in these interviews and discussions reinforced the risk realm model and many highlighted the growing importance of the "people" realm. Several referred to this as the "human realm." This notion of "people" and "human realm" is presented in this book as "social-human field."

Numerous interviewees pointed to two imperatives with their staffs, and EHSS risk management professionals in general. One had to do with gaining increased perspective. That is, to view their activities – such as risk assessment and control – more widely than regulatory compliance or consensus standard's needs. To think beyond immediate operational needs to consider, for example, value chain and other "outside the fence line" issues. The second imperative was characterized in terms of empathy and giving attention to risk profile impacts on people associated with the organization, including employees at all levels, service providers, contractors, etc. Each of these – increasing perspective and empathy – are threads throughout the book's tapestry. Both contributed to the development of the multidimensional risk management framework presented in this book, particularly with the contexts/drivers and actors/motive force dimensions.

At around the time that I executed the contract for this book, I was asked to represent CSHS in an initiative to develop a Culture of Health for Business framework sponsored by the Robert Wood Johnson Foundation (RWJF) and the Global Reporting Initiative (GRI). It was through this engagement that I saw and began to articulate, that organizational risk management platforms could serve as a foundation from which a culture of health could flourish.

From its inception in 2010, a central piece of the CSHS's mission was to impact sustainability and ESG (environmental, society, and corporate governance) frameworks, metrics, and reporting structures. We provided technical support to GRI in its evolution from its G3 to G4 standards. As it was sunsetting in 2019, with the leadership of Kathy Seabrook and Malcolm Staves, the center began to support the Capitals Coalition with their Human and Social Capitals framework and metrics development. The coalition provides an important anchor in considering the social dimensions of risk. Parallel to preparing this book, I was serving as a strategic advisor on the coalition's Worker Health and Safety Project Team and was the lead on its metrics workstream.

The coalition's work and "capitals thinking" orientation present a transformational perspective not only in the sustainability and ESG spaces but also in organizational governance norms, practices, and decision-making. The focus of this book is not on ESG, but ESG does have a role in EHSS risk management, impact (materiality and double materiality) reporting, and vice versa. I highlight the coalition's transformational thinking and processes in the book – with a focus on integrated thinking.

The organizational risk management landscape revolves around the two dominant frameworks and standards used by organizations. In 2017, the Committee of Sponsoring Organizations of the Treadway Commission (COSO) published the second edition of its enterprise risk management (ERM) framework; and, in 2018, the International Organization for Standardization (ISO) published the second edition of its risk management standard, ISO 31000. These second editions reflect an evolution in organizational risk management thinking and processes, including the integration of risk decision-making with activities ranging from daily boots-on-the ground routines, up to enterprise-wide strategic planning. This book provides increased dimensionality to these works.

The increasing complexity and tight coupling in risk profiles, of which social-human dynamics are a piece, necessitate, if not demand, fresh perspectives that transcend historic organizational risk management practices. With the integrated and multidimensional framework presented here, my goal is to provide you with tools and techniques to see, transform, and unleash organizational capacity that escalates impact.

September 2024

Charles Redinger, PhD, MPA, CIH
Institute for Advanced Risk Management
Harvard, MA, USA

Acknowledgments

I am thankful first and above all for my parents' role-modeling of intellectual curiosity and commitment to community service. Their unending attention and focus on educating and mentoring those whose lives they touched has provided both inspiration and a foundation for my life's work.

Building on this was accelerated by Thomas Turner, Richard Ridge, Maurine Saint-Gaudens, and others in their extended community. Their influence in my teens through the present set a learning-oriented trajectory toward critical thinking, science and empirical inquiry. I am thankful for the mentoring and character-building that Lynn Newcomb at Mt. Waterman Ski Lifts provided.

Larry Birkner was initially a client with ARCO and became a good friend and collaborator. He modeled grit, stick-to-itiveness, and charm. He connected me with Steve Levine at the University of Michigan and was an early collaborator in organizational improvement quests.

Professor Levine modeled intellectual courage and dogged determination to impact worker health and well-being. He brought together an amazing group of doctoral students who have provided international leadership and have been friends and collaborators over the years, including David Dyjack, Michael Brandt, and Doo Young Park.

In my doctoral work, I was very fortunate to have crossed paths with and received guidance from numerous people at the Occupational Safety and Health Administration (OSHA), including Marta Kent, Leo Cary, Steve Newell, Frank White, Cathy Oliver, Ed Stern, and Michael Seymour. They supported field work and learning along with Joe Wolfsberger, Marc Majewski, Mike Blotzer, Jere Ingram, and others who opened their organizational doors to offer assistance.

The American Industrial Hygiene Association (AIHA) has provided a strong learning community for decades. I am thankful for their leaders and staff who have provided opportunities and made contributions to my growth and bodies-of-work. AIHA CEOs Peter O'Neil and Larry Sloan have been particularly helpful, along with their staffs, including Manuel Gomez, Mary Ann Latko, and Alla Orlova. I am appreciative of friendships and collaborations with Zack Mansdorf, Vic Toy, Thea Dunmire, Alan Leibowitz, Kyle Dotson, Fred Boelter, Mary O'Reilly and Glenn Barbi.

Collaborators and co-founders at the Center for Safety and Sustainability have been valuable thought-partners in developing innovative EHSS and organizational risk management solutions. I am indebted to Kathy Seabrook, Tom Cecich, and Dennis Hudson for their contributions to my development in the sustainability standards arena. I am thankful for the many conversations with Natalie Nicholles and Malcolm Staves during work with the Capitals Coalition.

I met JoAnne Kellert in 1999, soon after completing my doctoral work. I reached out to her for support in my quest to create a new, post-regulatory compliance, context for EHSS professionals. As a business coach, mentor, and friend, she has provided invaluable guidance in the development of the ideas in this book. Her advice and coaching rigor has helped bring this project to fruition. Her recommendation to take Peter Senge's *Foundations for Leadership* course in 2005 began a valuable phase in my development.

There are many thought-leaders, scholars, academics, etc. whose work and teachings have made significant contributions to my growth and development. Peter Senge and others – including Otto Scharmer and Mette Miriam Boell from MIT and the Society for Organizational Learning, have had impact not just for me, but widely in society. Their quest to integrate research, methodological development, and boots-on-the-ground practice in pursuit of solving complex problems, is groundbreaking and noble.

Invaluable input was provided throughout the development of this book from Glenn Barbi, Fred Boelter, Michal Brandt, Thea Dunmire, JoAnne Kellert, Steven Lacey, Alan Leibowitz, Zack Mansdorf, Anthony Panepinto, and numerous other colleagues.

I am especially grateful to Thomas Turner, Richard Ridge, and SusanMary Redinger for their tireless efforts in proofing and editing throughout this effort.

My wife SusanMary and daughters Maggie and Ally have had amazing patience supporting me in seeing this project throught to the end. They have provided unending support and encouragement for which I am deeply grateful.

September 2024

Charles Redinger, PhD, MPA, CIH
Institute for Advanced Risk Management
Harvard, MA, USA

Business has become, in this last half-century, the most powerful institution on the planet; it is critical that the dominant institution in any society take responsibility for the whole, as the church did in the days of the Holy Roman Empire. But business has not had such a tradition.[1]

Willis Harman
Stanford University

---

[1] Harman, W. (1987). Why is there a world business academy? https://worldbusiness.org/why-a-world-business-academy/ (accessed 23 July 2019).

Acronyms

COH4B	Culture of Health for Business
CSR	Corporate social responsibility
CC	Capitals Coalition
CDP	Climate Disclosure Project
CDSB	Climate Disclosure Standards Board
CERES	Coalition for Environmentally Responsible Economies
CSHS	Center for Safety and Health Sustainability
DE&I	Diversity, equity, and inclusion
EHS	Environmental, health, and safety
EHSS	Environmental, health, safety, and sustainability
ERM	Enterprise risk management
ESG	Environmental, social, and governance
ESRS	European Sustainability Reporting Standards
GAAP	Generally accepted accounting practices
GRI	Global Reporting Initiative
GSF	Generative Social Field
HC	Human capital
HCM	Human capital management
HHRA	Human health risk assessment
IASB	International Accounting Standards Board
IFRS	International Financial Reporting Standards
IIRC	International Integrated Reporting Council
ILO	International Labor Organization
IRMS	Integrated risk management system
ISO	International Organization for Standardization
ISSB	International Sustainability Standards Board
MOC	Management of change
MS	Management system
MSS	Management system standard

NFRM	Non-financial risk management
NGO	Non-governmental organization
NIOSH	National Institute for Occupational Safety and Health
NOMS	National Occupational Mortality Surveillance
NRC	National Research Council
OEHS	Occupational and environmental health and safety
OHS	Occupational health and safety
OHSMS	Occupational health and safety management system
ORMS	Organizational risk management system
ORM	Organizational risk management
RDM	Risk decision-making
RMS	Risk management system
RWJF	Robert Wood Johnson Foundation
SASB	Sustainability Accounting Standards Board
SDG	Sustainable development goals
SEC	Securities and Exchange Commission
SHCP	Social and Human Capital Protocol
TCFD	Task Force on Climate-Related Financial Disclosures
TEEB	The Economics of Ecosystems and Biodiversity
TBL	Triple bottom line
VRF	Value Reporting Foundation

1

Introduction

Leverage to Create

CONTENTS

Imperatives, 3
Becoming Aware, 5
Postulates and Design Principles, 5
Who You Are, 7
What You Will Learn, 7
1.1 A New Era, 8
    1.1.1 Value and Purpose, 8
    1.1.2 Social–Human Dimension of Risk, 9
    1.1.3 Environmental, Social, and Governance, 10
    1.1.4 Diversity, Equity, and Inclusion, 11
    1.1.5 Health – Organizational and Human, 11
1.2 Leverage, 12
    1.2.1 Levine's Lever, 12
    1.2.2 A → B, Current Reality and Where You Want to Go, 14
    1.2.3 Thinking in Systems, 15
    1.2.4 Shifts, 15
    1.2.5 Frameworks and Structures, 19
    1.2.6 Metrics and Indicators, 20
1.3 Integration, 20
    1.3.1 Integrating What?, 21
    1.3.2 Integrated Thinking, 22
    1.3.3 Integrated Risk Management, 22
1.4 Culture of Health, 24
1.5 Leadership and Seat at the Table, 25
    1.5.1 Motive Force and Organizational Energy, 26
    1.5.2 The Table, 26
1.6 Finding Leverage, 27
    1.6.1 Your A → B, 27
    1.6.2 Developing a Playbook, 28
    1.6.3 Creating a Project, 28

Organizational Risk Management: An Integrated Framework for Environmental, Health, Safety, and Sustainability Professionals, and their C-Suites, First Edition. Charles F. Redinger.
© 2025 John Wiley & Sons, Inc. Published 2025 by John Wiley & Sons, Inc.

1.7 Book Logic and Chapter Summaries, 29
    1.7.1 Chapter Flow and Book Logic, 29
    1.7.2 Chapter Summaries, 30
Suggested Reading, 34

Call me Trim Tab.[1]

Buckminster Fuller

Fundamentally, this book addresses how to create new clearings[2] and capacities that increase the ability to generate and preserve value, resilience, and fulfillment for an organization and its stakeholders. The focus is on using organizational risk management (ORM) as a platform and environmental, health, safety, and sustainability (EHSS) management as a vehicle. Integrated thinking and decision-making are the fuel that generates organizational energy. This happens within a generative risk field. The possibility offered here is the transformation of ORM, and in turn, the transformation of organizational governance and purpose.

Few topics are more prominent in organizational governance than risk. Throughout the terrain of insurance and regulatory compliance, risk ideas and concepts have evolved. Things like COVID-19, environmental, social, and governance (ESG), and diversity, equity, and inclusion (DE&I), have brought to the fore new entries to risk profiles and a need for fresh thinking and leadership about how to interact with risk in organizational life. In this mix, there is increased attention to organizational purpose, value generation, and preservation. While concepts of value are not new in organizational governance, a relatively new evolution has been framing value in terms of capital, not just financial capital but also natural, social, and human capital.

Beyond-compliance thinking was a mantra when International Organization for Standardization (ISO) management systems began to appear in the 1980s and 1990s, first with quality and then environmental management. Along with other approaches, ISO management systems offered frameworks to organize activities that were aligned with organizational objectives, not just regulatory compliance. To say current ORM approaches are bankrupt may be too strong. But there is ample evidence that fresh thinking is needed, along with approaches and

---

1 Buckminster Fuller, *Playboy*, February 1972, p. 59 (v.19, No. 22).
2 Clearing is a term used throughout this book to mean, "The state and context where frameworks, structures, policies, procedures, and culture arise."

solutions that address changing risk profiles and the bundle of external challenges that organizations face, often characterized as volatile, uncertain, complex, and ambiguous (VUCA).[3]

A risk field framework is offered here as a fresh approach. Field ideas and theories are not new. They were found in the physical and social sciences as early as the 1860s with James Maxwell's identification of electromagnetic fields and in the 1940s with Kurt Lewin's field theory work in psychology. Numerous sociologists were developing organizational field theories in the 1970s and 1980s, which have developed into a burgeoning arena in institutional thinking.[4] In the 2015 timeframe, Peter Senge, Otto Scharmer, and Mette Boell worked on developing a lexicon to further ideas on generative social fields. I first learned about organizational field concepts in my master's degree work in public policy in the 1990s. In 2018, I was fortunate to attend a workshop at the Garrison Institute led by Peter, Otto, Mette, and others. The workshop focused on the application of generative social field concepts to impact large organizations and complex social systems. At that time, I was in the early stages of writing this book. The knowledge I gained there and in a subsequent workshop in 2019 provided a platform and language for my evolving ideas on how to transform ORM.

The risk field framework is operationally realized with a three-dimensional Risk Matrix that can be used as a tool to (1) foster integrated thinking, decision-making, and action; and (2) impact organizational culture, specifically in generating a learning culture, and a culture of health. The Matrix presents ORM drivers, EHSS management elements, and "actors" and entities that provide motive force that translates into organizational energy.

Book flow and overview follow at the end of this chapter. But first, let us look at some foundational things to set the stage.

Imperatives

It is an understatement to say that much changed in 2020 with the global pandemic and the host of things that cascaded from it. Norms were skewed, and weaknesses in human, public, community, and organizational health were revealed. The "social" aspects of ESG and DE&I were on the radar well before 2020 but took on

---

3 HBR Jan/Feb (2014) – Bennett and Lemoine, "What VUCA really means to you."
4 Fligstein, N. and McAdam, D. (2012). *A Theory of Fields*. Oxford University Press. This topic is addressed in Chapter 7, and other resources are offered at the end of this chapter.

new dimensions in the years that followed. Our understanding of workplace impacts was still playing out when this book went to press.[5] What is clear is that things like fatigue, burnout, and psychosocial issues are common in the post-2020 era; that rules of engagement between organizations and their workforce have shifted; and a general sense of things being "out of balance" has taken hold. Issues of health and safety have taken on a new dimension, along with concerns related to diminished engagement.

Sustainability, corporate social responsibility (CSR), and ESG – generally referred to in the aggregate as ESG – have also been added to the organizational agenda. As this book was in its final stages, there was significant consolidation of principles and standards in the ESG space.[6] The International Sustainability Standards Board (ISSB) was formed in 2021 and issued its first standards in 2023. ESG issues and organizational responses are woven throughout this book. This introduction's title, "Leverage to create" partially originates from a need for nimbleness in responding to ESG's impact on risk profiles, and in turn, how to respond. While there are many complexities in the ESG arena, possibly the most significant are associated with its "social" aspect.

Some of the things I have heard from EHSS professionals over the years that have inspired this book's development are:

- "I don't know how to handle/manage the diversity of things in front of me";
- "I don't know how to increase my staff's ability to look beyond compliance and audit checklists"; and,
- "I know I/we can contribute more. How do I/we get a 'seat at the table' to offer our skills beyond compliance things."

Throughout the book, I point to an ever expanding list of risks, their complexity, and the tight coupling between them. In my decades of practice in ORM and EHSS management, I have seen and experienced this evolution as an academic, an advisor to regulators, and as a consultant to Global 500 companies, as well as in my teaching to thousands of EHSS/ORM professionals where I have heard about their challenges. Through these travels and my exposure to practitioners and scholars from diverse academic disciplines, I have formed this view, and suggest to you that:

The historic ways we frame and practice ORM have diminished ability to meet post-2020 challenges.

---

5 I distinguish between workplace and company here to highlight granularly how work and the work environment occur to workers, employees, etc., as distinct from organizational culture and performance.
6 The term "space" is used to capture the totality of issues, dynamics, reporting, standards, expectations, organizations, etc. related to human capital and ESG.

Notions of urgency and imperatives seem to me to be overused at times. Nevertheless, I strongly suggest that with the increased prominence and presence of "the social" and "the human" in organizational life, there is an imperative, if not urgency, to explore fresh approaches to navigate in these new paradigms.

Becoming Aware

Becoming aware of awareness is the first step in change, transformation, and creating new clearings. Key distinctions in this book are awareness and becoming aware of new organizational risk contexts. And in turn, to expand and grow into these.[7] Becoming aware is framed here in terms of learning and transformational shifts, both individual and organizational. This involves increasing knowledge and perceptions of ORM – yours, your teams, and the structures and frameworks within which we, you, and they operate. The notion of "becoming aware" is the focus of a body of work expressed in the book *On Becoming Aware*,[8] and is expanded on in the book *Theory U*.[9] Aspects of these are touched on throughout this book.

Awareness of risk contexts is a necessary and fundamental first step in transcending the limits of historic ORM approaches. Shifting the context from one of compliance to value creation and preservation and focusing on capitals and their strengths/health, alters being and action. It alters internal states and perceptions. It provides a clearing whereby the outcome space represented in Figure 1.1 expands – and where outcomes such as increased resilience and fulfillment – for the organization and its stakeholders – can happen.

Postulates and Design Principles

Postulates that this work is organized around are:

1) Organizational risk profiles have grown in complexity and interconnectedness (tight coupling).
2) A social–human era is evolving.
3) Fresh ORM frameworks, orientations, and perspectives are needed to meet new challenges.

---

7 Become (v), "begin to be; grow to be; turn into; quality or be accepted as; acquire status of." New Oxford American Dictionary (in Apple OS 10.15.7).
8 Depraz, N., Varela, F., and Vermersch, P. (2003). *On Becoming Aware, A Pragmatics of Experiencing*. Johns Benjamins Publishing Co.
9 Scharmer, C.O. (2016). *Theory U – Leading from the Future as It Emerges*. Oakland, CA: Berrett-Koehler Publishers.

Figure 1.1 Design realms.

4) Framing ORM within an organizational field construct provides leverage.
5) Increased awareness of, and attention to value and purpose, generates organizational energy and motive force.
6) Increasing understanding and awareness of organizational energy[10] reduces uncertainty[11] and increases the power of diagnostics tools.
7) Embodying interiority is the secret sauce of generating organizational energy and shifting from a system to a generative field.[12]

Increasing awareness, fostering integrated thinking, and shifting attention to purpose and value- generation are primary threads in the fabric I weave with you in this book. My goal is to evoke new ways of thinking about risk and transforming ORM beyond risk-related outcomes.

ORM is framed here in terms of non-tangible and tangible realms, or if you like, nonphysical and physical realms. Their intersection is the space of decision-making and the generation of organizational energy, and where outcomes are produced. It is in this space that shifts happen and new possibilities arise (Figure 1.1). This is a key insight from my earliest days following graduate school. It has framed much of my work and thinking since the late 1990s. Examples of tangibles include systems, frameworks, policies, procedures, and "bricks and mortar." Examples of non-tangibles include human capital (people; "the human"), social capital, communication, culture, and community.

Regulatory and technical issues are ORM's historical focus. These are not addressed in depth here. Rather, the primary focus is on things in the non-tangible

---

10 Energy (n). "The strength and vitality required for sustained physical or mental activity." New Oxford American Dictionary, Apple OS 13.5.2.
11 Uncertainty, COSO 2017, p. 110 "The state of not knowing how or if potential events my manifest."
12 Interiority is a term used throughout this book to mean "The phenomena of individual, team, or organizational engagement and generation of motive force that produces outcomes and impacts."

realm, like culture, leadership, cognition, and ontology, and their intersection with systems and frameworks. Facets of leadership are reviewed in Chapter 6 (Field Leadership), along with the introduction of a fresh leadership perspective that I call *field leadership*. Cognitive dynamics are covered in Chapter 5 (Awareness in Risk Management) and Chapter 7 (Decision-Making).

Who You Are

This book is written with attention to people at multiple organizational levels and functions. The primary focus is on mid-to senior-level EHSS professionals; the executives to whom they report and who have ultimate accountability; and those who design, build, implement, operate, and evolve programs and systems in an organization. EHSS activities are framed within an ORM context because these activities, while EHSS-focused, are happening within an organizational context. It is my intention to assist you in your endeavors, increasing your professional confidence, personal resilience, and ability to impact your organization.

Non-EHSS ORM practitioners and executives will also find value here, as will early career professionals. There are clearly many levels within an organization, from production and support functions up to the C-suite (an organization's most senior executives) and board of directors, in addition to an array of stakeholders that include vendors, strategic partners, community members, customers, and those found throughout supply chains. All of these are connected by EHSS/ORM norms, practices, and goals which are also elements of an organization's risk field.[13]

What You Will Learn

My goal is to provide you with tools, skills, constructs, and places from which to orient that will help you navigate and lead in the post-2020 landscape. I am focused on helping you, your teams, and your organization increase resilience and fulfillment, both for the organization itself, and the people associated with it. While resilience and business continuity have been on the radar for decades, they took on increased salience in 2020, along with challenges in meeting basic goals

---

13 Organizational risk field is the relational and contextualized space that creates the interactions and collective behavior that in turn produce the organization's risk management governance, strategy, execution, and outcomes.

and objectives. I advocate elevating "fulfillment" from its current aspirational status to an intentional goal for everyone associated with the organization.

In this book, you will learn processes, tools, and constructs to:

1) Increase risk awareness and perspective;
2) See, characterize, and transform systems;
3) Develop and use new metrics and assessment schemes; and
4) Foster integrated thinking, decision-making, and action.

Uncertainty is possibly the most prominent concern in ORM. Understanding it and reducing it – if not eliminating it, is a paramount goal in all risk management endeavors. The risk field construct and its operationalization as a Risk Matrix (Chapter 8) help to accomplish this goal by identifying key risks and their interconnections within organizational systems and then developing strategies to minimize them. The Matrix provides a multidimensional framework that promotes integrated thinking and decision-making.

1.1 A New Era

As suggested, impacts from the 2020 pandemic were significant and have lingered. Social, organizational, and individual norms were skewed, and vulnerabilities in human, public, community, and organizational health became more visible. While the "social" aspects of ESG and DE&I were on the radar well before 2020, they took on new dimensions in the years that followed. It is bold, and maybe too strong, to suggest that we are in a new era, but what is clear is that there has been increased attention to social and human capital on numerous fronts.

It is timely to consider whether the historic ways we frame and practice ORM, and within it EHSS risk management, are sufficient in this new reality of increased complexity and tight coupling. Historic approaches are compliance-driven, tend to be siloed, and have narrow and often ill-defined decision-making processes.

As this book was going to press, ESG and DE&I had taken on a political/ideological dynamic. Regardless of one's views, these are topics that need to be considered in an organizational risk context.

1.1.1 Value and Purpose

Issues related to value generation and protection are woven throughout this book. These are key themes. Value protection is fundamental to what EHSS/ORM professionals have done for decades; it is foundational to our work.

I suggest at numerous junctures that we intentionally "flip the script" from value protection to value generation. Of course, protection cannot be ignored and must remain a priority. However, you will see as we progress that value generation is framed in terms of integrated capitals that are intricately interconnected, and on which EHSS/ORM professionals have both the opportunity and the ability to provide impact.

In like fashion, there is considerable emphasis from numerous perspectives related to organizational purpose. The integrated Risk Matrix offered here provides a framework for considering purpose not only from the enterprise/company perspective but also from the perspective of individuals, teams/departments, and the community. Determining purpose is unique for each organization and the other entities identified. As with value, I offer that EHSS/ORM professionals' historic attention to health – human, environmental, and organizational – provides us with unique perspectives and abilities to contribute to overall organizational purpose.

1.1.2 Social-Human Dimension of Risk

Values are at the heart of ORM, none more so than within its social–human dimension. The word "social" means different things to different people and in different contexts – much as risk does. The way you and your organization define this dimension is important. As suggested, it is valuable, if not critical, to be clear about your orientation, perspective, and awareness regarding the social–human dynamic. This topic bumps up against charged legacy tensions in the interactions of organizations, communities, and society. These are important considerations when upgrading ORM systems and structures in the social–human era.

In this book, the term social–human refers to people individually and collectively, both inside and outside an organization's "fence line", including those in supply chains, communities, consumers, and the population in general. Bedrock are the people who work for a company, historically referred to as workers or employees. Over time, work relationships have morphed with the use of contractors and the evolution of the gig economy. Also in this first layer are the supply/value chains and the people in them. Another layer includes the people in communities where the company resides (brick and mortar), as well as its customers (product stewardship). And finally, and more broadly, mainly from an environmental perspective, is the way externalities from the business impact human health beyond the immediate communities. There is a lot to unpack with these layers/levels. I do this throughout the book and offer guidance and suggestions on how to gain clarity about how the dynamics of these levels/layers impact your company's risk profile.

Social determinants of health (SDH) are well-developed and prominent concepts in public health.[14] They are defined by the World Health Organization (WHO) as "the conditions in which people are born, grow, work, live, and age, and the wider set of forces and systems shaping the conditions of daily life." Such forces and systems are relevant topics in the social–human era and are seen in ESG and DE&I. I suggest that it is valuable to have an understanding of SDH issues and their terrain. This understanding will be helpful in being clear on what you and your company mean, or will mean by "social" as you upgrade your ORM frameworks and approaches. More on this in Chapters 8, 9, and 10.

1.1.3 Environmental, Social, and Governance

Events of 2020 brought to the fore numerous social aspects of risk not historically considered in organizational risk profiles – namely the health and well-being not just of people but of the organizations themselves and the communities that support them and provide implicit and explicit license to operate. The social aspects of ORM have also been growing in the ESG reporting space. In the 2010s, ESG ideas and thinking – if not norms and "requirements" – evolved in several respects. Notions of "capitals" thinking began to be highlighted, along with integrated thinking and actions related to them. While the "social" of ESG has had historic attention, it increased as the 2010s decade progressed. My focus in this book is not on ESG reporting *per se*. However, given its (1) relation to ORM, (2) the evolving nature of its attention to social and human capital, and (3) its attention to integrated thinking and integrated RDM – it is touched on throughout.

Considering the social dimension of EHSS/ORM is not new. Historically, the domains of human resources, organizational development functions, and public relations – "social" aspects of organizational activities – have not been prominent in risk considerations. Numerous points in time can be identified where this began to change. In the United States, significant laws and regulations began to appear in the early 1970s, which are referred to as an era of "social regulations." These include the National Environmental Policy Act and the Occupational Safety and Health Act, which respectively established the EPA and OSHA. Within the sustainability space that grew in the 1980s, ideas, practices, norms, etc. for "corporate social responsibility" (CSR) began to accelerate. In addition, in the 1980s, the triple-bottom-line emerged, reflecting a shift from traditional accounting and reporting practices to ones that considered ESG.

---

14 World Health Organization (2008). Closing the Gap in a Generation: Health Equity Through Action on the Social Determinants of Health. Final report of the Commission on Social Determinants of Health. Geneva, Switzerland. https://www.who.int/publications/i/item/WHO-IER-CSDH-08.1.

1.1.4 Diversity, Equity, and Inclusion

In the first two decades of the 21st century, DE&I evolved into a significant aspect of human capital management in organizations, if not an important issue in risk management. The three terms are used differently in different circles, but they have been increasingly referred to simply as DE&I. Their roots trace back to the 1960 civil rights movement, the Equal Rights Amendment in the 1970s, and fairness in the workplace initiatives in the 1980s and 1990s.

The Greenlining Institute, a nonprofit policy and research thinktank, offers definitions of the DE&I components as: *"Diversity* refers to difference or variety of a particular identity; *Equity* refers to resources and the need to provide additional or alternative resources so that all groups can reach comparable, favorable outcomes; and *Inclusion* refers to internal practices, policies, and processes that shape an organization's culture."[15] Examples of common diversity factors are: age, disability, ethnicity/national origin, family status, sex, gender identity or expression, generation, language, life experiences, neurodiversity, organizational function levels, physical characteristics, race/color, religion, sexual orientation, and veteran status.[16]

The label DE&I is not familiar to many EHSS and ORM professionals. But from an ORM perspective, DEI&I issues have been in play for decades; it is just that the DE&I label is a relatively new term. The human resources department commonly takes the lead on issues related to policies and procedures. Framing DE&I within a risk management context – such as considering risk tolerance, acceptable risk, risk transfer, and risk owner – is critical in the social–human era and as ORM frameworks and approaches evolve and address social–human issues.

1.1.5 Health – Organizational and Human

> Health is a state of complete physical, mental and social well-being and not merely the absence of disease or infirmity.[17]
>
> *World Health Organization (WHO)*

---

15 Beavers, D. (2018). *Diversity, Equity, and Inclusion Framework*. Oakland, CA: Greenlining Institute. http://greenlining.org/wp-content/uploads/2018/05/Racial-Equity-Framework.pdf (accessed 9 May 2023).
16 Society for Human Resource Management (SHRM) (2023). Introduction to the human resources discipline of diversity, equity, and inclusion. https://www.shrm.org/resourcesandtools/tools-and-samples/toolkits/pages/introdiversity.aspx (accessed 9 May 2023).
17 World Health Organization (WHO) https://www.who.int/about/accountability/governance/constitution (accessed 19 May 2022).

Building on WHO's foundational definition related to human health, notions of health have been increasingly viewed from multiple perspectives. Numerous bodies of work have evolved in the 2000s focused on organizational health. The consulting firm McKinsey and Company has done extensive work on this topic that is summed up in an excellent book titled, *Beyond Performance: How Great Organizations Build Ultimate Competitive Advantage (2011)*. The Robert Wood Johnson Foundation (RWJF) has funded and participated in numerous projects related to community and workplace health. Later in this chapter, I touch on the Culture of Health for Business (COH4B) initiative that they co-sponsored with the Global Reporting Initiative (GRI).

In later chapters, I offer suggestions and ideas on considering WHO's definition in terms of ORM and generative fields. As we dive deeper into purpose and value issues, I propose that organizational and human health be included among the top purpose and value drivers.

1.2 Leverage

> Give me a lever long enough and a fulcrum on which to place it, and I shall move the world.
>
> *Archimedes*

Many, if not all, organizational activities seek to improve efficiencies and performance, to do more with less, and so on. Whether it is looking for leading performance indicators, processes, or systems, we seek ways to use less force or energy and accomplish more. This is a central theme in this book, with particular focus on the non-tangible realm depicted in Figure 1.1.

1.2.1 Levine's Lever

When I joined Professor Steven Levine's research group at the University of Michigan in 1994, he was transitioning his focus from analytical instrumentation used in exposure science to a management agenda that focused on management systems, audit tool development, and activity-based cost accounting.[18] It was his view that while traditional research on things like glove permutation, sampling

---

18 Two colleagues and fellow doctoral students were David Dyjack and Michael Brandt. David developed an ISO-9001-based occupational health and safety (OHS) management system framework, and Michael developed a robust activity-based accounting framework for OHS management.

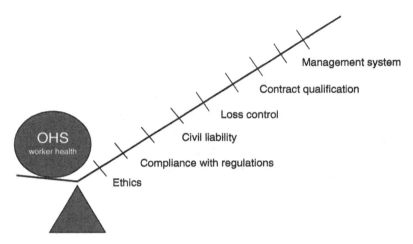

Figure 1.2 Levine's lever.

efficiency, and the like was of course important, it had limited impact on improving workplace and community health.

Early in my time with him, Levine showed me and other students a hand-drawn figure that had a circle, on a fulcrum, with a lever attached. In the circle was OHS, short for occupational health and safety (OHS), and along the lever were a series of words and phrases; these are depicted in Figure 1.2. He did this as part of his quest to characterize activities and their relative impact on OHS performance.

While the diagram is simple, it left a lasting impression on me. It grew over the years through a range of engagements that included developing management systems in large organizations, upgrading audit programs, and teaching. Through conducting many EHSS audits, I observed many efforts to determine upstream actions (causes) that impacted downstream outcomes (effects). Identifying upstream actions and their indicators is akin to creating a longer lever.

In this book, three questions I address are: (1) in the suggested social–human era, how might the circle be depicted; (2) what are the goal(s); and (3) what are the activities and actions that are expected to lead to greater leverage to accomplish it/them?

I leave this to you and your organization to identify the content of your circle, essentially that which you want/need to leverage, improve, transform, etc. Going forward in the book, this is depicted as "B" in Figure 1.3. You might say ESG. You might say EHSS. You could say, generally, resilience and fulfillment. We explore these questions together in later chapters, where I will prompt inquiries on them for you and your organization.

1 Introduction

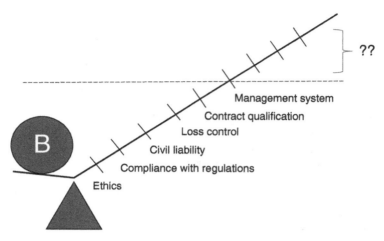

Figure 1.3 Leveraging your needs.

The futurist Buckminster Fuller contributed the metaphor that profoundly crystallized my vision of what is possible for individual actions in large and complex systems. It deserves repeating:

> Something hit me very hard once, thinking about what one little man could do. Think of the Queen Elizabeth – the whole ship goes by and then comes the rudder. And there's a tiny thing at the edge of the rudder called a trim tab. It's a miniature rudder. Just moving the little trim tab builds a low pressure that pulls the rudder around. It takes almost no effort at all. So I said that the individual can be a trim tab. Society thinks it's going right by you, that it's left you altogether. But if you're doing dynamic things mentally, the fact is that you can just put your foot out like that and the whole ship of state is going to turn around. So I said, 'call me Trim Tab.'[19]

At several points in this book, the trim tab metaphor is revisited in terms of you and your team(s) and the actions that generate organizational energy, motive forces, and engagement.

1.2.2 A → B, Current Reality and Where You Want to Go

While simplistic, "B" is used to depict a future state. This could be simple or complex. It could be the result of a visioning or futuring process or pegged to something like ESG. Throughout the book, I use the notation A → B as shorthand for a simple causal model for getting from point A to point B. Integrating

---

19 Buckminster Fuller, *Playboy*, February 1972, p. 59 (v.19, No. 22).

decision-making to identify relevant risks, develop goals and strategies for minimizing and mitigating these risks, implementing these strategies, and evaluating the results are all accomplished using the Risk Matrix as a framework to get you from point A to point B.

The arrow (→) represents an intervention. This could be implementation of an ORM management system, building a generative risk field, or upgrading an audit program. In important ways, many aspects of this book revolve around the simple A → B model. Where are you, where do you want/need to go, and how will you do it? Not illustrated in this model are the feedback loops from B to A so that any necessary mid-course corrections can be made. Numerous "Bs" are suggested, especially those related to creating generative fields (§10.3). Key B's to pay attention to as we progress together are purpose and value generation.

1.2.3 Thinking in Systems

This work has been influenced by many people, one of whom is Donnella Meadows. She was a pioneering thinker in the systems world (modeling, dynamics, thinking, etc.) that came out of Jay Forrester's System Dynamics group at MIT. Meadows was the lead author of *The Limits to Growth* in the early 1970s, and has "...helped usher in the notion that we have to make a major shift in the way we view the world and its systems in order to correct our course."[20]

In the late 1990s, she published an influential article titled "Leverage Points: Places to Intervene in a System." The 12 points she identified are listed in Table 1.1.[21]

Key points to note in her list are leverage points 1 and 2 – "the mindset or paradigm out of which the system – its goals, structure, delays, parameters – arise," and "the power to transcend paradigms." These are themes I weave with you throughout this book.

1.2.4 Shifts

The higher the leverage point, the more the system will resist change.[22]

Shifts can be small or large. They can happen at individual, team, or enterprise levels. They refer to moving from one place or state to another. In organizational development, terms related to shifts include change and transformation. Embedded in the A → B model are shifts, changes, or transformations.

---

20 Meadows died in 2001. This quote is from *Thinking in Systems, a Primer (2008)*. The book was in progress when she died. The manuscript was published posthumously by the Sustainability Institute. Quote is from page xi.
21 Meadows, D. (1999). *Leverage Points: Places to Intervene in a System*. Hartland, VT: The Sustainability Institute, p. 3.
22 Meadows, D. (2008). *Thinking in Systems, a Primer (2008)* (ed. W. Diana), Sustainability Institute, p. 165.

Table 1.1 Places to Intervene in a System (in Increasing Order of Effectiveness).

---

12) Constants, parameters, numbers (such as subsidies, taxes, standards).
11) The sizes of buffers and other stabilizing stocks, relative to their flows.
10) The structure of material stocks and flows (such as transport networks, population age structures)
9) The length of delays, relative to the rate of system change.
8) The strength of negative feedback loops, relative to the impacts they are trying to correct against.
7) The gain around driving positive feedback loops.
6) The structure of information flows (who does and does not have access to what kind of information).
5) The rules of the system (such as incentives, punishments, constraints)
4) The power to add, change, evolve, or self-organize system structure.
3) Goals of the system.
2) The mindset or paradigm out of which the system – its goals, structures, rules, delays, parameters – arises.
1) The power to transcend paradigms.

Figure 1.3 depicts a lever longer than the one in Figure 1.2, with question marks at end of the lever to prompt thinking about how more leverage can be obtained. There are trim tabs for increasing leverage: (1) expanded perspective and awareness; (2) language; and, (3) an integrated capitals perspective. These are addressed throughout the chapters that follow.

A central dimension of shifts is learning as well as the speed at which people and their organizations learn. A thought leader on this is Arie de Gues of Shell Oil who writes, "We understand that the only competitive advantage that the company of the future will have is its managers' ability to learn faster than their competitors. So that the companies that succeed will be those that continually nudge their managers toward revising their views of the world."[23]

1.2.4.1 Expanding Perspective and Awareness

We grow, evolve, and advance through a complex cycle of learning and acting on what we have learned. Einstein's scientific breakthroughs are well known. He was obviously an inquisitive person who thought deeply about many things – he famously observed, "We cannot solve our problems with the same level of thinking that created them."[24] Embedded in his seminal observation are the notions of

---

23 de Gues, Arie P. (1988). Planning as learning. *Harvard Business Review*, March-April, p. 6. (reprint, 88202).
24 https://www.brainyquote.com/quotes/albert_einstein_121993. Accessed, October 2, 2023.

awareness, paying attention to what is happening, considering our orientation to problems, and thinking about solutions. This is not a new idea to you, and I bet you can quickly come up with examples where you have had shifts in thinking personally and professionally. I highlight this idea to reinforce the message that there are numerous shifts in both orientation and thinking that are wise, and necessary, to meet post-2020 ORM challenges.

I have introduced you to several things that involve shifts, including: (1) suggesting that the historic ways we frame and practice ORM are archaic and have diminished ability to meet post-2020 needs; (2) viewing ORM as occurring within a larger risk field; and, (3) highlighting the social dimension of a risk field.

Embracing the social–human dimension of ORM requires shifts in orientation and thinking for many of us. As I have suggested, ORM is an enterprise full of paradoxes and tensions that are generated by a stew of personal, organizational, and social values, perspectives, and orientations. Keeping this notion of shifts in thinking in your toolbox will be valuable as we journey forward. It provides an opening for litmus tests along the way as we consider RDM, as well as when we begin to consider what it takes to bring a generative field into existence.

A question to ponder is, "what leads to shifts in orientation and thinking?" Numerous things are identified throughout the book that generate and support shifts. The role that organizational culture plays in this process is addressed in Chapter 5, in terms of integrated thinking and the extent to which principles of learning are embedded in the culture. Leadership provides a motive force within a culture. The practice of *Field Leadership* in Chapter 6 facilitates trust, engagement, and inclusion, which in turn supports a culture where shifts can occur. Techniques for creating such shifts are presented in Chapters 9 and 10.

1.2.4.2 Language as Currency

We know that words matter. This is taught in communication training and reinforced by people who critique our writing and speaking. Their importance is not reinforced as much in our technical and scientific endeavors. The regulatory/technical field tends to be driven by data (numbers) that are interval or ratio in nature, and its decision-making processes tend to be well defined and concrete. In organizational and social fields, the role of words takes on increased importance in the decision-making and communication processes. When we enter the domain of possibility – creating it, seeing it, standing in it, committing to it – the role of words, the awareness of them, along with vocabulary and language, is critical.

Significant contributions to decision-making theory and practices have recently been made in behavioral economics. Daniel Kahneman, Amos Tversky, and many of their colleagues have been pioneers in decision science, behavioral economics, cognitive psychology, and other disciplines that have contributed to risk management. Kahneman points to the role of language and vocabulary in organizational life:

> An organization is a factory that manufactures judgments and decisions. My aim for water cooler conversations is to improve the ability to identify and understand errors of judgment and choice, in others and eventually in ourselves, by providing a richer and more precise language to discuss them. Learning medicine consists in part of learning the language of medicine. A deeper understanding of judgments and choices also requires a richer vocabulary than is available in everyday language.[25]

Words, vocabulary, and language – whichever one resonates with you – are currency in the inquiries in this book. Their role in impacting orientations, perspectives, and decision-making is highlighted through out, as they are key tools in generating transformational shifts. New terms and concepts are offered to aid in applying the Risk Matrix to ORM upgrades, in strengthening your risk field, and in developing an integrated RMS.

1.2.4.3 Integrated Capitals Perspective

Ideas about capital have been around at least since the writings of the 18th-century economist Adam Smith, and are historically associated with things ("capital") related to business and finance.

The term "natural capital" was first used in the 1970s and adopted into environmental economics in the late 1980s. The application of natural capital thinking in the ESG space led to the formation of the TEEB[26] for Business Coalition in 2012, which became the Natural Capital Coalition in 2014. In 2018, the Social and Human Capital Coalition was founded, and in 2020, the two coalitions joined to form the Capitals Coalition (CC). The new entity characterizes itself as a "global collaboration transforming the way decisions are made by including value provided by nature, people, and society." The CC presents a common concept of capital as natural, social, human, and produced. Organizational decision-making and risk management are strengthened (more leverage) as capital assessment evolves beyond single-capital to multi-capital assessments.[27] This is depicted in Figure 1.4 and is expanded on in the chapters that follow.

---

25 Kahneman, D. (2011). *Thinking Fast and Slow*. New York: Farrar, Straus and Giroux, p. 418.
26 TEEB is the acryonym for The Economics of Ecosystems and Biodiversity.
27 Capitals Coalition (2021). Principles of integrated capitals assessments. p. 10. www.capitalscoalition.org.

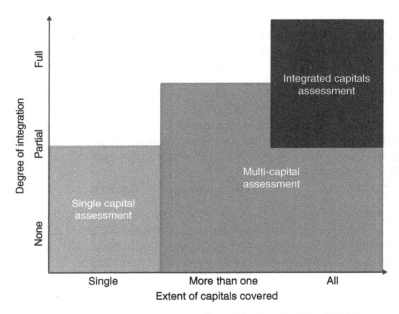

Figure 1.4 The capitals assessment spectrum. *Source:* Capitals Coalition (2021)/ with permission of Capitals Coalition.

1.2.5 Frameworks and Structures

Frameworks provide the "containers" in which EHSS/ORM activities and processes happen. These include processes, programs, and systems. The risk field introduced in this book can be considered a framework or container as well. This provides the scaffolding and defines boundaries and scope. It is important to pay attention to this notion of EHSS/ORM frameworks and containers, as we know that structure has power and influences culture and behavior. Strengthening these increases leverage.

We have all experienced in our professional endeavors – the power, if not impact, of organizational charts, management systems, and the design of physical spaces. Following the World War II bombing of the House of Commons, Winston Churchill famously pointed to the power of structure in addressing the British Parliament about rebuilding the Commons Chamber and whether to rebuild the rectangular pattern or change to a semi-circle or horseshoe design. He said, "we shape our buildings and afterwards our buildings shape us."[28]

---

28 Churchill, Winston (1943). https://www.parliament.uk/about/living-heritage/building/palace/architecture/palacestructure/churchill/ (accessed 12 October 2020).

This book provides guidance on how to strengthen your existing EHSS/ORM frameworks and structures, and how to evolve them in an integrated organizational risk field. We will be exploring aspects of the following frameworks and applying them through the Risk Matrix (Chapter 8) to meet social–human era challenges:

1) COSO's[29] Enterprise Risk Management (ERM) framework.
2) ISO's risk management standard (31000).
3) ISO's bundle of management system standards that are familiar to EHSS and risk management professionals are based on what is called ISO's "Annex SL," namely, ISO 14001 (environment), ISO 45001 (OHS), ISO 50001 (energy), and ISO 22301 (business continuity).
4) ISO 37001:2021, Governance of organizations – Guidance.

As a precursor to upgrading your ORM frameworks and structures, there is value in distinguishing how you characterize them, your orientation to them, and your understanding of how things operate within them. When I ask clients or students to describe these three dynamics (characterization, orientation, operation) within their EHSS/ORM frameworks/structures, they often begin by describing them in terms of their management systems (e.g. ISO MSS, COSO ERM). This makes sense due to the popularity of these dominant approaches. With a bit of dialogue, we discover that the tapestry is often richer than a single approach, especially in larger and multinational companies.

1.2.6 Metrics and Indicators

Any consideration or conversation about ORM and ways to improve it quickly touches on metrics and indicators. I have identified three trim tabs, along with frameworks and structures, as key to increasing leverage and generating shifts. Fundamental to any intervention are metrics and indicators that describe and measure their effectiveness, and tracking the progression of A → B. Current ORM metrics/indicators are addressed throughout the chapters that follow. Chapters 9 and 10 introduce a bundle of metrics that cascade from the Risk Matrix presented in Chapter 8.

1.3 Integration

Increasing ORM leverage in the social–human era requires upgrading and integrating decision-making processes. Core to this is integrated thinking, awareness of risk orientations, and expanding perspectives.

---

29 Committee of Sponsoring Organizations of the Treadway Commission.

> **Box 1.1 Definitions: Integration, Integrate, Whole, Wholeness[30]**
>
> *Integrate (verb)*: Combine one thing with another so that they become whole. Latin root (integrare) – to make whole.
> *Integration (noun)*: The act or process of integrating.
> *Whole (noun)*: A thing that is complete in itself. All of something.
> *Wholeness (noun)*: the state of forming a complete and harmonious whole; the state of being unbroken or damaged; good physical or mental health.

These are important concepts to understand and gain facility with. The use of the term "integrated thinking" in business management and ESG spaces is relatively new. Its application here is primarily in terms of RDM and risk field building. The "integration" aspect of risk field development relates to the space where three subfields (regulatory/technical, organizational, social–human) merge.

1.3.1 Integrating What?

Integration is a central phenomenon in decision-making; it is not new to EHSS/ORM professionals. When ISO management systems began to appear in the 1990s, methods for integrating common aspects of siloed programs were offered, if not required, when third-party certifications were pursued. Training, hazard and risk identification, and auditing for example, became commonplace. There was integration of regulatory compliance programs within systems and frameworks. Organizational science perspectives offered increased EHSS performance when these activities were integrated with processes from boots-on-the-ground operations up to design. With the evolution of sustainability pressures, focus expanded to include external impacts (e.g. externalities, double materiality) and things like climate change and biodiversity, whereupon the integratation of these additional dynamics became necessary.

In the 2010s, ideas of capitals thinking, and integrated capitals assessments gained currency. The CC, a Geneva-based think tank, has developed several protocols, including a Social and Human Capital Protocol (2019) along with Integrated Capitals Assessment Principles (2021). The coalition's work focuses on redefining value to transform decision-making. Their core principles are woven throughout this book.

---

30 New Oxford American Dictionary (in Apple OS 10.15.7).

1.3.2 Integrated Thinking

Integrated thinking is a common topic in ESG and business circles, and is often discussed in conjunction with integrated reporting. It is a topic I touch on throughout the book, particularly in relation to decision-making and subsequent action(s), and is addressed in Chapter 9 (§9.3.3). A logic model presented in Chapter 2 (Figure 2.1) contains what I am referring to as "outcome space," which includes a flow from integrated thinking to decision-making to action. The Risk Matrix presented in Chapter 8 provides an environment for this sequence to play out and be used in ORM.

In the ESG reporting space, practices of integrated reporting of financial, environmental, social, and corporate governance began to appear in the 2010 timeframe; the term "One Report" evolved to describe this phenomenon.[31] The International Integrated Reporting Council (IIRC) was formed in 2010 and developed an integrated reporting framework and associated protocols.[32] In the IIRC efforts, integrated reporting ideas, if not practices, began to focus on "integrated thinking" within their reporting schemes. In 2019, IIRC published an integrated reporting document that states:

> Integrated reporting is about more than just creating a report. At the heart of integrated reporting is a process founded on integrated thinking which intentionally joins how an organization's strategy, governance, performance and prospects, in the context of its external environment, lead to the creation of value in the short, medium and long term.[33]

In this report, as well as earlier ones by other entities, the term integration was presented in the context of integrating "capitals" associated with business; six capitals identified included financial, manufacturing, intellectual, social, and relationship capital, human capital, and natural capital. The IIRC has developed what it calls a "spring model" that focuses on optimizing value through a multi-capital approach. This concept and other approaches provide a foundation for the Risk Matrix presented in Chapter 8.

1.3.3 Integrated Risk Management

Integrated risk management is a central theme in COSO's ERM framework and ISO's risk management standard (ISO 31000). Each of these, in their second editions, which are referenced in this book, reflects increased emphasis on integration.

---

31 Eccles and Krazus (2010). One Report: Integrated Reporting for a Sustainability Strategy.
32 IIRC and SASB merged in 2021, to form the Value Reporting Foundation.
33 IIRC (2019). Integrated thinking and strategy, state of play report. p. 12.

1.3 Integration

In COSO's framework document, it is keenly observed that "For most entities, integrating enterprise risk management is an ongoing endeavor. Factors that influence integration are entity culture, size, complexity, and how long a risk-aware culture has been embraced."[34] Key here is acknowledging that integration is an ongoing endeavor, and is a function of the maturity of risk awareness understanding and acceptance in the culture. The framework document lists the benefits of integration:

> Integrating enterprise risk management with business activities and processes results in better information that supports improved decision-making and leads to enhanced performance. In addition, it helps organizations to: anticipate risks earlier or more explicitly, opening up more options for managing the risks and minimizing the potential for deviations in performance, losses, incidents, or failures; identify and pursue existing and new opportunities in accordance with the entity's risk appetite and strategy; understand and respond to deviations in performance more quickly and consistently; develop and report a more comprehensive and consistent portfolio view of risk, thereby allowing the organization to better allocate finite resources; and improve collaboration, trust, and information sharing across the organization.[35]

Viewing integration this way reflects the value of increasing anticipation, widing and deepening risk awareness (e.g. risk portfolio wider view), and increasing responsiveness to deviations.

Risk management integration is a central component of ISO 31000:2018. It states: "The purpose of the risk management framework is to assist the organization in integrating risk management into significant activities and functions. The effectiveness of risk management will depend on its integration into the governance of the organization, including decision-making. This requires support from stakeholders, particularly top management."[36] The standard continues, "Integrating risk management into an organization is a dynamic and iterative process, and should be customized to the organization's needs and culture. Risk management should be a part of, and not separate from, the organizational purpose, governance, leadership and commitment, strategy, objectives and operations." Key here is the observation that integration is a dynamic and iterative process and that there is essentially no "one-size fits all" approach. That is, customizing to an organization's needs and culture is part of the process.

---

34 COSO ERM (2017), p. 18.
35 COSO ERM (2017), p. 17.
36 ISO 31000:2018, p. 4.

1.4 Culture of Health

> If we get health right, everything else will follow.[37]
>
> *Dr. Alistair Fraser*
> *Vice President, Health at Royal Dutch Shell*

In the late 19th century, Nikola Tesla offered a visionary precursor to the future primacy of health, suggesting that "hygiene" (aka health) would be prominent in a future president's cabinet. He said, "The Secretary of Hygiene or Physical Culture will be far more important in the cabinet of the President of the United States who holds office in the year 2035 than the Secretary of War."[38]

You are familiar with the idea of organizational culture. It is talked about formally in training and informally in "water cooler" conversations with colleagues. Organizational culture considerations are at the center of the development and implementation of any new initiative of consequence. Many books have been written on organizational culture – how to characterize it, assess it, change it, and so on. While not always recognized as such, the ebb and flow of an organization's culture is a social phenomenon. It is impacted by many things, such as systems, "bricks and mortar" (the physical stuff), policies/procedures, and so on. It manifests, through people, and is reflected in the bundle of things I have lumped into the terms orientation, perspective, and mental models.

I touch on organizational culture at numerous points in numerous ways. Generally, in terms of "conditioning" a culture when upgrading ORM frameworks and approaches, bringing a risk field into existence, or implementing an RMS. And, specifically, in terms of infusing a culture with principles of health and learning, as expressed as *culture of health* and *learning culture.*

A relatively recent addition to the organizational culture dialogue is "culture of health." The idea of proactively developing a culture of health in a company started in the mid-2010s with an RWJF initiative called, "Culture of Health for Business" (COH4B). This effort evolved out of the foundation's broader efforts to promote cultures of health in communities. Related to business, the foundation partnered with the GRI on developing the COH4B framework and metrics that could be used in GRI's reporting scheme.[39]

Throughout this book, I point to aspects of the RWJF/GRI framework and show how you can incorporate them to strengthen EHSS/ORM initiatives. My

---

[37] Warner Lecture, 2015 IOHA/BOHS Conference, London, England; April 29, 2015. "Health – Who Cares?"
[38] Seifer, M. (2016). *Wizard: The Life and Times of Nikola Tesla: Biography of a Genius.*
[39] I supported the Advisory Council for his effort.

comment on SDH regarding the depth to which you need to know details applies here with COH4B. There are many parts to this framework, and for organizations that are ready and able, with respect to their culture, it provides a comprehensive approach.

There are ample trends that suggest health is a key consideration in fourth-generation risk management, comprising human, organizational, community, environmental, public, and global health.[40] There are a number of definitions for health depending on the area of focus. For human health, a common definition is WHO's "a state of complete physical, mental, and social well-being and not merely the absence of disease or infirmity." On the business side, McKinsey & Company has developed an organizational health framework that is expressed in Scott Keller's book *Beyond Performance*. Keller presents three attributes of healthy companies: internal alignment on direction, high quality of execution, and capacity for renewal. On the occupational health front, the National Institute for Occupational Safety and Health (NIOSH) has developed a *Total Worker Health* paradigm that is having impact and being embraced by many companies. Finally, Shell Oil and others have defined a culture of health as a key value and operating principle. Shell presents this as:

> Culture of health is when Shell's people are thriving, engaged, competent, and performing at their best with a deep sense of purpose. People will readily demonstrate behaviors like autonomy, agility, and empowerment to make meaningful decisions; they will feel energized to be the best that they can be, even if they are ill or have limitations. In a Culture of Health, we are a true Business Enabler, committed to being outcome focused and ensuring that everything we do drives business performance.[41]

1.5 Leadership and Seat at the Table

Much has been written on leadership. Many of us have had formal leadership training, and it is not uncommon for large companies to have embedded leadership institutes. Similar to my assertion that "The historic ways we frame and practice ORM have diminished ability to meet post-2020 challenges," we find the same diminished capacity with leadership in organizations, and that fresh thinking is needed. I suggest – and explore in this book – an evolving imperative in the post-2020 era that necessitates a new kind of leadership and that few

---

40 Fourth generation risk management is introduced in Chapter 2 (2.1.1).
41 Warner Lecture, 2015 IOHA/BOHS Conference, London, England; April 29, 2015. "Health – Who Cares?"

organizational functions are as uniquely positioned as EHSS teams and professionals to provide it.

1.5.1 Motive Force and Organizational Energy

I have touched on awareness of the relationship between motive force and organizational energy. Central questions in this book are: Where does motive force come from? Who generates it? How is it generated? Over the years, I have had colleagues and clients who have been daunted by these questions. They have pointed to an already full plate of responsibilities and said things like, "I can't take on any more," when confronting the expanding complexity of their risk profiles and risk objects. In response, I have advocated for decades that advances in areas such as management and decision science provide tools to help them expand. These tools are what are being presented here.

Elucidating the phenomenon of organizational energy is relatively new in organizational science. Professors Heike Bruch and Bernd Vogel in their seminal book, *Fully Charged: How Great Leaders Boost Their Organization's Energy and Ignite High Performance*, define organizational energy as, "The extent to which an organization, division, or team has mobilized its emotional, cognitive, and behavioral potential to pursue its goals. Simply put, it is the force with which a company (or division or team) works." They identify three attributes of it as:

1) "Organizational energy comprises the organizations' activated emotional, cognitive, and behavioral potential;
2) Organizational energy is a collective attribute – it comprises the shared human potential of a company (unit or team); and
3) Organizational energy is malleable."[42]

The risk field and Matrix provide a platform to generate organizational energy through the integration of the social–human dimension of ORM. The risk field construct and Field Leadership (Chapter 6) within it provide the motive force for this.

1.5.2 The Table

Having a "Seat at the table" is a common metaphor in EHSS circles. It points to questions about the degree of influence EHSS professionals have in their organizations beyond technical and regulatory issues.

---

42 Bruch, H. and Vogel, B. (2011). *Fully charged: How great leaders boost their organization's energy and ignite high performance*. Boston: Harvard Business Review Press.

> **Box 1.2 Definitions: Influence[43]**
>
> *Influence (noun)*: the capacity to have an effect on the character, development, or behavior of someone or something, or the effect itself. The power to shape policy or ensure favorable treatment from someone, especially through status, contacts, or wealth. (verb) have an influence on.

Over time, particularly with sustainability issues and the rise in ESG, EHSS professionals have been increasingly engaged by the C-suite in creating general organizational strategy. The degree of influence they have varies and has numerous dimensions.

The table metaphor raises the question, "whose table?" In the social–human era, this metaphor is possibly no longer relevant, particularly within a risk field context and increased demands for integrated thinking, decision-making, and action. Rather, we might ask: "Who has a seat at (or in) the Matrix? That is, who has influence in Risk Matrix decision-making and management? I will discuss this issue with you in later chapters when considering leadership, engagement, motive force, and organizational energy.

1.6 Finding Leverage

We are handicapped in our search for leverage. Ensuring regulatory compliance and solid risk management (assessment, control, elimination, etc.) are foundational. It is a challenge to evolve EHSS programs/systems and begin to address the non-tangible things addressed in this book. An EHSS function/department maturity model is presented in Chapter 2 (Risk Logics), which provides a construct for evolving from a purely regulatory/technical orientation. The EHSS management elements in the Risk Matrix (Chapter 8) are presented in three groups to distinguish among elements that need 24/7/365 attention (Foundational Five), those that provide leverage (Trim Tabs), and those that are operationally oriented.

1.6.1 Your A → B

A valuable question to ask is, "what is your future pull," and how well is this distinguished? Flipping this question is to ask, "are you being pushed by the past, or are you being pulled by the future?" It could be that simply meeting regulatory requirements is all that is needed. However, more likely than not, there are

---

43 New Oxford American Dictionary (in Apple OS 10.15.7).

pressures to address the rise of risk profile complexities. Identifying your "B" is addressed in Chapter 9 (Matrix in Action), and ways to accomplish it are addressed in Chapter 10 (Escalate Impact).

1.6.2 Developing a Playbook

I use the term "building a playbook," and suggest you create one to capture insights along the way. This practice will help in tailoring the book's content to your specific dynamics. There are times when we need to have already developed procedures and frameworks and be able to use them without any modifications in our organization. This is common and understandable. However, while there are certainly similarities among organizations, there are also differences ranging from nuanced to significant. Tailoring the book's content to your needs is discussed in Chapters 9 and 10. But to get started, it would be valuable to think about and track a few things related to your organization. These are:

- *EHSS/ORM processes*: How are they characterized, and what defines them? Look for interconnections between elements and activities.
- *Risk decision-making (RDM)*: Begin to pay attention to RDM processes. Are expected or desired outcomes happening?
- *Integrated thinking*: Is this happening, and if yes, how is it impacting RDM?
- *Social–human dynamics*: Pay attention to how these are being characterized, defined, and integrated in your organization's field.
- *Orientations, perspectives, and awareness*: Begin to observe and characterize: orientations (individual, team, enterprise) to risk; abilities to integrate RDM inputs beyond regulatory and technical perspectives; and awareness-related practices.

Think about baselines for each of these queries. Make written notes of initial thoughts. Engage your teams. Ponder these questions:

1) "What's happening now?" (A)
2) "Where do we want or need to go, or what do we want to accomplish?" (B)
3) "How can it be done?" ($\rightarrow$)

1.6.3 Creating a Project

To maximize your value in reading this book, I recommend that you consider a project to work on as you proceed through the chapters, and, if you do this, augment your playbook with insights and discoveries as you progress. The project should be relevant now, and be one that addresses a current challenge. This can be a personal or team-related professional challenge. It can be an organization-wide challenge, such as upgrading existing frameworks to meet some of the challenges identified above. Some ideas are:

Personal/Individual:

1) Improve integrated thinking and decision-making skills.
2) Increase perspectives and levels of awareness.
3) Develop Field Leadership skills.

Team/department:

4) Develop, strengthen, and/or bring value generation and purpose to ESG reporting activities.
5) Establish uniformity in MOCs across the enterprise.
6) Shift the EHSS/ORM culture to one that is generative.

Enterprise/company:

7) Develop/strengthen RDM processes embedded with the principles of integrated thinking.
8) Develop an integrated risk management system using the Risk Matrix framework.
9) Build a generative risk field.

Get started with the following actions and considerations:

1) Write a short description of the challenge, fewer than 25 words.
2) What will be different when it is achieved? You can think of goals here.
3) What's the time horizon? When can it be realized?
4) What resources are needed?
5) Are shifts in the orientations, perspectives, mental models bundle needed?

1.7 Book Logic and Chapter Summaries

1.7.1 Chapter Flow and Book Logic

There are 10 chapters, nine following this introduction, that are grouped into three parts, as depicted in Figure 1.5

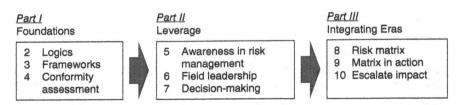

Figure 1.5 Book flow.

1.7.2 Chapter Summaries

Part I – Foundations
Chapters 2, 3, and 4 provide the foundations for ORM. This includes logics, structures, metrics, and performance assessment.

1.7.2.1 Chapter 2 – Risk Logics

Few topics are more important in organizational life than ones related to risk. With roots in finance, insurance, and regulatory compliance, ORM has evolved into a dominant organizational governance theme with impacts well beyond its traditional roots. The term "risk logics" is used to describe the range of approaches, principles, ideas, and contexts that establish the foundation for an organization's risk-related systems, frameworks, decision-making processes, and subsequent actions. Operationalization of the tangible/non-tangible model in Chapter 1 (Figure 1.1) depicts an ORM outcome space where risk logics interact (Figure 2.1). A range of risk logics are reviewed along with foundational ones that include uncertainty, acceptable risk, risk tolerance, risk transfer, and a systems perspective. The role of conformity assessment in ORM is introduced along with the benefits of applying program evaluation concepts and methods to performance improvement. A historical overview of environmental, health, safety, and sustainability (EHSS) and ORM drivers, and their evolution is provided. The risk field construct is introduced.

1.7.2.2 Chapter 3 – Frameworks

Frameworks define much of what is done in EHSS/ORM. Programs evolve into systems, and systems evolve into interconnected fields. Framework is a catch-all term for all of these. We can also think of them as structures, scaffolding or even containers, that establish the context within which the multitude of EHSS and ORM processes and practices occur. Frameworks establish the parameters for processes, systems, and norms within which ORM is executed and assessed. An overview of aspects of risk science that affect ORM is presented along with foundational National Academy of Sciences (NAS) and EPA risk models/frameworks that have impacted risk management's evolution. ISO management systems frameworks are reviewed along with ISO's risk management standard, ISO 31000. COSO's Enterprise Risk Management (ERM) framework is reviewed, and aspects that relate to the Risk Matrix (Chapter 8) are highlighted. An overview of ESG frameworks is presented. This is followed by three initiatives (Total Worker Health®, Culture of Health for Business, and the Social and Human Capitals Protocol) that are framed as "transcending paradigms," and that along with the Risk Matrix, help build bridges between the three organizational risk management dimensions, regulatory/technical, organizational, and social-human.

1.7.2.3 Chapter 4 – Conformity Assessment and Measurement

Conformity assessment is an important topic in organizational governance and risk management. While there are numerous aspects to this topic, the most important one relates to the decision-making process regarding whether an organizational function, activity, process, or system conforms with a defined internal or external standard. Internal and external auditing are formal activities that perform the measurement and evaluation activities that are integral to this process. In the context of this book, a central conformity assessment question is "who's the judge?" to refer to a decision-maker. A host of entities can be identified as the judge with the number of judges increasing with the proliferation of stakeholders linked to organizational governance. Conformity assessment practices that have evolved with ISO management systems are used as a guide to consider assessment and decision-making issues related to the Risk Matrix. Topics introduced in this chapter set the stage for Chapters 7 (Decision-Making) and 9 (Matrix in Action). Internal auditing and internal audit departments are suggested as unrecognized resources for organizational learning which can provide leverage in increasing impact and evolving generative fields.

> *Part II – Leverage*
>
> Chapters 5, 6, and 7 address motive forces and the processes that use them. Building on Part I, these chapters set the stage for developing integrated frameworks for ORM. The focus of these chapters is on people and decision-making, which are at the core of generating organizational energy.

1.7.2.4 Chapter 5 – Awareness in Risk Management

Increasing awareness in ORM is key to creating new clearings and capacities that increase the ability to generate and preserve value, resilience, and fulfillment for an organization and its stakeholders. Awareness of risk contexts is a necessary and fundamental first step in transcending the limits of historic ORM approaches. Shifting the context to value creation and preservation and focusing on capitals and their strength/health alters being and action. It alters internal states and perceptions. It allows the outcome space to expand, and fosters outcomes such as increased resilience and fulfillment for the organization and its stakeholders. Seven risk awareness elements are presented, along with framing language as currency in impacting change. Steps for shifting risk perceptions, mindsets, and perceptions that center around a learning context are offered.

1.7.2.5 Chapter 6 – Field Leadership

Fulfillment of organizational purpose and the creation of value are not stand-alone activities of a singular function. They necessitate the alignment of organizational functions, engagement at all levels, and an "all hands on deck" perspective. The phenomenon of generating integrated context, motive force, and engagement that drives the organization toward the fulfillment of its purpose and value creation is framed here as field leadership. Field leaders have a wide, multi-capital perspective that drives integrated thinking, decision-making, and action. Integrated thinking, decision-making, and action approaches and practices are posited as fundamental to effective leadership, along with skills, competencies, practices, and capacities such as creating context, seeing systems, awareness, empathy, listening, and fostering trust. Readers are invited to consider their leadership perspectives, legacies, and Buckminster Fuller's profound insight – "Call me Trim Tab."[44]

1.7.2.6 Chapter 7 – Decision-Making

While organizational decision-making has been discussed in the organizational theory and behavior literature for a long time, its role in ORM is evolving. A brief overview of decision science and theory and the impact that cognitive science has had on them is provided. General ideas and concepts on how these have been applied and implemented in risk management areas are also addressed. The chapter begins with the question, "How aware are you and your team of your risk decision-making process(es)?" Topics addressed in previous chapters and applied to decision-making include awareness, heuristics, and purpose. A systems perspective is offered to frame organizational risk decision-making (RDM). Stocks and flows that travel through impact-dependency pathways are presented as key aspects of the decision-making process. Key decision-making considerations include carriers of meaning, decision-making currency, and the RDM kernel.

Part III – Integrating Eras

Parts I and II set the stage for transforming ORM from a silos- and systems-based orientation to an integrated organizational field. The social–human dimension is highlighted for generating organizational energy and integrated RDM. The Risk Matrix is presented, along with the first steps in putting it into action. Ways to escalate impact with the Risk Matrix are offered in the final chapter.

---

44 Buckminster Fuller, *Playboy*, February 1972, p. 59 (v.19, No. 22).

1.7.2.7 Chapter 8 – Risk Matrix – An Integrated Framework

A multidimensional ORM framework is presented in this chapter. It reflects an operationalization of the risk field construct introduced in Chapter 2; it takes the construct from the conceptual, to a boots-on-the-ground application to generate integrated thinking, decision-making, and action. This integrated framework is represented as a three-dimensional matrix. The dimensions are (1) Contexts/Drivers (y-axis); (2) Actors/Motive Force (z-axis); and, (3) Risk Management Elements (x-axis). The Risk Matrix is offered for use in impacting EHSS/ORM, such as upgrading existing processes, programs, and systems; upgrading measurement activities such as audit programs and metrics; and strengthening external reporting/disclosure needs. It serves as a foundation for evolving generative organizational fields and escalating organizational impact.

1.7.2.8 Chapter 9 – Matrix in Action

The Risk Matrix represents an environment within which organizational risk management happens – it represents the relational and contextualized space that creates the interactions and collective behavior, that in turn produce the organization's risk management governance, strategy, execution, and outcomes. The Matrix provides a wider multifaceted lens of increased dimensionality through which to view risks. The dimensionality of the Risk Matrix is increased by making explicit the social–human as an ORM Drivers/Contexts, including the organizational field element, called "actors," and identifying them as motive forces – individual, team/department, enterprise, and community. The organizational risk management taxonomy depicted in the Risk Matrix offers a valuable tool to begin to identify the complexity and tight coupling of all relevant risk management factors.

1.7.2.9 Chapter 10 – Escalate Impact

Consideration is given to how ORM outcomes can lead to impacts, and how these can escalate, either in terms of rate of impact, its magnitude, or both. ORM goals and purposes are revisited in terms of both outcomes and impacts, not only for the enterprise/company but also within the other three actors/motive force risk dimension's elements: individual, team/department, and community. The generative field construct is introduced, along with ways to create it. An evolution of the outcome/impact space is offered as a purpose- space where the risk field/matrix fosters integrated decision-making and escalated impact. The assertion is made that value generation is central to what EHSS/ORM professionals have done for decades. Interiority is characterized and linked with engagement and participation. Generative field leverage within the Risk Matrix is associated with the social–human engagement (E7) and leadership (E8) Risk Management Elements. The carriers of meaning phenomenon in field theory has evolved to carriers of a field. Shifting EHSS/ORM assessments (e.g. mainly auditing) to a pedagogical context offers access to transcending historic EHSS/ORM paradigms. It is asserted that EHSS/ORM professionals are carriers of a field committed to improving organizations, and the health and well-being of people, communities, and the environment.

Suggested Reading

Bruch, H. and Vogel, B. (2011). *Fully Charged: How Great Leaders Boost their Organization's Energy and Ignite High Performance*. Boston: Harvard Business Review Press.

Cameron, K.S., Dutton, J.E., and Quinn, R.E. (2003). *Positive Organizational Scholarship, Foundations of a New Discipline*. San Francisco, California: Berrett-Koehler Publishing, Inc.

Capitals Coalition (2019). *Social and human capital protocol*. https: //capitalscoalition. org/capitals-ap- proach/social-human-capital-protocol/ (accessed 14 August 2021).

Capitals Coalition (2021). *Principles of integrated capitals assessments*. https://capitalscoalition.org/wp-content/uploads/2021/01/Principles-of-integrated-capitals-assessments_v362.pdf. (accessed 23 February 2023).

Edmondson, A.C. (2019). *The Fearless Organization – Creating Psychological Safety in the Workplace for Learning, Innovation, and Growth*. Wiley.

Fligstein, N. and McAdam, D. (2012). *A Theory of Fields*. Oxford University Press.

Fritz, R. (1999). *The Path of Least Resistance for Managers – Designing Organizations to Succeed*. San Francisco, CA: Berrett-Koehler Publishers.

Howard-Grenville, J., Lahneman, B., Pek, S. et al. (2020). Organizational culture as a tool for change. *Stanford Social Innovation Review* 2020: 29–33.

Kahneman, D. (2011). *Thinking Fast and Slow*. New York: Farrar, Straus and Giroux.

Keller, S. and Price, C. (2011). *Beyond Performance: How Great Organizations Build Ultimate Competitive Advantage*. New York: Wiley.

Kotter, J. and Rathgeber, H. (2005). *Our Iceberg is Melting – Changing and Succeeding Under any Conditions*. New York: St. Martins Press.

Meadows, D. (1999). *Leverage Points, Places to Intervene in a System*. Hartland, VT: Sustainability Institute.

Meadows, D. (2008). *Thinking in Systems, A Primer* (ed. D. Wright). White River Junction, Vermont: Chelsea Green Publishing.

Scharmer, C.O. (2016). *Theory U – Leading from the Future as it Emerges*. Oakland, California: Berrett-Koehler Publishers.

Scott, W.R. (2014). *Institutions and Organizations, Ideas, Interests and Identities*, 4e. Thousand Oaks, California: Sage Publications, Inc.

Wooten, M. and Hoffman, A. (2017). Organizational fields: past, present and future. In: *The Sage Handbook of Organizational Institutionalism* (ed. R. Greenwood et al.), 55–74. Thousand Oaks, CA: Sage Publications.

Part I

Foundations

2
Risk Logics

CONTENTS
Defining Risk Logics, 39
Core Logic, 39
Premises, 41
Uncertainty, 41
2.1 Contexts, Drivers, Orientations, 42
2.1.1 Evolutions, 43
2.1.2 Corporate Governance – Purpose and Value Creation/Protection, 47
2.1.3 ESG, 51
2.1.4 Social and Human Capital, 52
2.1.5 Culture of Health, 53
2.2 Defining Risk, 54
2.2.1 Risk to Whom?, 54
2.2.2 Definitions, 55
2.2.3 Risk-Reward and Opportunity, 56
2.2.4 Fourth-Generation Risk Management, 57
2.3 Core Concepts, 58
2.3.1 Analysis, Assessment, Communication, and Management, 58
2.3.2 Risk Profile, 59
2.3.3 Owner – Risk Owner, Generator, and Source, 59
2.3.4 Acceptability – Acceptable Risk and Risk Appetite, 60
2.3.5 Tolerance – Risk Tolerance, 60
2.3.6 Transfer – Risk Transfer, 62
2.3.7 Risk-Based Thinking, 62
2.4 Conformity Assessment, 63
2.5 Systems Perspective, 64
2.5.1 Systems Thinking, 65
2.5.2 System Dynamics Iceberg, 66
2.5.3 Deeper Levels, 68
2.6 Risk Field, 68
2.6.1 Field Background, 70
2.6.2 Characterizing an Organizational Field, 71
2.6.3 Operationalization – Risk Field → Risk Matrix, 74
Suggested Reading, 74

Organizational Risk Management: An Integrated Framework for Environmental, Health, Safety, and Sustainability Professionals, and their C-Suites, First Edition. Charles F. Redinger.
© 2025 John Wiley & Sons, Inc. Published 2025 by John Wiley & Sons, Inc.

Being alters action; context shapes thinking and perception. When you fundamentally alter the context, the foundation on which people construct their understanding of the world, actions are altered accordingly. Context sets the stage; being pertains to whether the actor lives the part or merely goes through the motions.[1]

Tracy Goss et al.

It has become an empirical fact that the concept of risk in its raw form has acquired social, political, and organizational significance as never before, and this needs to be explained even if, as it seems likely, that risk itself is an essentially contested concept.[2]

Michael Power (2007)
London School of Economics

The purpose of risk management is the creation and protection of value. It improves performance, encourages innovation, and supports the achievement of objectives.[3]

ISO 31000:2018

Aspects of risk are found throughout the operation and governance of organizations. Organizational risk management's (ORM) historical focus on finance and insurance has evolved to address compliance, operational, strategic, environmental, and reputational risks. The advent of the Chief Risk Officer (CRO) reflects risk's growth as a primary organizational governance theme. Risk issues and risk objects[4] have increased in quantity and complexity, along with tighter coupling and interconnectedness between them that gives rise to paradox and tension in risk decision-making (RDM). Power's observation of risk's "acquired, political, and organizational significance" highlights the context within which environmental, health, safety, and sustainability (EHSS) professionals operate, and one for which their classical training and orientations do not prepare them.

The EHSS function is historically driven by regulatory compliance and the avoidance of events that injure people, property, or create liabilities (e.g. legal,

---

1 Goss, T. et al. (1993). The reinvention roller coaster: risking the present for a powerful future. *Harvard Business Review*. November-December 1993. p. 101. Goss et al. acknowledge being indebted to numerous philosophers, scholars, and thinkers who inquired into the nature of being, especially Werner Erhard, Martin Heidegger, and Ludwig Wittgenstein. p. 108.
2 Power, M. (2007). Organized Uncertainty, Designing a World of Risk Management. Oxford University Press. p. 3.
3 ISO 31000:2018, p. 2.
4 "Risk object" refers to a specific risk, such as debt, a new acquisition, a regulation, or a chemical exposure.

reputational). The focus here broadens this, frames organizational EHSS management within a larger ORM context, and examines how EHSS operates within and in concert with this larger context. Core questions in this inquiry are "how are decisions made" and "by whom?" That is, by individuals, by teams, within teams/departments collectively, or within the larger enterprise. Another question is "how does RDM impact resilience, fulfillment, and conditions to flourish?"

The risk landscape reaches far and wide and is full of paradox and tension. It touches us personally from the seemingly mundane question of what to eat to the complex act of buying a house. It is front and center in public policy decision-making, whether with things like exchange rate issues or regulating greenhouse gases. Many of the risk management decision-making endeavors EHSS professionals deal with involve human health.

Why frame what EHSS professionals do in terms of ORM? The rationale is simple, EHSS happens in organizations. Yes, there are things like regulatory compliance, technologies, and assessment techniques that drive the "doing" of EHSS management in general and risk management more specifically, but these happen within (1) organizational contexts, and (2) ORM structures and thinking.

Defining Risk Logics

The term risk logics is used to reflect ORM's multifaceted nature that includes bundles of individual logic constructs and models. In this book, it is used to describe *the range of approaches, principles, ideas, and contexts that establish the foundation for an organization's risk-related systems, frameworks, decision-making processes, and subsequent actions.*

Core Logic

ORM outcomes are a function of the integrated thinking, decision-making, and actions that happen within systems and frameworks driven by people and teams in an organization. This is reflected in an operationalization of Figure 1.1 as depicted in Figure 2.1. The "people" driver can also be viewed in terms of human and social capitals.[5] These two capitals are bundled with natural capital at numerous junctures in the book.

"Outcome space" is a focal point throughout the book. As ideas progress and build through the chapters, so does this space and how it is characterized. This is particularly the case when the Risk Matrix presented in Chapter 8, is put into

---

5 The term "people" is used as shorthand to refer to the human element of social and human capital.

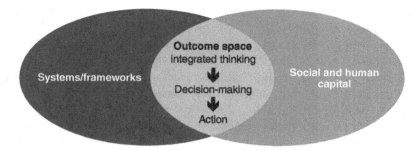

Figure 2.1 Risk management outcomes – core logic.

action in Chapter 9, and again in Chapter 10 when our attention turns to escalating impact and creating generative fields.

In examining the Deepwater Horizon oil rig explosion in the Gulf of Mexico (2010), Robert S. Kaplan and Anette Mikes observed that, "A rules-based risk-management system may work well to align values and control employee behavior, but it is unsuitable for managing risks inherent in a company's strategic choices or the risks posed by major disruptions or changes in the external environment. These types of risks require systems aimed at generating discussion and debate."[6] The limitations of the "rules-based" approach they identify are minimized by focusing on integrated thinking, decision-making, and action that consider drivers beyond regulatory requirements; namely, by considering the impacts and dependencies associated with the capitals bundle. Integrated thinking is a core concept in the evolution of sustainability as seen in things like the One Report[7] concept, the work of the Capitals Coalition, and the International Integrated Reporting Council (IIRC).[8] Definitions of these capitals are as follows:

Human capital[9]
The knowledge, skills, competencies, and attributes embodied in individuals who facilitate the creation of personal, social, and economic well-being.
Social capital
Networks together with shared norms, values, and understanding that facilitate cooperation within and among groups.

---

[6] Kaplan, Robert and Mikes, Anette (2012): "Managing Risks: A New Framework, Smart companies match their approach to the nature of the threats they face." Harvard Business Review, June, 2012.
[7] For example, as expressed in – Eccles and Krazus (2010), *One Report: Integrated Reporting for a Sustainability Strategy*.
[8] IIRC and SASB merged in 2021 to form the Value Reporting Foundation, which subsequently merged with the International Sustainability Standards Board (ISSB).
[9] Capitals Coalition (2019). Social & Human Capital Protocol. p. 11.

***Natural capital*[10]**
The stock of renewable and nonrenewable natural resources (e.g. plants, animals, air, water, soils, and minerals) that combine to yield a flow of benefits to people.

Premises

The key concepts woven throughout the book are:

Managing risks is a journey, not a destination. Risk management is not static, in a silo, or a checklist activity, rather it is a dynamic and iterative balancing of bundles of interconnections and needs within an organization and among its stakeholders.

Risk management is full of paradox and tension. Understanding scope, purpose, and context is as important as understanding the technical mechanics. The Risk Matrix in Chapter 8 aids in distinguishing these.

Risk is not absolute. Risk is relative and context-dependent.

Residual risk is ever present. In EHSS parlance, elimination and substitution may create new risks (transfer) – some degree of risk may remain – transferred elsewhere – even after controls are implemented, i.e. residual risk. All other choices in the hierarchy of controls have residual risk. Whether the transfer or residual risks are acceptable is central to the decision-making process.

Uncertainty

> People live under uncertainty whether they like it or not.[11]
>
> Amos Tversky

Reducing uncertainty is the overriding objective of ORM. Like residual risk considerations, uncertainty is at the core of risk science. Uncertainty refers to "... imperfect knowledge or lack of precise knowledge of the real world, either for specific values of interest or in the description of the system. Although numerous schemes for classifying uncertainty have been proposed, most focus on two broad categories: parameter uncertainty and model uncertainty."[12] In the

---

10 Capitals Coalition (2016). Natural Capital Protocol. p. 2.
11 Lewis, M. (2017). *The Undoing Project*. New York: W. W. Norton & Company, p. 197.
12 Environmental Protection Agency, *Framework for Human Health Risk Assessment to Inform Decision Making*, EPA/100/R-14/001 (2014), p. 34.

COSO enterprise risk management (ERM) and ISO 45001 frameworks, uncertainty is defined as:

> *COSO ERM*: "The state of not knowing how or if potential events may manifest."[13]
>
> *ISO 45001*: "Uncertainty is the state, even partial, of deficiency of information related to, understanding or knowledge of an event, its consequence, or likelihood."[14]

The historic collection of risk management tools, approaches, and ideas can all be framed as endeavors to reduce uncertainty. However, with the increasing complexity of ORM and the tight coupling within it, consideration of new ORM approaches and tools is clearly needed.

A goal of this book – through the use of the Risk Matrix – is to make the invisible visible and to consider the perspectives of a larger number of relevant organizational actors regarding a range of potential risks. That is, to increase the field of vision and that which we see. To do this, one must increase the awareness of risks and the ability to see various systems and their interconnections. One must pay attention to the awareness of orientations and contexts, including ORM drivers, decision-making units in an organization (e.g. individual, team, department, c-suite, and board), and systems/frameworks.

Again, most, if not all, risk-related activities are attempts to eliminate or decrease uncertainty. EHSS/ORM risk-related decision-making has become more challenging as risk profiles have become more complex and tightly coupled, along with increased stakeholder expectations. RDM endeavors are probabilistic in nature. That is,

> There are multiple outcomes, each having varying degrees of certainty or uncertainty of its occurrence. Probabilistic is often taken to be synonymous with stochastic but strictly speaking, stochastic conveys the idea of (actual or apparent) randomness whereas probabilistic is directly related to probabilities and therefore is indirectly associated with randomness. Thus, it might be more accurate to describe a natural event or process as stochastic, and to describe its mathematical analysis (and that of its consequences) as probabilistic.[15]

2.1 Contexts, Drivers, Orientations

There are many dimensions to ORM. Insurance and regulatory compliance considerations have been historical drivers. In one way or another, all ORM issues are linked to finance, money, and an organization's bottom line. As organizational

---

13 COSO ERM (2017), p. 110.
14 *45001 (note 2) to risk definition, 3.20, p. 5.*
15 Environmental Protection Agency (2014). *Framework for Human Health Risk Assessment to Inform Decision Making*, EPA/100/R-14/001, p. 34.

governance systems, structures, strategic thinking, and complexities have evolved, so have ORM risk categories, including operational, strategic, cybersecurity, market, and reputational risks. These are further parsed into areas such as environmental, sustainability (climate change), ESG, occupational health and safety (OHS), supply chain, and diversity, equity, and inclusion (DE&I).

The focus here is on nonfinancial risk management decision-making, without a particular focus on a particular risk, such as regulatory requirements, ESG, DE&I, or OHS. While financial considerations are not ignored, they are not the focus, other than in terms of "value" as a driver in decision-making processes, and with issues related to capitals considerations in decision-making.

In ORM decision-making, identification and understanding of risk-related contexts, drivers, and orientations is essential. These play an important role in defining decision-making scope and boundaries and in the human components (individual and team) of perspective and sensemaking. Given their importance here and throughout the book, let us refresh on their definitions.

Box 2.1   Definitions – Context, Driver, Orientation[16]

Context (noun): "The circumstances that form the setting for an event, statement, or idea, and in terms of which it can be fully understood and assessed."
Driver (noun): "A factor which causes a particular phenomenon to happen or develop."
Orientation (noun): "The determination of the relative position of something or someone (especially oneself). A person's basic attitude, belief, or feelings in relation to a particular subject or issue."

Gaining an appreciation of ORM's contexts, drivers, and orientations – and the risk logics embedded in them – sets a foundation from which we will join together to build integrated thinking, decision-making, and action frameworks and processes to increase resilience and fulfillment in your organization – for the organization itself and the people associated with it. I say "together" to reflect that as ideas, concepts, tools, and the Risk Matrix are presented, I periodically turn your attention to the application of these to your situation(s). And, with that, encourage tailoring these to your needs.

2.1.1   Evolutions

As with any aspect of an area of focus, field of interest, or social practice, there is an evolution of norms, laws, regulations, and ways of responding. Five

---

16 New Oxford American Dictionary (in Apple OS 10.15.7).

EHSS/ORM evolution sequences are identified here. There are certainly any number of ways these can be conceived and organized. How I have done this is based on what I have observed in my organizational, public policy, and standards-development endeavors. Admittedly, I struggled to organize them. The lines of demarcation between the five blurs, there is some redundancy among them, and the progression (evolution) within each is not necessarily linear. The key point I will circle back to throughout the book is the end point of each (e.g. impact, field, capitals, stakeholders, value and purpose). It is this bundle that sets the stage for the final chapter (§10) "Escalate Impact." The five areas are:

1) *EHSS management*: compliance, performance, impact
2) *Frameworks*: process, program, system, field
3) *Sustainability*: ESG, materiality, double materiality, value, capitals
4) *Object/foci*: shareholder, people/workers, stakeholders
5) *ORM*: four generations – insurance, regulatory compliance, consensus standards, and value and purpose

2.1.1.1 EHSS Management: Compliance, Performance, Impact

It is valuable to distinguish EHSS/ORM evolutionary phases and the drivers and orientations in them. Awareness of this helps in understanding – and improving – RDM processes. I have simplified the scope to three orientations and evolutionary phases, which are (1) compliance, (2) performance, and (3) impact. There are distinct processes, drivers, and actions occurring in each of these. Regulatory compliance is the bedrock and a predominant orientation. It is common that basic regulatory compliance is achieved; and from that there is a progression to focus on organizational performance and the use of things like consensus standards, such as ISO management systems, and metrics. Over time, orientations expand and consider an organization's impacts outside the fence line, as demonstrated with a focus on things like sustainability, social responsibility, and ESG. These orientations are depicted in Figure 2.2 and provide a foundation for the risk field framework presented later in this chapter.

RDM is multifaceted with many drivers. In organizational science parlance, decision-making can be characterized as being reactive, proactive, or generative. When operating primarily within a compliance orientation, decision-making, and actions are generally reactive – they are done in response to regulations and regulatory compliance. When operating primarily within a performance orientation, actions are generally proactive; the focus is on meeting organizational objectives, say, through formal management systems. And, when operating primarily within an impact orientation, actions are generally generative, with awareness of, say, externalities and improving the commons.

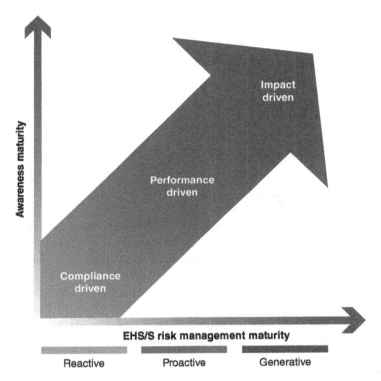

Figure 2.2 Evolution of EHSS/ORM.

2.1.1.2 Frameworks: Process, Program, System, Field

Regulatory compliance is foundational in EHSS management and ORM. Historically, compliance is defined in terms of processes and programs. In the 1980s and 1990s, commonalities between processes and programs began to be viewed in terms of systems thinking. While "systems approaches" could be identified in nongovernmental standards before this, they began to proliferate in the 1990s with the development of ISO 14001 (environmental management), and various OHS management systems, namely Britain's BS 8800. A predominant driver in the management system space has been the market – that is, to meet customer and supply chain demands. Roots of the systems phenomenon can be traced back to the quality assurance arena and the ISO 9000 family of standards.

A central component in this book is framing EHSS management and ORM within a field context and advocating an evolution from a system to a field orientation. And with this, asserting that this evolution offers greater leverage to achieve greater impact. This is depicted in terms of Levine's Lever in Figure 2.3.

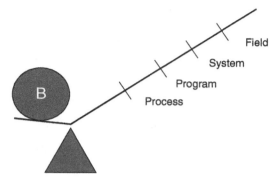

Figure 2.3 Framework leverage.

2.1.1.3 Sustainability: ESG, Materiality, Double Materiality, Value, Capitals

The *sustainability* evolution sequence is an example of the cautionary note I made above about blurred lines and redundancies. The point here is to recognize that over time, a variety of terms and nuances have emerged since sustainability momentum accelerated in the 1980s, with increasing impacts on organizational governance and risk management. Important concepts here, and again teased out as we go along, are ESG, materiality (single and double), and attention to value and capitals. I also include value in the fifth evolution sequence that I have identified.

The elements in this area are addressed at numerous points in the chapters that follow, namely in Chapter 3 (§3.6) where the bundle of terms associated with sustainability and ESG is presented. Corporate social responsibility (CSR) could also be included here and is addressed later.

2.1.1.4 Object/Foci: Shareholder, People/Workers, Stakeholders

The notion of object/foci is rooted in the foundational dynamic in regulations and consensus standards referred to as scope. Consideration of the scope of developing or using a regulation or standard is a key first step. This is fundamental, for instance, in ISO standards. In each of their versions, ISO 9001 is focused on meeting customer specifications; ISO 14001 on environmental aspects; and, ISO 45001 on workers.

In ORM, there has been a historical concentration on meeting shareholders expectations. Over time, this has morphed into a broader conception of stakeholders, which can include workers, customers, communities, and regulators.

I have included this ORM evolution sequence because of the increasing complexity of risks that organizations face. This notion of object or foci of risk outcomes is important to consider in decision-making processes and is covered in depth in Chapter 7 (Decision-Making).

2.1.1.5 Organizational Risk Management: Four Generations – Insurance, Regulatory Compliance, Consensus Standards, and Value and Purpose

Certainly, at a macro-level, there are a number of ways in which risk management's evolution can be characterized. I suggest four eras, or generations, and their foci.

First-generation: financial and insurance
Second-generation: governmental regulations/compliance
Third-generation: nongovernmental standards, and professional practices
Fourth-generation: value generation and preservation, and purpose

The Risk Matrix (Chapter 8) and the Awareness-Based Risk Management™ concept (Chapter 5) represent fourth-generation tools and guidance.

2.1.2 Corporate Governance – Purpose and Value Creation/Protection

Corporate governance is a large topic that is extensively covered in business literature. My purpose here is to highlight three governance-related drivers that are central in organizational decision-making in general, and specifically ORM. These drivers are purpose, shareholder versus stakeholder, and value creation/protection.

For this discussion, the distinction is not made between public and private companies, and the terms "corporate" and "organizational" are used interchangeably. For example, public corporations are subject to a myriad of financial and governance regulations that highlight the distinction between shareholder and stakeholder through what we call corporate governance.

Increased attention has been given to organizational purposes with the growth in ESG reporting, DE&I, and the COVID-19 pandemic in the early 2020s. Embedded here is the question of "purpose to whom?" Social responsibility is a concept that dates back to at least the 1950s and is addressed a bit later in this chapter along with ESG. In 1970, the noble laureate economist Milton Freedman penned an article in the *New York Times Magazine* that is mentioned often in discussions on organizational purpose. He framed this (purpose) in terms of social responsibility, which was a prominent topic at that time. Freedman essentially asserted that a business's purpose is to provide profits to its stockholders. He stated,

> There is one and only one social responsibility [purpose] of business – to use its resources and engage in activities designed to increase its profits so long as it stays within the rules of the game, which is to say, engages in open and free competition without deception or fraud.[17]

---

17 Freedman, M. (1970). The Social Responsibility of Business is to Increase its Profits. *New York Times Magazine* (13 September), p. 122.

The bundle of ideas that Freedman offers in this seminal article had become central in ESG debates as this book was going to press, and are addressed further in Chapter 9 (Matrix in Action) and Chapter 10 (Escalate Impact).

Harvard professors Joseph Bower and Lynn Paine offer an alternative to Freedman's shareholder-centered governance model, which they call a "company-centered" model, based on entity theory as opposed to agency theory. They state,

> A better model (than agency theory), we submit, would have at its core the *health of the enterprise* rather than the near-term returns to its shareholders. Such a model would start by recognizing that corporations are independent entities endowed by law with the potential for indefinite life. With the right leadership, they can be managed to serve markets and society over long periods of time. Agency theory largely ignores these distinctive and socially valuable features of the corporation, and the associated challenges of managing for the long term, on the grounds that corporations are 'legal fictions.'[18]

The notion of "health of the enterprise" is visited and expanded throughout this book in terms of "organizational health" and "culture of health."

Bower and Lynn offer eight propositions that provide a "...radically different, and [they believe], more realistic foundation for corporate governance and shareholder engagement." These are:[19]

1) "Corporations are complex organizations whose effective functioning depends on talented leaders and managers.
2) Corporations can prosper over the long term only if they are able to learn, adapt, and regularly transform themselves.
3) Corporations perform many functions in society.
4) Corporations have differing objectives and differing strategies for achieving them.
5) Corporations must create value for multiple constituencies.
6) Corporations must have ethical standards to guide interactions with all their constituencies, including shareholders and society at large.
7) Corporations are embedded in a political and socioeconomic system whose health is vital to their sustainability.
8) The interests of the corporation are distinct from the interests of any particular shareholder or constituency group."

---

18 Bower, J. and Paine, L. (2017). The error at the heart of corporate leadership – most CEOS and boards believe their main duty is to maximize shareholder value – it's not. *Harvard Business Review*, p. 57.
19 Bower, J. and Paine, L. (2017), pp. 57–59.

2.1 Contexts, Drivers, Orientations | 49

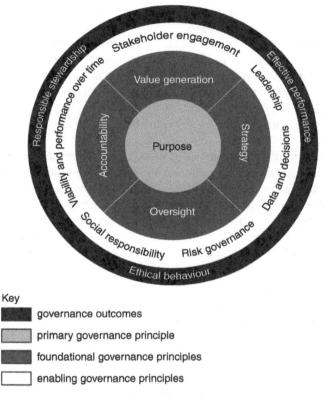

Figure 2.4 Governance of organizations – principles and outcomes. *Source:* ISO 37000:2021/ with permission of International Organization for Standardization.

Expanded notions of organizational purpose proliferated in the 1980s and 1990s, and into the 2000s. In 2021, ISO published a guidance document on "Governance of Organizations," in which it defines organizational purpose as an "organization's meaningful reason to exist."[20] Numerous "purpose" examples are offered by ISO, along with suggested governance principles and outcomes (Figure 2.4)[21]; 11 "principles of governance" are offered in ISO 37000:2021 and depicted in this figure. These are purpose (6.1), value generation (6.2), strategy (6.3), oversite (6.4), accountability (6.5), stakeholder engagement (6.6), leadership (6.7), data and decisions (6.8), risk governance (6.9), social responsibility (6.10), and viability of performance over time (6.11).

---

20 ISO 37000:2021. *Governance of Organizations – Guidance*. Definition 3.2.10, p. 4. Geneva, Switzerland.
21 ISO 37000:2021. *Governance of Organizations – Guidance*, p.10. Geneva, Switzerland.

ISO 37000:2021's first principle of governance – purpose (6.1) – states,

> The [organization's] governing body should ensure that the organization's reason for existence is clearly defined as an organizational purpose. This organizational purpose should define the organization's intentions towards the natural environment, society, and the organization's stakeholders. The governing body should also ensure that an associated set of organizational values is clearly defined.[22]

Numerous entities – standard developers (e.g. ISO), academics, and business associations (e.g. Business Roundtable) – have developed ideas and guidance on the topic of organizational purpose. One of these is The Enabling Purpose Initiative (EPI),[23] which considers purpose as a strategy as opposed to purpose as culture. Considering purpose as a strategy provides the foundation for organizational decision-making. It is suggested that "...there is confusion and disagreement about what purpose is and how it differs from values, mission, and vision." EPI offers clarity on this as follows:

- Purpose articulates why an organization exists.
- Values articulate how the organization behaves.
- Mission sets out what the organization does.
- Vision describes where the organization intends to have impact.[24]

Evolving notions of organizational purpose increase attention on "to whom" questions. Freedman, Bower, and Lynn touch on this in terms of responsibility/purpose to whom and value creation to whom? Later in this chapter, I pose the question, risk to whom? Shifting from a solely stock/shareholder perspective to a broader stakeholder perspective is an important consideration when navigating in the social–human era.

When upgrading ORM – evolving from systems to fields – it is important to consider who your stakeholders are. ISO 37000 offers a stakeholder definition as a "person or *organization* that can affect, be affected by, or perceive itself to be affected by a decision or activity."[25] Stakeholder examples include customers, regulators, suppliers, and employees. In addition, two types of stakeholder categories

---

22 ISO 37000:2021. *Governance of Organizations – Guidance*, p. 13. Geneva, Switzerland.
23 EPI is a multi-institution partnership between the University of Oxford, the University of California Berkeley, BCG BrightHouse, EOS at Federated Hermes, and the British Academy. www.enactingpurpose.org.
24 Enabling Purpose Initiative (2020). Enacting Purpose within the Modern Corporation: A Framework for Directors. pp. 4–5. https://enactingpurpose.org/assets/enacting-purpose-initiative---eu-report-august-2020.pdf (accessed 10 January 2022).
25 ISO 37000:2021, p. 5.

are defined as "member stakeholder" and "reference stakeholder." Member stakeholder is defined as a *"stakeholder* who has a legal obligation or defined right to make decisions in relation to the *governing body* and to whom the governing body is to account." And, reference stakeholder is defined as a *"stakeholder* to whom the *governing body* has decided to account to when making decisions pertaining to the *organizational purpose."* Examples of reference stakeholders include "scientific advisory board to a research organization, parents of the pupils in a school, and community advisory boards for companies."[26]

Risk management's role in value creation and protection is the senior principle presented in ISO 31000, as well as COSO's ERM framework. These frameworks are addressed in Chapter 3 and inform the application of the Risk Matrix (Chapter 8) addressed in Chapters 9 and 10. Value and valuation are also central to the Capitals Coalition protocols that are referenced throughout the book.

ISO 37000:2021's second principle of governance – value generation (6.2) – states, "The governing body should define the organization's value generation objectives such that they fulfil the organizational purpose in accordance with the organizational values and the natural environment, social and economic context within which it operates."[27] Rationale (6.2.2) is offered to support this governance principle.

> The focus for all organizations should be to fulfil their organizational purpose by generating value over time. To achieve this, organizations need to generate value which represents something of worth to its stakeholders. The ultimate value an organization is trying to generate (articulated in the organizational purpose) can only be achieved through collaboration with stakeholders. Appropriate value needs to be generated for stakeholders so that they are willing and able, to support the organization in fulfilling its organizational purpose over time. The value stakeholders expect can take different forms and can impact the natural environment and society, as well as the stakeholders themselves. The governing body's function in this value generation includes stewardship – to ensure the organization not only creates but also protects value over time.[28]

2.1.3 ESG

ESG is addressed in greater length in Chapter 3 (Frameworks), and in later chapters that address the application of the Risk Matrix in organizations. Addressing it here in this chapter as a part of risk logics is done because of its far-reaching nature,

---

26 ISO 37000:2021, p. 5.
27 ISO 37000:2021, p. 15.
28 ISO 37000:2021, p. 15.

and impact on organizations. The "ESG space" is fluid with numerous dimensions and needs to be considered in ORM activities. The Risk Matrix is designed with an eye toward navigating in this space, namely, offering a tool for the integrated thinking, decision-making, and action suggested – if not required – in ESG approaches. To use "required" is tricky and may not be the best term. A question is, "required by whom?" The nuances in this question are addressed in later chapters.

While ESG has roots in numerous areas, such as social responsibility, triple bottom line (TBL), and sustainability in general, its formalization is pegged to a 2004 United Nations-led initiative to "better integrate environmental, social, and corporate governance issues in asset management, securities brokerage services, and associated research functions."[29]

ESG's historical focus has been on financial asset decision-making, specifically where investments that were aligned with sustainability and TBL goals could be placed. While its finance orientation is still dominant, there has been increased pressure to impact corporate governance through ESG; the term "stakeholder capitalism" is associated with this. There has been increasing emphasis on social and human capital (HC) within this context. There is a wide range of nongovernmental organizations (NGOs), financial advisories, and third-party rating agencies that have prepared HC/ESG frameworks and reporting schemes.

It is important to gain awareness of and understand the tensions that have evolved in the ESG space from the increased emphasis on human and social capital.

It is often asked, "what is the difference between 'sustainability' and 'ESG'?" In some ways, it depends on one's perspective. There is a general consensus that "sustainability," usually thought of in terms of climate change, is senior, and that ESG is a subset of it.

2.1.4 Social and Human Capital

Considering a social–human dimension of ORM is not new. Historically, the domain of human resources, organizational development functions, and public relations – "social" aspects of organizational activities have not been prominent in risk considerations. Numerous points in time can be identified where this began to change. In the United States, significant laws and regulations began to appear in the early 1970s, referred to as an era of "social regulations." These include the National Environmental Policy Act and the Occupational Safety and Health Act

---

29 The Global Compact (2004). *Who Cares Wins: Connecting Financial Markets to a Changing World*. Recommendations by the financial industry to better integrate environmental, social, and governance issues in analysis, assessment management, and securities brokerage. Swiss Federal Department of Foreign Affairs (Bern) and the United Nations (New York). https://www.unepfi.org/fileadmin/events/2004/stocks/who_cares_wins_global_compact_2004.pdf.

2.1 Contexts, Drivers, Orientations | 53

that respectively established the EPA and OSHA. Within the sustainability space that expanded dramatically in the 1980s, ideas, practices, and norms for CSR began to emerge. In 1980s, the TBL emerged as well, which reflected a shift from traditional accounting and reporting practices to ones that considered ESG.

A social dimension of ORM has emerged that will have an ongoing impact on how organizations characterize their risk profiles, strengthen resilience, and define objectives. Many norms were skewed by the pandemic, the resulting economic fallout, a host of social issues, and political tumult. Facets of public, community, and personal health were amplified as not seen in generations, including socioeconomic issues related to health. The confluence of the pandemic's socioeconomic health issues and social unrest pushed DE&I forward in social attention. Adding to this complexity is a political dimension that many attribute to diminished civility and trust. In the ESG space, events of 2020 have driven and validated increased emphasis on the "S."

This is an issue you want to stay ahead of. A common metaphor used during the development of ISO management system standards that addressed social issues in the 1990s (e.g. 14000 – environment and 45001 – OHS[30]) was "the train has left the station." I suggest that this is apropos here. When I was in the final stages of finishing this book, I was participating in several standard development activities in the human capital and ESG spaces. A common observation from participants, including ones with large multinational companies was, "this is happening." That is, social–human issues/dynamics that coalesced in 2020 are on their radars and will be addressed well into the next decade.

2.1.5 Culture of Health

This topic was introduced in Chapter 1 and is addressed in greater depth in the chapters that follow. It is included here as part of risk logics for forward-looking purposes. This is not a historic topic in ORM thinking, at least to any significant degree. I suggest this is a topic that deserves attention as ORM evolves. Aspects of this can be valuable for navigating in fourth-generation ORM, and in developing a generative risk field.

There are three impacts that a culture of health has in an organization. The first is related to its ability to impact workplace efficiency and productivity. The RWJF/GRI framework research offers robust evidence for this. The second is that "health" – of the organization and the people associated with it – can serve as a foundation for organizational purpose, and in applying the Risk Matrix to your organization. Third, health's role in the GRI framework will support ESG

---

30 ISO 45001 was published in 2018. Initial consideration of an ISO OHSMS began in the mid-1990s around the time ISO 14001 was first published.

reporting efforts, in particular those related to human capital management reports – for instance to meet Security and Exchange Commission (SEC) requirements in the United States.

2.2 Defining Risk

> Risk does not exist out there, independent of our minds and culture, waiting to be measured. Human beings have invented the concept of risk to help them understand and cope with the dangers and uncertainties of life. Although these dangers are real, there is no such thing as real risk or objective risk.[31]
>
> Paul Slovic, University of Oregon

Risk is not monolithic – it is fluid and context dependent. Paul Slovic, a luminary[32] in the risk analysis and decision-making field, has pioneered numerous fundamental ideas and methods in a wide range of risk science issues. He succinctly points to the importance of having awareness of one's (and one's company's) orientation to risk. This chapter's opening quote from Michael Power underscores this idea. For EHSS/ORM professionals Slovic's assertion is a head-turner. Key objectives of this book are to encourage you to think about risk contexts and orientations in your decision-making process, and provide tips for integrated thinking that will help you navigate the inherent tensions and paradoxes when there are multiple contexts in play.

2.2.1 Risk to Whom?

RDM is rife with complexity, tension, and paradox, which arise from the values-dependent, and subjective nature suggested in Slovic's quote. Tension is present when numerous values are in play – for example, consider the often competing and potentially adversarial interests of labor and management or plaintiff and defendant or producer and consumer. Paradox is present when something appears to be other than what it is. Compliance with regulations and consensus standards can introduce a paradox because compliance/conformance does not necessarily mean risks have been eliminated. Central here is the question, "to whom or what does something pose a risk?" Is it the organization, the worker, the community, the environment, or endangered salamanders? Depending upon the context and the "who or what," a risk may be acceptable or not. There is always some level of risk unless a decimal limitation is set; computer calculations by design cannot yield a result of "zero risk."

---

31 Kahneman, D. (2011). *Thinking Fast and Slow*, 141. New York: Farrar, Straus and Giroux.
32 Professor Slovic has been a leader in the development of decision theory. His work has been seminal in the field's evolution.

2.2 Defining Risk | 55

This question – to whom does something pose a risk? – is at the heart of the integrated RDM processes addressed in later chapters. And, embedded in this question – touched on above – is "value to whom?" as well as "purpose to whom?"

2.2.2 Definitions

Notions of risk have evolved for centuries. Aspects of risk can be identified in all corners of society. Here, the focus is on organizational risk. Examples of risk-related distinctions can be found in education, religion, public health, and national security. In public health fields, risk is equated with the severity and probabilities associated with hazards. More recently, ISO standards and guidelines, and COSO's ERM framework, offer several definitions related to organizational risk, which is commonly framed in terms of uncertainty and objectives. For example, in ISO 31000, risk is defined as the "effect of uncertainty on objectives."[33] Within ISO management system standards, risk is defined in Annex SL, as[34] "effect of uncertainty." "Notes" to this definition state, "An effect is a deviation from the expected – positive or negative. Uncertainty is the state, even partial, of deficiency of information related to, understanding or knowledge of, an event, its consequence, or likelihood." And "Risk is often characterized by reference to potential *'events'* (as defined in ISO Guide 73:2009, 3.5.1.3) and *'consequences'* (as defined in ISO Guide 73:2009, 3.6.1.3), or a combination of these."

A note is included that points to a risk definition common to OHS, and public health professionals as well. That is, "Risk is often expressed in terms of a combination of the consequences of an event (including changes in circumstances) and the associated *'likelihood'* (as defined in ISO Guide 73:2009, 3.6.1.1) of occurrence."

Given its ubiquitous and context dependent nature, it is important to understand the nuanced differences in how risk is defined, and who is affected by the array of contexts in which you operate and make decisions. This may sound obvious, and in some ways it is. However, especially in the EHSS and the human health spaces, notions of risk are not the same as in the ORM space. With human health and EHSS management, risk is commonly defined in terms of the likelihood and severity that a hazard presents to a person or group of people. The key here is distinguishing between hazard and risk, where hazards are precursors to risks; hazards can exist that pose little or no risk.

ISO 45001:2018 includes a definition specific to OHS risk that states that this is the "combination of the likelihood of occurrence of a work-related hazardous event(s) or exposure(s) and the severity of *injury and ill health* (3.18) that can be

---

33 ISO 31000 (2018), p. 1.
34 Annex SL (2013), p. 138.

caused by the event(s) or exposure(s)."[35] This standard defines *injury and ill health* as "adverse effect on the physical, mental, or cognitive condition of a person." This definition includes two "notes" that state, "These adverse effects include occupational disease, illness, and death; [and] the term 'injury and ill health' implies the presence of injury or ill health, either on their own or in combination."[36]

In the COSO ERM framework, risk is defined as

> The possibility that events will occur and affect the achievement of strategy and business objectives. Note: 'Risks' (plural) refers to one or more potential events that may affect the achievement of objectives. 'Risk' (singular) refers to all potential events collectively that may affect the achievement of objectives.[37]

More broadly, in the social sciences, the notion of "risk society" was introduced by sociologists Ulrich Beck and Anthony Giddens in the 1980s. This concept refers to the ways society responds to risk, more specifically the way society responds to what they refer to as "modernity." Giddens defines a risk society as, "a society increasingly preoccupied with the future (and also with safety), which generates the notion of risk."[38] And Beck defines it as, "a systematic way of dealing with hazards and insecurities induced and introduced by modernization itself."[39]

Throughout the book, I refer to this range of risk definitions and their contexts. The Beck/Giddens thread offers insights when considering integrated decision-making issues addressed in Chapters 7 and 9.

2.2.3 Risk-Reward and Opportunity

The idea and practice of coupling "risk and opportunity" can be an anathema to public health professionals. When I served on the US TAG to ISO 31000, there were colleagues who pushed back on this coupling even though it had become common in the business community. I have also seen colleagues challenged by ISO's definition of risk – "The effect of uncertainty on objectives" – which in a generic sense, suggests that managed risk can be welcome, as in insurance with a deductible, or finance where a loan is collateralized, or even when taking on a personal risky activity that might endanger one's health.

---

35 ISO 45001:2018, clause 3.21, p. 5.
36 ISO 45001:2018, clause 3:18, p. 4.
37 COSO (2017), p. 110.
38 Giddens, A. and Pierson, C. (1998). *Conversations with Anthony Giddens, Making Sense of Modernity*. Stanford University Press, p. 209.
39 Beck, U. (1986). *Risk Society: Toward a New Modernity*, p. 21. Sage Publications, Inc.

Early ISO management approaches focused only on risk (if not by name, by idea, or context) identification, assessment, control, and mitigation; requirements to consider opportunities for improvement were not explicitly addressed. Consideration of opportunities began to be mandated, as seen in ISO's high-level MSS, and, as such, in ISO 45001:2018 as "OH&S opportunities." ISO's risk management activities (ISO 31000:2009) supported an expanded view when considering opportunities, as it states (§5.4.2) "It is important to identify the risks associated with not pursuing an opportunity."[40]

ISO 45001:2018 defines OH&S opportunity (3.22) as "circumstance or set of circumstances that can lead to improvement of *OH&S performance,*"[41] and, in its Annex (A.6.1.1), it presents a robust bundle of opportunities to improve OH&S performance.

From an integrated thinking, decision-making, and action perspective, the inclusion of risk-related opportunities into the ORM vernacular is a significant evolution as it nudges EHSS/ORM professionals to think more broadly than simply in terms of hazards, risks, and their immediate controls.

2.2.4 Fourth-Generation Risk Management[42]

New vocabulary and concepts that allow new ways of thinking are abundant in the social arenas of sustainability, corporate citizenship, and social responsibility. Examples of words and distinctions that are impacting organizations include transparency, license-to-operate, health and wellness. Words and distinctions shown to be important to millennials are collaboration, civic-minded, diversity, global, and green.

Reframing the "objectives" and "uncertainty" as stated in ISO's definition of risk, in terms of "what matters and is important," and "paradox," is key to understanding the evolution to fourth-generation risk management, Awareness-Based Risk Management™, and the Risk Matrix. This reframing increases engagement and vitality – it fundamentally shifts the relationship to risk, for individuals and teams. ISO's risk definition can be restated as:

> *Effect of tension and paradox on what matters and is important.*

At the International Occupational Hygiene Association (IOHA) meeting in London in 2015, Dr. Allister Fraser, who was at that time the Vice-President of

---

40 ISO 31000:2009, p. 17.
41 ISO 45001:2018, p. 5.
42 *Fourth-generation risk management* is addressed in greater depth in Chapter 5. I highlight here aspects of it that relate to thinking about and defining risk.

Health at Royal Dutch Shell, offered in his plenary speech a novel definition of risk that provided the foundation for Shell's worker health and safety programs. This definition incorporated engagement, stating:

$$\text{Risk} = (\text{Hazard}) \times (\text{Exposure} / \text{Engagement})$$

2.3 Core Concepts

Risk management is often parsed into four categories: analysis, assessment, communication, and management. There is variation among organizations regarding high-level risk topographies, such as defining the term to describe risk-related activities. For instance, is risk analysis the senior label to describe the field, or, say, is it risk management? These nuances are not important to dive into here, but rather to acknowledge that there are variations and to understand the sequence and relationships between these risk categories.

2.3.1 Analysis, Assessment, Communication, and Management

Brief introductory definitions are provided here to help with understanding risk logics. The Society for Risk Analysis (SRA) is a preeminent multidisciplinary international society that provides a forum for risk analysis scholars, practitioners, and policymakers. SRA states that "risk analysis is broadly defined to include risk assessment, risk characterization, risk communication, risk management, and policy relating to risk, in the context of concerns to individuals, to public and private sector organizations, and to society at a local, regional, national, or global level."[43] This society has published numerous valuable documents, one of which is a glossary of terms. For our purpose here, four definitions to help us along the way are:[44]

Risk analysis: "Systematic process to comprehend the nature of risk and to express the risk, with the available knowledge."
Risk assessment: "Systematic process to comprehend the nature of risk, express and evaluate risk, with the available knowledge."
Risk communication: "Exchange or sharing of risk-related data, information, and knowledge between and among different target groups (such as regulators, stakeholders, consumers, media, general public)."

---

43 https://www.sra.org/about-sra/ (accessed 23 August 2022).
44 SRA (2018). *Society for Risk Analysis Glossary*, 8. Society for Risk Analysis.

Risk management: "Activities to handle risk such as prevention, mitigation, adaptation, or sharing. It often includes trade-offs between costs and benefits of risk reduction and choice of a level of tolerable risk."

2.3.2 Risk Profile

This term is used to characterize the entirety of an organization's risk inventory. COSO defines this as "a composite view of the risk assumed at a particular level of the entity, or aspect of the business that positions management to consider the types, severity, and interdependencies of risks, and how they may affect performance relative to the strategy and business objectives."[45]

EHSS professionals are familiar with concepts embedded in risk profiles and risk inventories. It is common, if not essential, to have a process for maintaining a risk inventory and assessing the EHSS risk profiles. The inventory and profile processes are formalized in EHSS management systems. With chemicals, for instance, it is common to maintain a list (inventory), as well as characterize individual and HEG/SEG (homogeneous/similar exposure group) exposure profiles.

With an expanded risk context, EHSS professionals are asked to consider risks associated with things like supply chains; work with strategic partners, neighbors, and community; reputation; fraud; and business continuity. With their experience in risk thinking, EHSS professionals can provide guidance and leadership to those responsible for the organization-wide risk management process.

2.3.3 Owner – Risk Owner, Generator, and Source

Embedded in identifying risk inventories and profiles is a risk's source or generator. For example, chemical processes, regulations, workers, policies, audits, and community groups can generate risks. It is also necessary to gain awareness of a risk's owner; that is, within an organization, identifying the person(s) who is/are accountable for managing individual risks in a risk profile. Risk source is defined in ISO 31000:2018 as an "element which alone or in combination has the potential to give rise to risk."[46]

Maintaining awareness of these terms is important in the RDM process, especially with integrated thinking, decision-making, and action. A topic that will be

---

45 COSO (2017), p. 110.
46 31000: 2018, p. 1.

introduced later is integrated capitals assessment, and with such assessments, identifying these items is central. This is also a precursor to and helps with, root cause analysis in addressing deviations found in audits.

2.3.4 Acceptability – Acceptable Risk and Risk Appetite

Notions of acceptability vary. In many cases related to EHSS management, there are governmentally prescribed limits, such as with wastewater effluents, air emissions, and occupational exposure limits that provide "acceptability" guidance, or at least regulatory guidance. In the National Resource Council's (NRC) 1983 Red Book, these are referred to as inference guidelines.[47] These guidelines are discussed in greater depth in Chapter 3 (Frameworks) and Chapter 7 (Decision-Making).

In ORM parlance, the term risk appetite is more commonly used versus the term acceptable risk, which is more common in public health circles. SRA defines risk appetite as "Amount and type of risk an organization is willing to take on risky activities in pursuit of values or interests."[48] And, COSO provides a similar definition as "The types and amount of risk, on a board level, an organization is willing to accept in pursuit of value."[49] COSO's ERM document (2017) provides guidance on determining ORM risk appetite(s). It states, "There is no standard or 'right' appetite that applies to all entities." This has to be determined on an entity-by-entity basis. The Risk Matrix (Chapter 8), and its operationalization (Chapter 9) offer guidance in determining appetite, especially when considering social–human drivers.

2.3.5 Tolerance – Risk Tolerance

Risk tolerance in ORM is historically viewed from an organizational perspective. With the increased prominence of social–human perspectives, there is increased consideration more broadly. A challenge addressed in later chapters with integrated thinking, decision-making, and action is how to "integrate" different notions of acceptable risk and tolerance held among individuals, teams/departments, the enterprise as a whole, and at the community level.

Definitions of tolerate, tolerance, risk tolerance, and risk capacity are offered in Box 2.2.

---

47 Red Book.
48 SRA glossary, p. 8.
49 COSO, p. 110.

2.3 Core Concepts | 61

> **Box 2.2 Definitions: Tolerate, Tolerance, Risk Tolerance, and Risk Capacity**
>
> *Tolerate*: "Allow the existence, occurrence, or practice of (something that one does not necessarily like or agree with) without interference. Accept or endure (someone or something unpleasant or disliked) with forbearance. Be capable of continued subjection to (a drub, toxin, or environmental condition) without adverse reaction."[50]
>
> *Tolerance*: "The boundaries of acceptable variation in performance related to achieving business objectives."[51]
>
> *Risk tolerance*: "An attitude expressing that the risk is judged tolerable."[52]
>
> *Risk capacity*: "The maximum amount of risk that an entity is able to absorb in the pursuit of strategy and business objectives."[53]

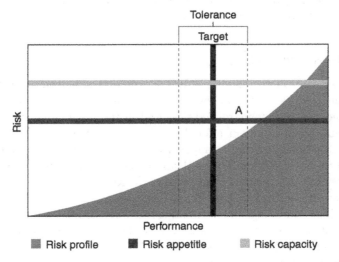

Figure 2.5 Relationship between risk profile, risk appetite, and risk capacity. *Source:* Adapted from COSO.

COSO's ERM framework offers a valuable diagram that depicts the relationship between risk profile, risk appetite, and risk capacity.[54] These relationships are examined in Chapter 7 (Decision-Making), and in their application in the Risk Matrix, in Chapters 8, 9, and 10 (Figure 2.5).

---

50 Apple OS dictionary. New Oxford American Dictionary.
51 COSO, p. 110.
52 SRA glossary, p. 9.
53 COSO, p. 110.
54 COSO, p. 62 (Figure 7.5).

2.3.6 Transfer – Risk Transfer

Risk transfer is a risk management and control strategy that shifts risk from a risk-averse party to a risk-tolerant party. Formal risk transfer is often done contractually such as through insurance. However, risk transfer has also occurred informally through globalization with resulting impact on markets, workers, consumers, and the environment. Related to the concept of risk transfer is the economic notion of externalities, that is, the positive or negative consequences of an economic activity experienced by third parties. SRA defines risk transfer as "Sharing with another party the benefit of gain, or burden of loss, from the risk. Passing a risk to another party."[55]

Historically, EHSS professionals have not taken risk transfer, whether from one person to another, to the environment, or elsewhere, into account. The increasing complexity of risk profiles and portfolios suggests that transferring risk is important to understand in the context of RDM.

2.3.7 Risk-Based Thinking

In some ways, this is a tricky topic. It is a topic that gained steam in ISO with that organization's push into the risk management arena with its array of management system standards. While sound, critics point to what is addressed in the Paul Slovic comment offered earlier in this chapter. Namely, "risk to whom?" Critics suggest that a danger with this concept is that it is historically framed in terms of risk to the organization (e.g. the materiality issue in ESG). I go deeper into this issue when addressing the Risk Matrix (Chapter 8) and its application (Chapter 9).

Identifying, controlling, and eliminating risk when possible has been a central EHSS activity from the field's earliest days. The term "risk-based thinking" began to appear in ISO MSS activities, both as a concept and a performance criterion, in some standards, as the high-level MSS was applied to specific areas, such as quality (ISO 9001) and environment (ISO 14001).

ISO 14001:2015 uses the term "risk-based thinking" in the introduction, embedded in 0.5, where it states, "This International Standard does not include requirements specific to other management systems, such as those for quality, OHS, energy, or financial management. However, this International Standard enables an organization to use a common approach and *risk-based* thinking to integrate its environmental management system with the requirements of other

---

55 SRA, p. 9.

management systems."[56] ISO 9001:2015 states in its introduction (0.3.3), "Risk-based thinking (see Clause A.4) is essential for achieving an effective quality management system. The concept of risk-based thinking has been implicit in previous editions of this International Standard including, for example, carrying out preventive action to eliminate potential nonconformities, analyzing any nonconformities that do occur, and taking action to prevent recurrence that is appropriate for the effects of the nonconformity."[57]

While the term "risk-based thinking" is not used in ISO 45001:2018, the concept is implicit throughout the standard, in much greater detail than earlier OHSMS standards (e.g. OHSAS 18001:2007). For instance, risk management and analysis are key elements included throughout the standard, with the main requirements outlined in Clause 6.1 – actions to address risks and opportunities. This clause addresses the requirement that an effective OHSMS must assess not only issues that may present risks, which the standard defines as the effects of uncertainty, but also opportunities, defined as circumstances that can lead to improvement. OH&S risk is further defined as a combination of the likelihood of occurrence and the potential severity of the event. These risks are often represented by a matrix.

2.4 Conformity Assessment

Conformity assessment is an important topic in ORM and is embedded throughout the risk logics addressed here. The primary activity in the conformity assessment bundle is auditing, and measurement in general, at a more granular level. Beyond auditing and measurement, in common lay terms, conformity assessment can be thought of as "who is the judge?" in determining acceptable performance. In regulatory settings, it is governmental agencies, in determining conformance with consensus standards, it is third-party registrars, and in ESG, it is investors. The National Research Council (NRC) defines conformity assessment as:

> The determination of whether a product or process conforms to particular standards or specifications. Activities associated with conformity assessment include testing, certification, and quality assurance system registration.[58]

---

56 International Organization for Standardization (2015). *Environmental Management Systems – Requirements with Guidance for Use, ISO 14001*, viii. Geneva, Switzerland.
57 International Organization for Standardization (2015). *Quality management systems – Requirements, ISO 9001*, ix. Geneva, Switzerland.
58 National Research Council (1995), p. 206.

Activities and distinctions associated with conformity assessment are certification; auditing; first, second, and third parties; registrar; and accreditation. Conformity assessment frameworks commonly have three levels:

- Primary level – Assessment (e.g. auditing)
- Secondary level – Accreditation
- Tertiary level – Recognition

The primary level represents measurement and auditing activities. Stack testing, soil sampling, perception surveys, workplace air sampling, and safety surveys, are examples of assessment activities, as are management system audits. The secondary level addresses the formal qualifications of the entities performing primary level activities and the bodies that provide confirmation of these qualifications. The tertiary level addresses either the formal codification of auditor/assessor qualifications – and the entity that provides them – or less formally, when recognition is provided by the marketplace.

Conformity assessment elements that are embedded in ORM include:

1) Existence of a standard against which assessments are made.
2) A way to perform the assessment using an agreed-upon measurement method (e.g. validated tools, and protocols).
3) An accreditation mechanism whereby first, second, and third parties can be certified to perform assessments.
4) A QA/QC mechanism whereby assessor performance is evaluated and modified as needed – a means by which affected parties can register complaints.

These topics are covered in greater depth in Chapters 4 (Conformity Assessment) and 7 (Decision-Making).

2.5 Systems Perspective

Decision-making is at the core of ORM and is addressed throughout the book. Chapter 7 is devoted to this topic, and numerous aspects of RDM processes are addressed, within which a systems perspective is highlighted. In that chapter, and the ones that follow it, I frame RDM in a systems context. I briefly start this thread here to reinforce its role as a risk logic. That is, the logic of your RDM process generally, and more specifically, the bundle of decision-making processes within your risk management endeavors.

I offer several systems-perspective anchors as a prequel, along with four questions that we'll explore in Chapter 7:

1) How aware are you and your team of your RDM process(es)?
2) Is this something you and your team think or talk about?

3) To what extent, if at all, does your process reflect integrated thinking?
4) How would you describe or characterize the process if asked?

2.5.1 Systems Thinking

Seminal work on the development of systems thinking originated within Jay Forrester's System Dynamics group at MIT. Daniel Kim, among others associated with the group, has been a leader in developing systems thinking and organizational learning ideas and concepts. Kim offers broadly that:

> Systems thinking is a way of seeing and talking about reality that helps us better understand and work with systems to influence the quality of our lives. In this sense, systems thinking can be seen as a perspective. It also involves a unique vocabulary for describing systemic behavior, and so can be thought of as a language as well. And, because it offers a range of techniques and devices for visually capturing and communicating about systems, it is a set of tools.[59]

Thinking from a systems perspective has roots in numerous disciplines, and is commonly found in such fields as operations research, system dynamics, and ecology, for example. A system can be defined as:

> Any group of interacting, interrelated, or interdependent parts that form a complex and unified whole that has a specific purpose; and is almost always within a larger system.[60]

In some disciplines, parts or elements of systems are identified as inputs, processes, outputs, and feedback, and their whole is characterized as open or closed. Most systems familiar to EHSS and ORM professionals are open systems, that is, where the system interacts with its external environment. For instance, this is the case with an EHSS/ORM program or department which, of course, interacts throughout an organization and its external environment.

Systems thinking is a critical skill to develop in order to create a learning organization. In fact, it is the "fifth discipline" referred to in the landmark book *The Fifth Discipline* by Peter Senge. He identifies an ensemble of five disciplines – systems thinking, personal mastery, mental models, building shared values, and team learning – where each provides a vital dimension in learning organizations, "It is this discipline [systems thinking] that integrates the disciplines,

---

59 Kim, D.H. (1999). *Introduction to Systems Thinking*, p. 2. Waltham, MA: Pegasus Communications.
60 Kim, D.H. (1999). *Introduction to Systems Thinking*, p. 2. Waltham, MA: Pegasus Communications.

fusing them into a coherent body of theory and practice."[61] Senge links systems thinking with the learning organization. He states

> Systems thinking makes understandable the subtlest aspect of the learning organization – the new way individuals perceive themselves and their world. At the heart of a learning organization is a shift in mind – from seeing ourselves as separate from the world to connected to the world, from seeing problems as caused by someone or something 'out there' to seeing how our own actions create the problems we experience.[62]

Key systems thinking concepts include causal loops, understanding system behavior over time, and deconstructing complex systems within common system archetypes.[63]

Management systems such as ISO 14001 and ISO 45001 promote systems thinking. After systems such as these are developed and implemented, through their ongoing maintenance and continual improvement, their users begin to have an increased ability to see interconnections between their immediate focus – such as OHS – and other organizational functions, extending to supply chains, and beyond.

2.5.2 System Dynamics Iceberg

An iceberg model, as illustrated in Figure 2.6 has commonly been used in the system dynamics and systems thinking disciplines. This model frames a system's cause–effect relationships within several levels that are progressively more difficult to see and/or define, with events being above the waterline, and then with a progression of levels below the waterline that are less obvious. Daniel Kim refers to these levels as "levels of perspective."[64]

Events: This level is above the water line and points to what is observed day in and day out and are often trailing indicators, such as an accident.

Patterns/trends: Under events, is a level that points to what is observed over time with an intention toward understanding what is causing an event, or events. Patterns and trends begin to emerge when data are collected and analyzed.

---

61 Senge, P. (2006). *The Fifth Discipline: The Art and Practice of the Learning Organization*, 2e, pp. 11–12. New York: Currency Doubleday.
62 Senge, P. (2006). *The Fifth Discipline: The Art and Practice of the Learning Organization*, 2e, p. 12. New York: Currency Doubleday.
63 System archetypes are "one of the tools used in systems thinking. Systems archetypes are classic stories in systems thinking – common patterns and structures that occur repeatedly in different settings." Definition from *Systems Thinking Basics*, by V. Anderson and L. Johnson, Pegasus Communications, Waltham, Massachusetts 1997, p. 130.
64 Kim, D.H. (1999). *Introduction to Systems Thinking*, 2, p. 17. Waltham, MA: Pegasus Communications.

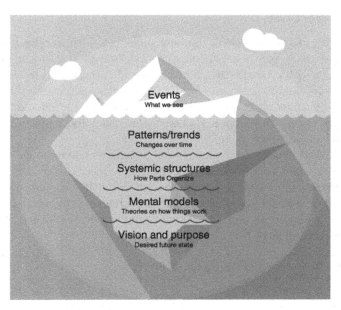

Figure 2.6 System dynamics iceberg.

Systems/structures: Under patterns/trends are the programs and systems that affect the patterns; causal connections are made. ISO management systems provide a way to strengthen this level; when missing, accident prevention programs are put in place. Many believe that this level holds the key to lasting, high-leverage change.[65]

Mental models: Under systems/structures are "deeply ingrained assumptions, generalizations, or even pictures or images that influence how we understand the world and how we take action; they determine not only how we make sense of the world, but how we take action."[66] The base level points to issues raised in Chapters 5 (Awareness in Risk Management) and 7 (Decision-Making) on the human operating system's role in decision-making, and the observation by Paul Slovic.

Gaining an understanding of the iceberg and how the levels interact is an important step in improving decision-making skills. Each level offers a distinctive mode of action, with each successive level (moving toward the iceberg's bottom) providing increased leverage. Being able to distinguish in real time the level (perspective) at which an issue originates increases the ability to select the most effective intervention.

---

65 Anderson, V. and Johnson, L. (1997). *Systems Thinking Basics*, p. 8. Waltham, MA: Pegasus Communications.
66 Senge, P. (2006). *The Fifth Discipline: The Art and Practice of the Learning Organization*, 2e, p. 8. New York: Currency Doubleday.

2.5.3 Deeper Levels

An intriguing inquiry in systems thinking circles is to identify what phenomena might be, or are, in play below the mental model's level. Daniel Kim and others identify a deeper level that is driven by vision and purpose. This level may prove to be the most difficult for an organization to articulate and can be thought of as the why (or purpose) at the core of the what and the how. From a different, but related perspective, Otto Scharmer's *Theory U* model points to deeper levels that are driven by heart and sensing.[67]

In the 2015-18 timeframe, Peter Senge, Otto Scharmer, Mette Bolle, and others began to formalize a body of work called "generative social fields (GSF)."[68] Similar to Kahneman's aim to develop more precise and richer language in decision science, Scharmer, Senge, *et al.* have been trailblazing the development of a lexicon for investigating GSF. The following definition for GSF was offered at a follow-up event in 2019.[69]

Box 2.3   Definition: Generative Social Field

"Generative social fields [are where] families, schools, businesses, governments, and the larger society in which we live can create relational interactions that support the nurturance of compassion, connection, curiosity, and well-being. Within such generative fields, a sense of belonging, meaning, and effective action emerges."

In Chapter 10 (§10.2), I provide more detail on the generative field construct.

2.6   Risk Field

> No concept is more vitally connected to the agenda of understanding institutional processes and organizations than that of organization field.[70]
> W. Richard Scott, Stanford University

> I think of the fields that surround my work as an educator as learning fields. If I were a physician working in a hospital, I would be talking about

---

67 Scharmer, C.O. (2016). *Theory U – Leading from the Future as it Emerges*. Oakland, CA: Berrett-Koehler Publishers.
68 Generative Social Fields Conference, at the Garrison Institute, Garrison, New York, 1–3 October 2018.
69 April 26, 2019. 2019 IPNB Pre-Conference Workshop Awareness-Based Systems Leadership: Cultivating Generative Social Fields and Systems Change, Marina Del Rey, CA.
70 W. Richard Scott – Institutions and Organizations – Ideas, Interests, and Identities (2014). 4e, 219.

healing fields, reflecting the repeated acts of healing that take place every day in hospitals. If I were a coach, I would be speaking about team fields.[71]

Christopher Bache (2008)

Risk field is a central construct in this book. The idea of organizational fields and subfields within organizations is a burgeoning area in the social sciences – namely in sociology and institutional theory. There are numerous ways field constructs are conceived, presented, and defined. Earlier in this chapter (§ 2.1.1.2), I included risk field as an element of ORM framework evolution. I gave some basic background in Chapter 1 and provide more detail here.

Richard Scott has been a thought leader and developer of organization field concepts and ideas for decades. He viewed the "field approach" as being central to understanding institutional and organizational dynamics and provided a way to explain and analyze behavior within them. In his work, he pointed to a long history of field ideas and concepts in the physical and social sciences. He writes,

> What [is] common to these, and related approaches is that the behavior of the objects under study is explained not by their internal attributes, but by their location in some physical or socially defined space. The objects, or actors, are subject to vectors of force (influences) depending on their location in the field and their relation with other actors as well as the larger structure within which these relations are embedded.[72]

To help queue up the discussion on risk fields, and later, on generative fields, let us look at definitions of field, organization field, and organizational risk field.

Box 2.4   Definitions: Field, Organizational Field, Organizational Risk Field, and Generative Risk Field

Field (noun): "An area of open land, especially one planted with crops or pasture, typically bound by hedges and fences. A space or range within which objects are visible from a particular viewpoint or through a piece of apparatus. The region in which a particular condition prevails, especially one in which a force or influence is effective regardless of the presence or absence of a material medium."[73]

Organizational field: "Those organizations that, in the aggregate, constitute an area of institutional life: key suppliers, resource and product consumers,

---

71 Bache, C. (2008). *The Living Classroom, Teaching and Collective Consciousness*, 53.
72 W. Richard Scott – Institutions and Organizations – Ideas, Interests, and Identities (2014), 4e, 220.
73 Apple OS dictionary. New Oxford American Dictionary.

> regulatory agencies, and other organizations that produce similar services or products."[74] // "A community of organizations that partakes of a common meaning system and whose participants interact more frequently and fatefully with one another than with actors outside of the field."[75]
>
> *Organizational risk field*: The relational and contextualized space that creates the interactions and collective behavior, that in turn produce the organization's risk management governance, strategy, execution, and outcomes.[76]
>
> *Generative organizational risk field*: An organizational risk field is generative when there is collective energy and leadership creating and preserving value that includes social, human, and natural capital.

2.6.1 Field Background

Field ideas, concepts, and theories are common in the natural and social sciences. Universal to these is consideration of phenomena not in isolation, but more broadly, within the boundaries of a field and in relation to other phenomena. In agriculture, there are fields where plants grow and are harvested; there are many necessary things that must happen in order to produce a bountiful harvest: seeds, water, and sunlight – and before planting, preparing the soil.

In physics, electromagnetic fields were identified by Michael Faraday in the 19th century. In psychology, Kurt Lewin characterized human behavior as occurring within a complex social field. He asserted that analysis could not be done in the absence of considering a person's whole environment. The educator Christopher Bache has evolved theories about the "living classroom" where he frames teaching and learning in a field context. In a learning field, Bache distinguishes two layers. One layer is a legacy or historic one created by a professor over time with a course's curriculum. The second is present in real time with a current class and is a function of what the class cohort brings with it in forming the field. In this book, I speak about organizational risk fields in terms of both of the layers identified by Bache. In fact, Figure 2.1 can be viewed in terms of Bache's two layers. That is, systems/frameworks are akin to what Bache refers to as the "legacy or historic" field, and social and human capital are akin to the "class cohort" field.

---

74 DiMaggio and Powell, 1983: 148.
75 Richard, S.W. (1994), pp. 207–208.
76 Redinger, C. (2019). Organizational Risk as a Generative Social Field – New Promise for Increasing Risk Awareness, Resilience, and Improving Decision-Making. Poster presented at the Society for Risk Analysis Annual Meeting, 9–11 December 2019.

2.6.2 Characterizing an Organizational Field

In the *Sage Handbook for Organizational Institutionalism*,[77] Professors Melissa Wooten and Andrew Hoffman provide a chapter that offers a robust overview of organizational fields. Several of their key concepts follow that serve as a foundation for developing and operationalizing the risk field construct.[78]

Box 2.5 Characterizing an Organizational Field

"Fields must be seen not as containers for the community of organizations, but instead as *relational spaces* that provide an organization with the opportunity to involve itself with other actors. Fields are richly contextualized spaces where disparate organizations involve themselves with one another in an effort to develop collective understandings regarding matters that are consequential for organizational and field activities."

"Fields are spaces that produce cultural and material products ranging from definitions of efficiency to organizational archetypes."

"Fields matter not only because of their investigative power but because actual people must deal with the consequences of their outcomes on a daily basis."

(*Source:* Wooten et al. (2017)/with permission of Sage Publications.)

There is a wide range of nomenclature, terms, and ideas used to characterize and describe fields in the literature. Some of the common ones are:

1) Fields have boundaries.
2) Fields contain systems – multiple systems – that are interconnected.
3) Fields are dynamic and fluid. There is energy within them.
4) Fields contain elements.

Elements identified in the field literature and augmented here, include people, actors (beyond individuals there are groups of people and teams), physical things (bricks and mortar), artifacts (policies/procedures), processes, programs, systems, organization logic, culture, rules, norms/practices, carriers of meaning, and their pathways/movement (movement of ideas and information, and their morphing). And in a risk field, there are obviously the risks (risk profiles/inventories). Definitions for field elements that might be new to you are highlighted in Box 2.6.

---

77 https://sk.sagepub.com/reference/hdbk_orginstitution.
78 Wooten, M. and Hoffman, A. (2017). Organizational fields: past, present and future. In: *The Sage Handbook of Organizational Institutionalism* (eds. Greenwood, R. et al.), 55–74. Thousand Oaks, CA: Sage Publications.

> **Box 2.6 Definitions: Actors, Artifacts, Carriers of Meaning, and Institutional Logic**
>
> *Actors*: These "include individuals, associations of individuals, populations of individuals, organizations, associations of organizations, and populations of organizations."[79]
>
> *Artifacts*: "An artifact is a discrete material object, consciously produced or transformed by human activity, under the influence of the physical and/or cultural environment."[80]
>
> *Carriers of meaning*: "These are fundamental mechanisms that allow us to account for how ideas move through space and time, who or what is transporting them, and how they may be transformed by their journey."[81]
>
> *Institutional logic*: "A set of material practices and symbolic constructions which constitutes its organizing principles, and which is available to organizations and individuals to elaborate."[82]

The term "institutional logic" might be new to you; it contains seeds that will be important when we dive into RDM (Chapter 7) and apply the Risk Matrix in an organization (Chapter 9). For our efforts with risk fields, this refers to the "logics" and dynamics – in common speech, I would say "what is going on" – in the three subfields identified in Figure 2.7 (regulatory/technical, organizational, social-human), and at their confluence – the risk field.

These logics are the source of tensions and challenges in ORM and distinguishing them is an important aspect of improving your RDM processes. Richard Scott characterizes these as:

> Many of the most important tensions and change dynamics observed in contemporary organizations and organization fields can be fruitfully examined by competition and struggle among various categories of actors committed to contrasting institutional logics. An important source of institutional tension and change experienced by both organizations and individuals in everyday life involves jurisdictional disputes among the various institutional logics.[83]

---

79 Scott, W.R. (2014). *Institutions and Organizations*. Sage Publications, Inc., 228.
80 Suchman (2003). The contract as social artifact. *Law and Society Review* 37: 91–142, p. 98. In Scott, W.R. (2014). *Institutions and Organizations*. Sage Publications, Inc., p. 102.
81 Scott, W.R. (2014). *Institutions and Organizations*. Sage Publications, Inc., 95.
82 Feidland and Alford (1991). Bringing society back in symbols, practices, and institutional contradictions. In: *The New Institutionalism in Organizational Analysis* (eds. Powell, W. and DiMaggio, P.), 248. Chicago University Press.
83 Scott (2014), p. 91.

2.6 Risk Field

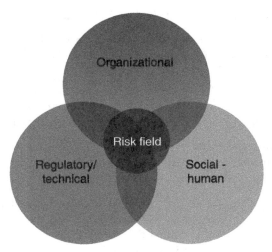

© 2023, Institute for Advanced Risk Management

Figure 2.7 Organizational risk field.

I have pointed toward numerous aspects of ORM and its complexity and noted that there are any number of ways to characterize it, with a plethora of narratives lurking about that describe it depending on one's perspective. Numerous ORM drivers can be identified. Within a "field" or "fields" perspective, drivers can be viewed as fields unto themselves. To keep this relatively straightforward, I am identifying three subfields from which an organization's risk field occurs: (1) regulatory/technical; (2) organizational; and, (3) social–human. These are depicted in Figure 2.7.

Each of the subfields contains drivers and elements. Some of these are depicted in Table 2.1 and are expanded on in Chapter 8 when addressing risk field operationalization as a risk matrix.

Table 2.1 Risk field, subfield elements.

Regulatory/technical	Organizational	Social–human
Laws and regulations	Mission, vision, values	Workers
Consensus standards	Objectives and goals	Community (society)
Risk science	Governance and structure	Consumers
Technology	Operations	Commons (environment)
	Frameworks and systems	
	Risk (risk profile)	

2.6.3 Operationalization – Risk Field → Risk Matrix

Matrix (n): "An environment or material in which something develops, a surrounding medium or structure."[84]

Operationalizing the risk field construct is needed in order to use it as an integrated thinking, decision-making, and action tool. This is done in Chapter 8 where a three-dimensional risk matrix is built. The three dimensions are:

- Contexts and drivers (y-axis)
- Actors and motive forces (z-axis)
- Risk management elements (x-axis)

I introduce you to aspects of the matrix in the chapters that follow. A foundation for ORM drivers and context has been offered in this chapter. This continues in Chapter 3 with a review of common frameworks used in ORM, such as ISO management systems, ISO's risk management standard (31000), and COSO's ERM framework. Core issues in risk science that impact ORM and RDM are addressed in Chapter 3, along with newer frameworks such as capitals approaches and culture of health for business. The content of Chapter 3 lays the foundation for the risk management elements dimension of the Matrix. The foundation for the matrix's actors and motive force elements is done in Chapter 5 (Awareness in Risk Management) and Chapter 6 (Field Leadership).

Suggested Reading

Anderson, V. and Johnson, L. (1997). *Systems Thinking Basics*. Waltham, Massachusetts: Pegasus Communications.

Bache, C. (2008). *The Living Classroom*. Teaching and Collective Consciousness: State University of New York Press, Albany.

Bower, J. and Paine, L. (2017). The Error at the Heart of Corporate Leadership – Most CEO and Boards Believe their Main Duty is to Maximize Shareholder Value – It's Not. *Harvard Business Review* May-June 2017.

Capitals Coalition (2019). *Social and human capital protocol*. https://capitalscoalition.org/capitals-ap-proach/social-human-capital-protocol/ (accessed 14 August 2021).

Christensen, K. (2018). Thought Leader Interview: Michael Porter – the founder of modern strategy explains why social progress should be on every business leader's agenda. *Rotman Management* 2018: 2–17.

---

84 Apple OS dictionary. New Oxford American Dictionary.

Committee of Sponsoring Organizations of the Treadway Commission (2017). *Enterprise Risk Management – Integrating with Strategy and Performance*. Association of International Certified Professional Accountants.

COSO and WBSCD (2018). *Enterprise Risk Management – Applying enterprise risk management to environmental, social and governance-related risks*. The Committee of Sponsoring Organizations of the Treadway Commission (COSO) and World Business Council for Sustainable Development. https://docs.wbcsd.org/2018/10/COSO_WBCSD_ESGERM_Guidance.pdf (accessed 2 November 2019).

Fligstein, N. and McAdam, D. (2012). *A Theory of Fields*. Oxford University Press.

Goss, T., Pascale, R., and Athos, A. (1993). The reinvention roller coaster: risking the present for a powerful future. *Harvard Business Review* 1993: 97.

International Organization for Standardization (2018). *Risk Management - Guidelines, ISO 31000*. Geneva: Switzerland.

International Organization for Standardization (2021). *Governance of organizations - Guidance, ISO 37000*. Geneva: Switzerland.

Kaplan, R. and Mikes, A. (2012). Managing risks: a new framework, smart companies match their approach to the nature of the threats they face. *Harvard Business Review* 90 (6): 48–60.

Kim, D.H. (1999). *Introduction to Systems Thinking*. Waltham, Massachusetts: Pegasus Communications.

Porter, M. and Kramer, M. (2006). Strategy & society: the link between competitive advantage and corporate social responsibility. *Harvard Business Review* 84 (12): 78–92.

Scharmer, C.O. (2016). *Theory U – Leading from the Future as it Emerges*. Oakland, California: Berrett-Koehler Publishers.

Scott, W.R. (2014). *Institutions and Organizations, Ideas, Interests and Identities*, 4e. Thousand Oaks, California: Sage Publications, Inc.

Senge, P. (2006). *The Fifth Discipline: The Art and Practice of the Learning Organization*, 2e. New York: Currency Doubleday.

3

Frameworks

CONTENTS
Power of Structure, 79
Expanding Perspective → Expanding Awareness, 79
Learning Context, 80
"Next Generation" Frameworks – Evolution and Integration, 80
3.1 Types of Frameworks, 80
3.1.1 Regulatory, 81
3.1.2 Consensus Standards, 81
3.1.3 Evolved Organizational and Professional Practices, 82
3.1.4 Tailoring, 82
3.2 National Academy of Sciences and EPA: Risk Decision-Making Anchors, 83
3.3 International Organization for Standardization (ISO), 86
3.3.1 Background, 86
3.3.2 Risk Management Evolution at ISO, 86
3.3.3 ISO 31000 Overview, 87
3.3.4 ISO 37000 – Risk Governance, Principle 6.9, 89
3.4 ISO Management System Standards, 90
3.4.1 High-Level Structure, 91
3.4.2 ISO MSS Demonstrative – Occupational Health and Safety (ISO 45001:2018), 92
3.5 COSO Enterprise Risk Management Framework, 113
3.5.1 Evolution, 113
3.5.2 Overview – 2017 Version, 114
3.6 Environmental, Social, and Governance, 117
3.6.1 Overview and Terminology, 117
3.6.2 Human Capital, 121
3.6.3 Reporting and Performance Criteria, 123
3.6.4 Global Reporting Initiative (GRI), 124
3.6.5 International Sustainability Standards Board, 125
3.6.6 Value Reporting Foundation, IIRC, SASB, 126

Organizational Risk Management: An Integrated Framework for Environmental, Health, Safety, and Sustainability Professionals, and their C-Suites, First Edition. Charles F. Redinger.
© 2025 John Wiley & Sons, Inc. Published 2025 by John Wiley & Sons, Inc.

3.7 Transcending Paradigms, 128
 3.7.1 NIOSH Total Worker Health, 129
 3.7.2 Culture Health for Business (COH4B), 130
 3.7.3 Capitals Coalition, 131
Appendix 3.A   ISO 3100:2018 Principles, 133
Appendix 3.B   COSO ERM (2017) Principles, 134
Suggested Reading, 136

We shape our buildings and afterwards our buildings shape us.[1]
Winston Churchill (1943)

Framework (n): A basic structure underlying a system, concept or text.
New American Oxford Dictionary

Frameworks define rules and provide guidelines and structure. Many of us started our EHSS careers in the regulatory/technical area. We studied and learned governmental regulations, and how to comply with them. As nongovernmental (NGO) consensus standards proliferated, we studied and learned these as well – particularly with management system standards. Frameworks can be thought of as scaffolding or containers that define boundaries and scope. It is valuable to pay attention to the extent to which frameworks impact performance, effectiveness, and power. We have all experienced this in our personal and professional endeavors – think of marriage, org charts, management systems, or the design of physical spaces, etc.

This chapter introduces a handful of EHSS/ORM frameworks. Predominant in this space are frameworks put forth by the International Organization for Standardization (ISO) and the Committee of Sponsoring Organizations of the Treadway Commission (COSO). Brief backgrounds are provided on governmental regulations, and foundational work by the National Academy of Sciences (NAS), published by the National Research Council (NRC). With the ongoing growth of ESG (environment, society, and governance) reporting schemes, and their influence beyond reporting, an overview is provided of the Global Reporting Initiative (GRI), the Value Reporting Foundation (VRF), and their frameworks.

The Risk Matrix presented in Chapter 8 is a framework. While it is informed by the frameworks covered in this chapter, it is distinct with its emphasis on (1) a multidimensional risk field construct; (2) integrated thinking, decision-making, and action; and, (3) the social-human aspects of risk management. I also use the

---

1 Churchill, W. (1943). https://www.parliament.uk/about/living-heritage/building/palace/architecture/palacestructure/churchill/ (accessed 12 October 2020).

term "container" to refer to the bundle comprising all of the EHSS/ORM activities, processes, policies, procedures, etc.

Frameworks define much of what is done in EHSS/ORM. Programs evolve into systems. Systems evolve into interconnected fields. We can also think of them as structures, or even containers, that establish the context within which the multitude of EHSS and ORM processes and practices occur. Frameworks establish the parameters, processes, systems, and norms within which ORM is executed and assessed. These are the structures we point to when asked about how we do EHSS. They reflect underlying organizational contexts. Often, we describe these in terms of regulations and consensus standards. Activities associated with the assessment of regulatory compliance and consensus standards conformance represent a framework called conformity assessment.

Before designing and building a risk field, it's important to characterize your EHSS/ORM container, your orientation to it, and your understanding of how things operate within it. When I ask clients or students to describe these three dynamics (characterize, orientation, operate) about their EHSS/ORM container, they often begin by describing them in terms of their management system (e.g. ISO MSS, COSO ERM). This makes sense due to the dominance of these approaches. With a bit of dialogue, we discover that the tapestry is often richer than a singular approach, especially in larger and multinational companies.

Power of Structure

Structures have power. The systems and frameworks of organizations have an enormous impact on their culture, effectiveness, and performance. Winston Churchill famously pointed to this power when he addressed the British Parliament about rebuilding the Commons Chamber after its bombing in World War II, and whether to rebuild the rectangular pattern or change to a semi-circle or horseshoe design.

Expanding Perspective → Expanding Awareness

There is value in thinking about this notion of the power of structure, and the relationship between your current EHSS/ORM frameworks and structures, and how they are impacting performance and risk outcomes. A core piece of this book is bringing and expanding awareness to your, and your company's risk logics.[2] And, as

---

2 Refreshing from Chapter 2, risk logics refers to – the approaches, principles, ideas, and contexts that establish foundations for risk-related systems and frameworks, decision-making processes, and subsequent actions.

suggested, embedded within this is awareness of risk decision-making processes. As you proceed, think about and characterize your risk logics and the ones in play in your company. Looking forward to Chapters 8 (Risk Matrix), 9 (Matrix in Action), and 10 (Escalate Impact), think about the logics' impact on ORM performance.

Learning Context

A learning context, or a culture of learning, is foundational in fourth-generation ORM. Frameworks provide the structure within which learning occurs. From a systems perspective, feedback loops provide the mechanism(s) through which information flows, thus making corrections possible. Learning context is addressed in later chapters but is mentioned here to reinforce the importance of frameworks.

"Next Generation" Frameworks – Evolution and Integration

As organizational governance approaches, and standards that influence them have evolved, there has been a growing emphasis on the concept of integrated thinking. This is reflected in ISO standards that were published in the 2010s, and the second edition of COSO's ERM framework in 2017. The roots of this idea are found in early sustainability frameworks such as the Triple Bottom Line and Integrated Reporting. In the early 2020s, numerous ESG standards-development entities placed more emphasis on integrated thinking. This chapter provides background on frameworks that will prepare you for the Risk Matrix presentation in Chapter 8.

3.1 Types of Frameworks

There are many types of frameworks. For this discussion, they are generally classified as regulations, consensus standards, and and evolved organizational and professional practices. Frameworks are developed and promulgated by a range of entities. These include governmental agencies (e.g. EPA, OSHA, SEC), NGOs (e.g. ISO, GRI), and professional, and sector-specific organizations (e.g. COSO, ISSB). And of course, individual organizations develop tailored frameworks for their internal use.

3.1 Types of Frameworks

Process → Program → System → Field

Figure 3.1 Framework maturity sequence.

Within all of these, there are processes and programs, which, in turn, can also be considered frameworks. The simple model in Figure 3.1 illustrates a typical maturity sequence.

An important distinction for all of these is between requirement and recommendation. Requirements are framed in in terms of "shall" clauses or elements, while recommendations are framed in terms of "should" clauses or elements. Regulations (laws) and standards are generally, if not always, framed in terms of "shall," while guidelines are framed in terms of "should." This distinction, shall versus should, is an important consideration for auditing and conformity assessment in general. It is difficult, if not impossible to have consistency in auditing should-based frameworks.

From an auditing and decision-making perspective, another important distinction is between compliance versus conformance. The use of the term compliance is generally reserved for regulatory or legal adherence, while the term conformance is reserved for consensus-standard adherence.

3.1.1 Regulatory

For EHSS/ORM purposes, regulations are predominantly developed at the national level. In addition, there are regional political and economic unions (e.g. European Union) that develop regulations that have the force of law. Within countries, there are also state and municipal agencies that develop laws, regulations, and guidelines.

3.1.2 Consensus Standards

There is a range of consensus standards-development entities. For the purposes here, the predominant one is ISO. However, there are national-level entities that impact EHSS/ORM performance such as the British Standards Institute (BSI), the American National Standards Institute (ANSI), and the Singapore Standards Council (SSC). The International Labor Organization (ILO) is an agency within the United Nations that develops standards related to social justice and labor rights.

An important consideration from a stakeholder perspective with consensus standards is how the standard-developing entity structures the "consensus" part of their activities. There is generally a goal to have "balance" among the people involved in the development process. However, challenges to the extent to which there was/is balance for a given standard, are not uncommon.

3.1.3 Evolved Organizational and Professional Practices

Professional and sector-specific organizations develop standards and guidelines. Those most frequently addressed throughout this book are the Committee of Sponsoring Organizations' (COSO), the Capitals Coalition, the Robert Wood Johnson Foundation, and a handful of ESG-related entities.

Beyond organization-wide frameworks that present high-level structures, there are process-related frameworks embedded within them. This includes the plan-do-check-act and systems thinking processes. In OHSMS, the anticipation-recognition-evaluation-control process is an example of this.

3.1.4 Tailoring

A theme I introduced in Chapter 1 was developing a playbook unique to your needs. In like fashion, it is not uncommon for organizations to develop frameworks unique to their needs. There are of course different levels to these as reflected in Figure 3.2.[3] From the process level up to the enterprise-wide level, it makes sense to take "off-the-shelf" standards and guidelines, and depending on conformity assessment needs or requirements, tailor them to match the needs of your organization.

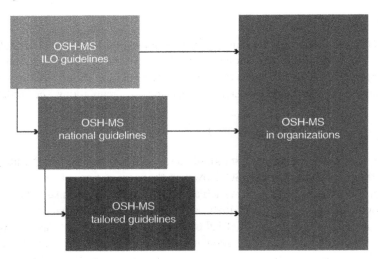

Figure 3.2 ILO's Elements of the national framework for OSH management systems. *Source:* International Labour Office (2001)/International Labour Organization CC BY 4.0.

---

3 International Labour Office (2001). *Guidelines on Occupational Safety and Health Management Systems ILO-OSH 2001*, 4. Geneva, Switzerland.

Tailoring is inspired by the ILO's occupational safety and health management system guidelines[4] that I was involved in developing in the late 1990s. A goal of that effort was to develop a high-level framework that could then be tailored as needed for the development of national standards in countries as well as by organizations. The guidelines state:

> 2.3.1. Tailored guidelines, reflecting the overall objectives of the ILO guidelines, should contain the generic elements of the national guidelines and should be designed to reflect the specific conditions and needs of *organizations* or groups of *organizations,* taking into consideration particularly:
>
> (a) their size (large, medium and small) and infrastructure; and
> (b) the types of hazards and degree of risks.
>
> 2.3.2. The links between the national framework for OSH management systems (OSH-MS) and its essential elements are illustrated in figure 1 (this book's Figure 3.2).[5]

Tailoring is addressed in greater length in later chapters within the context of developing and implementing the Risk Matrix in your organization. "Frameworks" is included as one of the 14 risk management elements (See Table 2.1) contained in the Matrix. These later chapters also address the extent to which an "off-the-self" framework needs to be tailored to meet an organization's needs with respect to integrated thinking, decision-making, and action needs.

3.2 National Academy of Sciences and EPA: Risk Decision-Making Anchors

This book's focus is on enterprise-wide risk management activities, embedded in which are many dimensions that impact the enterprise's risk profile. Within the EHSS component, fundamental aspects of risk decision-making, namely risk assessment and management, have been established by the NAS, and the Environmental Protection Agency (EPA). A quick overview of their contributions to EHSS/ORM follows.

The NAS was established in 1863 by the United States Congress as a private nongovernmental institution. Over time, it has grown and is referred to now as

---

4 International Labour Office (2001). *Guidelines on Occupational Safety and Health Management Systems ILO-OSH 2001*. Geneva, Switzerland.
5 International Labour Office (2001). *Guidelines on Occupational Safety and Health Management Systems ILO-OSH 2001*, 4. Geneva, Switzerland.

the National Academies of Sciences, Engineering, and Medicine (NASEM); its operating arm and brand for its publications is the NRC. The NRC's work and publications have played an important role in the development of risk-related structures in a range of public health arenas. Many, if not all, regulatory agencies have used and tailored the NRC risk frameworks to meet their specific mandates.

The NRC's work in the public health and safety risk arena originated from a request from the Food and Drug Administration (FDA) to help them fulfill a congressional mandate to conduct a "study to strengthen the reliability and objectivity of scientific assessment that forms the basis for federal regulatory policies applicable to carcinogens and other public health hazards; and to investigate the institutional means for risk assessment." The committee formed to do this viewed its mandate to be that of codifying practices of the 1970s in use at that time in federal agencies, as opposed to assessing their efficacy, or to creating new approaches. The NRC model is commonly referred to as the *Red Book* and characterized regulatory activities as having "two distinct elements," risk assessment and risk management.[6] Four steps in risk assessment were identified as: hazard identification, dose-response assessment, exposure assessment, and risk characterization. While risk management was identified as one of the elements of "regulatory actions," practices in use at that time were not the focus of the document.[7]

The *Red Book* served an important role in establishing a foundation for improving public health. Its four-step risk assessment process remains central in the field. The same cannot be said for the book's finding of, or characterization of, a distinct separation between "risk assessment" and "risk management." Problems with this separation of assessment and management began to be identified, and as time progressed, support grew to rethink its impact on achieving desired outcomes. Debate on this led to an evolution in risk-thinking that was published by the NRC in 2009 at the request of the Environmental Protection Agency (EPA); this document is often referred to as the *Silver Book*.[8]

A clear shift in thinking is observed in the *Silver Book* which recommended that risk assessments should not be done without a known purpose or independent of interpretation and management considerations. It states that, "...risk assessment should be viewed as a method for evaluating the relative merits of various options for managing risk rather than as an end in itself." The *Silver Book* also

---

6 National Research Council (1983). *Risk Assessment in the Federal Government: Managing the Progress.* Washington, DC: National Academy Press.
7 National Research Council (1983). *Risk Assessment in the Federal Government: Managing the Progress.* Washington, DC: National Academy Press.
8 National Research Council (2009) *Science and Decisions: Advancing Risk Assessment.* Washington, DC: National Academy Press.

recommended that problem formation and scoping should be conducted prior to conducting a risk assessment. Related to this recommendation was an additional recommendation to improve risk assessment utility, where utility entails making risk assessment more relevant to and useful for risk management decisions. The *Silver Book* summarized these points as:

> Risk assessment in EPA is not an end in itself but a means to develop policies that make the best use of resources to protect the health of the public and of ecosystems .... By focusing on early and careful problem formulation and on the options for managing the problem, implementation of the framework can do much to improve the utility of risk assessment. Indeed, without such a framework, risk assessments may be addressing the wrong questions and yielding results that fail to address the needs of risk managers.[9]

An additional important concept reinforced in the *Silver Book* was cumulative risk assessment (CRA). EPA's work on the development of CRA ideas and frameworks began in the 1990s, and was first formally introduced in 2003. It is defined as "...an analysis, characterization, and possible quantification of the combined risks to human health or the environment from multiple agents or stressors."[10] CRA's roots are clearly in Human Health Risk Assessment (HHRA) and environmental/ecological health. It is highlighted here because it provides a frame of reference and tool for EHSS/ORM professionals when making risk management decisions.

In 2014, the EPA published *Framework for Human Health Risk Assessment to Inform Decision-Making*.[11] This publication was the output of a 2010 colloquium on HHRA and is often referred to as the *Purple Book*. In it, key points from the *Silver Book* are reinforced and expanded. The *Purple Book* also introduced the concept of "fit for purpose," which is based on quality assurance principles. In the publication, fit for purpose is defined as "...the development of risk assessments and associated products that are suitable and useful for their intended purpose(s), particularly for informing risk management decisions."[12]

---

9 National Research Council (2009). *Science and Decisions: Advancing Risk Assessment*, 240, 244. Washington, DC: National Academy Press.
10 Environmental Protection Agency (2003). *Framework for Cumulative Risk Assessment*, EPA/600/P-02/001F.
11 Environmental Protection Agency (2014). *Framework for Human Health Risk Assessment to Inform Decision Making*, EPA/100/R-14/001.
12 Environmental Protection Agency (2014) *Framework for Human Health Risk Assessment to Inform Decision Making*, EPA/100/R-14/001, p. 4.

3.3 International Organization for Standardization (ISO)

3.3.1 Background

The International Organization for Standardization was formed in 1947 as part of post-WWII reconstruction efforts. ISO has grown into the world's largest voluntary consensus standards-development organization and publishes a wide range of standards and guidelines. The development of ISO 9001 on quality assurance in the 1980s was the organization's initial step into the arena of formal and organization-wide management systems. This was followed by other management system frameworks. Noteworthy for EHSS/ORM professionals are ISO 14001 (environment), and ISO 45001 (occupational health and safety). ISO's management system standards (e.g. 9001, 14001, and 45001) can be viewed as robust risk management tools.

3.3.2 Risk Management Evolution at ISO

In the early 2000s, ISO began to infuse risk-based thinking broadly into its endeavors. In 2002, it published Guide 73, titled *Risk Management – Vocabulary*. With risk management standards and approaches developing globally, ISO's working group on risk management began the development of what would be known as ISO 31000 "Risk Management Principles and Guidelines," which was first published in 2009, and revised in 2018. ISO 31000 presents a generic risk management framework and process and includes eight risk management principles.

Parallel to the development of Guide 73, an Ad Hoc Group on Management System Standards formed the Joint Technical Coordination Group (JTCG) to work on establishing consistency between ISO's various management system standards (MSS). The JTCG developed and issued in December 2011 ISO Guide 83, *High Level Structure, Identical Core Text and Common Terms and Core Definitions for use in Management Systems Standards*. However, while ISO Guide 83 was never formally adopted, it recommended the establishment of an ISO "high level MSS structure" that was subsequently adopted and published in 2013 in Annex SL of ISO's *Directives* (also referred to as the *ISO Supplement*). Annex SL presented the high-level generic MSS that all future ISO MSSs were required to follow, and with this infusion into their MSSs, the idea of risk-based thinking beyond traditional EHSS compliance-based orientations became more known to EHSS practitioners. For example, with ISO 45001:2018, the distinction is made between OHS risks and "other risks to the OHS management system" (6.1.2.2). OHS risks refer to what could be considered traditional risks, such as chemicals, slips, trips, and falls. Risks to the OHS management system refer to things that can affect OHS performance, such as day-to-day operations and decision-making; regulatory changes; the organizational culture; or changes in resources. However, a methodology for assessing OHS risks is required and needs to "be defined with respect to their scope, nature, and timing to ensure they are proactive rather than reactive."[13]

As of 2024, ISO had not developed an organizational risk management system (ORMS) standard. Its risk management guidelines (31000) document states it is not a management system, rather, it recommends that its contents be integrated into existing management systems. In 2013, ISO published a Technical Report (ISO/TR 31004) which provides guidance integrating ISO 31000 with ISO MSSs.[14] There is plenty to work with in the ISO family of standards for organizations that want to develop a tailored ORMS. ISO's business continuity management system (22301) provides a good starting point for the development of an ORMS as does the Risk Matrix presented in Chapters 8 and 9.

3.3.3 ISO 31000 Overview

As mentioned, ISO published its risk management guidelines (ISO 31000) in 2009, and its second edition in 2018. In the 2018 version, it states that "This document is for use by people who create and protect value in organizations by managing risks, making decisions, setting and achieving objectives and improving performance."[15] It also identifies four main changes it contains compared to the original 2009 document:

1) "review of the principles of risk management, which are the key criteria for its success,
2) highlighting of the leadership by top management and the integration of risk management, starting with the governance of the organization,
3) greater emphasis on the iterative nature of risk management, noting that new experiences, knowledge and analysis can lead to a revision of process elements, actions and controls at each stage of the process, and
4) streamlining of the content with greater focus on sustaining an open systems model to fit multiple needs and contexts."[16]

These guidelines offer risk management principles, a framework, and a process. These are shown in Figure 3.3[17] and are expanded on in later chapters related to the design and building of the Risk Matrix presented in Chapter 8. Along with risk management principles from COSO's ERM, ISO 31000:2018's eight principles are presented in appendix 3.A.

---

13 International Organization for Standardization (2018). *Occupational Health and Safety Management Systems – Requirements with Guidance for use; ISO 45001*, 13. Geneva, Switzerland.
14 ISO/TR 31004, "Risk management – Guidance for the implementation of ISO 31000."
15 ISO 31000:2018, p. v.
16 ISO 31000:2018, p. iv.
17 ISO 31000:2018, p. v.

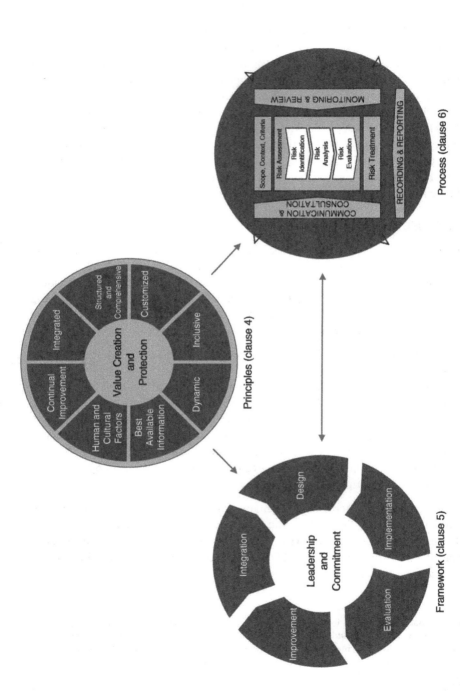

Figure 3.3 ISO 31000:2018, principles, framework and process. *Source:* ISO 31000:2018/with permission of International Organization for Standardization.

3.3.4 ISO 37000 – Risk Governance, Principle 6.9

In 2016, ISO formed a technical committee (TC) "to address standardization of the field of governance relating to aspects of direction, control and accountability of organizations."[18] The committee published a report mentioned in Chapter 1, titled "Governance of organizations – Guidance," referred to as ISO 37000:2021. The document states that it "gives guidance on the governance of organizations. It provides principles and key aspects of practices to guide governing bodies and governing groups on how to meet their responsibilities so that the organizations they govern can fulfill their purpose. It is also intended for stakeholders involved in, or impacted by, the organization and its governance."[19] Governance of organizations is defined as a: "human-based system by which an organization is directed, overseen and held accountable for achieving its defined purpose."

Eleven "principles of governance" are offered, and grouped as primary, foundational, and enabling. The primary principle – purpose (6.1) – was discussed in Chapter 1. Principle 6.9 – risk governance – is offered as one of six "enabling" principles. These are described as "principles that address the governance responsibilities pertinent to today's organizations – to meet evolving stakeholder expectations and the changing natural environment, social and economic context."[20]

The risk governance principle (6.9) is stated as: "The governing body should ensure that it considers the effect of uncertainty on the organizational purpose and associated strategic outcomes."[21] Recommendations for setting "the tone for management of risk," practices of "effective" risk management and risk management oversight are included.[22] Aspects of these recommendations are identified in later chapters as resources for consideration in designing, building, and navigating the Risk Matrix (Chapter 8).

Important concepts that are reinforced include stakeholder engagement; proactive and anticipatory actions; risk management integration across the organization; and, attention to new and emerging risks. Of particular interest here is the reinforcement of a holistic view of ORM; the document states: "The governing body should oversee the organization's management of risk, ensuring that a holistic view is taken by the organization, including consideration of all relevant types of risk."[23] Later chapters explore this notion of "all relevant types of risk," specifically within the evolving Social–Human Era.

---

18 https://committee.iso.org/home/tc309 (accessed 14 September 2023).
19 37000, p. 1.
20 37000, p. 12.
21 37000, p. 30.
22 37000, pp. 30–32.
23 37000, page 32, §6.9.3.4.a.

3.4 ISO Management System Standards

ISO's first MSS was published in 1987 and addressed quality management. The primary standard in the ISO 9000 family of quality management standards was 9001. Since that time, four subsequent editions have been published, with the most recent being in 2015.

Since 9001:1987's publication, there have been many subsequent MSSs published by ISO. Relevant to this discussion are their MSSs related to:

- Environmental management (ISO 14001)
- Occupational health and safety (ISO 45001)
- Security and resilience – Business continuity (ISO 22301)

ISO 14001 and 45001 are highlighted given their direct applicability to EHSS management. As suggested earlier, ISO 22301 is highlighted given its closeness to ORM in general, and its ability to serve as a platform to develop ORMS.

ISO's guidelines on the development of MSS defines a management system as a "system to establish policy and objectives and to achieve those objectives"[24] This ISO guideline states that:

> Management systems are used by organizations to develop their policies and to put these into effect via objectives and targets, using: an organizational structure with roles, responsibilities, authorities, etc., of people defined; systematic processes and associated resources to achieve the objectives and targets; measurement and evaluation methodology to assess performance against the objectives and targets, with feedback of results used to plan improvements to the system; and, a review process to ensure problems are corrected and opportunities for improvement recognized and implemented when justified.[25]

ISO 45001:2018 defines a management system (3.10) as a "...set of interrelated or interaction elements of an organization to establish policies and objectives and processes to achieve those objectives."[26] Comments related to this state that ... "a management system can address a single discipline or several disciplines; the system elements include the organization's structure, roles, and responsibilities,

---

[24] International Organization for Standardization (2001). *Guidelines for the Justification and Development of Management System Standards*, Guide 72, Geneva.
[25] International Organization for Standardization (2001). *Guidelines for the Justification and Development of Management System Standards*, Guide 72, Geneva.
[26] International Organization for Standardization (2018). *Occupational Health and Safety Management Systems – Requirements with Guidance for use; ISO 45001*, Geneva, Switzerland.

planning, operation, performance evaluation and improvement; and, the scope of the management system may include the whole of the organization, specific and identified functions of the organization, specific and identified sections of the organization, or one or more functions across a group of the organization." Given that ISO 45001 refers to an Occupational Health and Safety Management System, the developers of this standard refined it by adding a definition of an OH&S management system as a "management system or part of a management system used to achieve the OH&S Policy" as set by the organization.

3.4.1 High-Level Structure

Shortly after ISO 14001:1996 was published, ISO published *Guidelines for the Justification and Development of Management System Standards* (ISO Guide 72) in 2001.[27] This guide presented common MS elements as:

1) policy,
2) planning,
3) implementation and operation,
4) performance assessment,
5) improvement, and
6) management review.

These elements followed the structure of ISO 14001:1996 and were found in many nation-specific approaches at that time.

ISO's high-level MSS structure mentioned earlier (3.3.2) has 10 sections, these are:

1) Scope,
2) Normative references,
3) Terms and definitions,
4) Context of the organization,
5) Leadership,
6) Planning,
7) Support,
8) Operation,
9) Performance evaluation, and
10) Improvement.

This outline is expanded in Table 3.1 to include elements within sections 4–10.

---

27 International Organization for Standardization (2001). *Guidelines for the Justification and Development of Management System Standards*, Guide 72, Geneva.

Table 3.1 ISO's MSS high-level structure. (Note: XXX refers to a specific topic area, e.g. OH&S).

---

4) Context of the organization
    4.1 Understanding the organization and its context
    4.2 Understanding the needs and expectations of interested parties
    4.3 Determining the scope of XXX management system
    4.4 XXX management system

5) Leadership
    5.1 Leadership and commitment
    5.2 Policy
    5.3 Organizational roles, responsibilities, and authorities

6) Planning
    6.1 Actions to address risks and opportunities
    6.2 XXX objectives and planning to achieve them

7) Support
    7.1 Resources
    7.2 Competence
    7.3 Awareness
    7.4 Communication
    7.5 Documented information
        7.5.1 General
        7.5.2 Creating and updating
        7.5.3 Control of documented information

8) Operation
    8.1 Operational planning and control

9) Performance Evaluation
    9.1 Monitoring, measurement, analysis and evaluation
    9.2 Internal audit
    9.3 Management review

10) Improvement
    10.1 Nonconformity and corrective action
    10.2 Continual improvement

3.4.2 ISO MSS Demonstrative – Occupational Health and Safety (ISO 45001:2018)

ISO 45001:2018 is used here to provide examples of key ISO MSS issues. In later chapters, aspects of ISO 14001:2015 will be referenced for application of the Risk Matrix (Chapter 8). *With this example, note that the standard's section numbers are included, please do not confuse these section numbers with this book's numbering system.*

As indicated, ISO's high-level MSS dictated ISO 45001's overall structure and many of the terms and requirements contained within it. A summary of how the high-level MSS's 10 sections[28] were applied to OH&S in 45001:2018 follows in Sections 3.4.2.1 through 3.4.2.9.

ISO 45001's requirements are contained in its Sections 4 through 10. Scope and definitions are provided in Sections 1 and 3. Section 2 is titled "Normative references" as required by ISO's high-level MSS. However, ISO 45001 states that "there are no normative references in this document." ISO 45001's Annex A – "Guidance on the use of this document" – provides "explanatory information [with the intent] to prevent misinterpretation of the requirements contained in" the standard.[29] The Annex also provides some guidance on interpretation of the requirements that can help with system development and implementation.

ISO 45001's introduction contains background information and rationale for the standard, as well as the OHSMS approach in general. It provides the following description of an OHSMS's "aim" (0.2):

> The purpose of an OH&S management system is to provide a framework for managing OH&S risks and opportunities. The aim and intended outcomes of the OH&S management system are to prevent work-related injury and ill health to workers and to provide safe and healthy workplaces; consequently, it is critically important for the organization to eliminate hazards and minimize OH&S risks by taking effective preventive and protective measures.
>
> When these measures are applied by the organization through its OH&S management system, they improve its OH&S performance. An OH&S management system can be more effective and efficient when taking early action to address opportunities for improvement of OH&S performance.
>
> Implementing an OH&S management system conforming to this document enables an organization to manage its OH&S risks and improve its OH&S performance. An OH&S management system can assist an organization to fulfill its legal requirements and other requirements.[30]

---

28 Excluding §2 "normative references" which is not included because ISO 45001 contains none.
29 ISO 45001:2018, p. 24.
30 ISO 45001:2018, p. vi.

Implementation success factors are identified in section 0.3 where it is stated that:

> ...the implementation of an OH&S management system is a strategic and operational decision for an organization. The success of the OH&S management system depends on leadership, commitment and participation from all levels and functions of the organization.[31]

The plan-do-check-act (PDCA) concept has been a central to OHSMS since the earliest approaches. ISO 45001's Figure 1 (presented here as Figure 3.4)[32] shows the relationship between PDCA and the standard's framework.

3.4.2.1 Scope (1)

Every management system standard includes a statement of scope that defines the area it covers. It is not uncommon for standard-users to skip looking at the scope and go directly to its content – which is often thought of as "the auditable part." However, it is important to consider the standard's scope before working with its content, especially if system certification will be pursued, or if an integrated (e.g. OH&S, environment, and quality) system is being developed.

ISO 45001's scope states "This document specifies requirements for an Occupational Health and Safety (OH&S) management system, and gives guidance for its use, to enable organizations to provide safe and healthy workplaces by preventing work-related injury and ill health as well as by proactively improving its OH&S performance."[33] An important point in the scope of this standard is that it "does not state specific criteria for OH&S performance, nor is it prescriptive about the design of an OH&S management system."[34] Rather, it points to achieving performance outcomes that are consistent with the OH&S policy. The importance of this is that it reflects the intent for the use of the standard to be a risk management tool rather than a prescriptive instrument such as is found in governmental regulations.

Determining the scope (or coverage) of an organization's OHSMS's is a critical first step in the development and implementation of its management system. These considerations include clarity on operations covered by the OHSMS, such as a single plant, multiple plants, or corporate-wide, and, the activities covered. As determined by the organization, environmental, sustainability, or product safety can be added to the scope of the OHSMS, or vice versa; but integration is not a

---

31 ISO 45001:2018, p. vi.
32 ISO 45001:2018, p. viii.
33 ISO 45001:2018, p. 1.
34 ISO 45001:2018, p. 1.

3.4 ISO Management System Standards

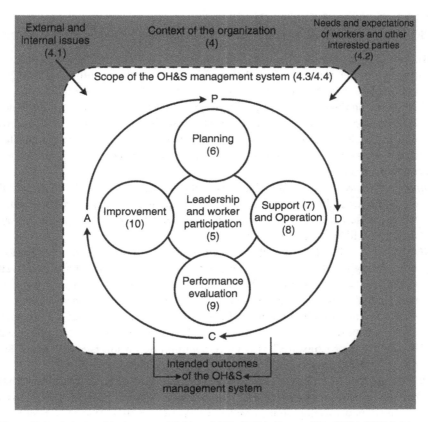

Figure 3.4 Relationship between PDCA and ISO 45001. *Source:* ISO 45001:2018/with permission of International Organization for Standardization.

requirement of ISO 45001:2018. If an OHSMS's scope is expanded beyond OH&S, consideration should be given to potential complications if certification will be sought only for the OHSMS (e.g. 45001) component.

Guidance and requirements for determining the OHSMS's scope are provided in section 4.3, where it states "the organization shall determine the boundaries and applicability of the OH&S management system to establish its scope. When determining this scope, the organization shall: consider the external and internal issues referred to in 4.1; take into account the requirements referred to in 4.2; and, take into account the planned or performed work-related activities." The scope needs also to "...be available as documented information."[35]

---

35 ISO 45001:2018, p. 8.

3.4.2.2 Terms and Definitions (3)

ISO 45001 references thirty-seven terms and definitions. Key definitions include:[36]

Worker: (3.3) is defined as a "person performing work or work-related activities that are under the control of the organization." In the evolution of OHSMS approaches, there has been movement toward this characterization as (worker), versus using the term "employee", to reflect that there are contract and temporary workers, who are conducting activities with health and safety implications that are under an organization's "control." Note 2 for this definition indicates that workers include top management, as well as managerial and nonmanagerial personnel.

Injury and ill health: (3.18) is defined as an "adverse effect on the physical, mental or cognitive condition of a person." New here is the term "cognitive condition." When implementing ISO 45001:2018, consideration should be given to how this term will be interpreted/defined.

Risk (3.20) is defined as the "effect of uncertainty." This is a generic definition used throughout ISO's MSS. It is meant to include direct risk to hazards as well as risks to the performance of the management system. To OH&S professionals, this at first seems like an odd definition because the common definition in OH&S is the "combination of the likelihood of an occurrence of a hazardous event or exposure(s) and the severity of injury or ill health that can be caused by the event or exposure." This more common definition is included in ISO 45001 as "OH&S risk" (3.21).

OH&S opportunity: (3.22) is defined as "circumstance or set of circumstances that can lead to improvement of OH&S performance." As indicated earlier, this is a relatively new development in MSSs in general as well as OHSMSs. It reflects an evolution in OHSMS approaches to focus not only on risk-related issues, but also to expand perspective and proactively look for, and act on, opportunities to improve OH&S performance.

3.4.2.3 Context of the Organization (4)

Examining, clarifying, and understanding organizational context establishes the foundation upon which the OHSMS is developed. The first step is to "determine external and internal issues that are relevant to [the organization's] purpose and that affect its ability to achieve the intended outcome(s) of its OH&S management system."[37] This includes gaining an understanding of the needs and expectations of workers and other interested parties including regulators and contractors.

---

36 ISO 45001:2018, pp. 2–8.
37 ISO 45001:2018, p. 8.

Consideration of organizational context is required by ISO's high-level MSS and represents an important advancement in MS parlance as it requires thinking more broadly (e.g. systems thinking) beyond traditional OH&S technical issues such as governance. The benefits of this approach include better harmonization and integration of OH&S activities with an organization's business processes.

ISO 9001:2015 and ISO 14001:2015 were the first to include this concept. They differ from earlier generations by including an ongoing assessment of changing circumstances along with the traditional elements of current conditions and practices. The primary goal of this approach is to ensure that the management system is tailored to the needs of the organization rather than a generic template that could overlook key local issues.

New to ISO 45001:2018 is its consideration of "interested parties." While ISO standards have traditionally included references to "interested parties" more recent versions have focused on a broader universe by requiring that the holistic "context of the organization" be considered when evaluating organizational impacts. Aside from interested parties in determining the context of the organization, OH&S professionals should consider possible implications and impacts from external as well as internal issues. In fact, OH&S professionals have always been concerned with internal issues such as health and safety skills, activities, and chemicals as well as biological and radiological materials. What is different, is the consideration of other types of external issues, such as cultures, competencies, and resources that have an impact on the, organization and subsequently, the health and safety of its workforce and workplaces.

3.4.2.4 Leadership and Worker Participation (5)

These two pieces – leadership and worker participation – are historically considered linchpins of the OHSMS approach. Detailed requirements are included for each in ISO 45001.

3.4.2.4.1 Leadership and Commitment ISO 45001:2018 defines (3.12) top management as a "person or group of people who directs and controls an organization at the highest level."[38] Notes to this definition state that "top management has the power to delegate authority and provide resources within the organization, provided ultimate responsibility for the OH&S management system is retained; and, if the scope of the management system covers only part of an organization, then top management refers to those who direct and control that part of the organization."[39]

---

38 ISO 45001:2018, p. 4.
39 ISO 45001:2018, p. 4.

This section (5.1) contains guidance and requirements related to top management "demonstrate[ing] leadership and commitment with respect to the OH&S management system."[40] These include taking responsibility and accountability for the prevention of work-related injury and ill health; ensuring the integration of the OHSMS into business processes; and, ensuring that there are sufficient resources for the OHSMS's functioning.

Three clauses that demonstrate important advancements from earlier OHSMS approaches are 5.1.j, 5.1.k, and 5.1.m:

Clause j: requires that "top management shall demonstrate leadership and commitment with respect to the OH&S management system by developing, leading, and promoting a culture in the organization that supports the intended outcomes of the OH&S management system."[41] This requirement reinforces the role that "culture" has in OH&S performance, and while directed at top management, it provides industrial hygiene and OH&S professionals with an important tool.

Clause k: requires that "top management shall demonstrate leadership and commitment with respect to the OH&S management system by protecting workers from reprisals when reporting incidents, hazards, risks, and opportunities."[42] This requirement supports greater worker engagement.

Clause m: requires that "top management shall demonstrate leadership and commitment with respect to the OH&S management system by supporting the establishment and functioning of health and safety committees."[43] This requirement helps promote greater participation and transparency.

3.4.2.4.2 Consultation and Participation of Workers

Worker consultation and participation guidance and requirements are nested throughout ISO 45001:2018. The standard defines consultation (3.5) as "seeking views before making a decision," and participation (3.4) as "involvement in decision-making."[44] In this section (5.4), consultation and participation requirements are broadly summed up as "the organization shall establish, implement, and maintain a process(es) for consultation and participation of workers at all applicable levels and functions, and, where they exist, workers' representatives, in the development, planning, implementation, performance evaluation, and actions for improvement of the

---

40 ISO 45001:2018, p. 9.
41 ISO 45001:2018, p. 9.
42 ISO 45001:2018, p. 9.
43 ISO 45001:2018, p. 9.
44 ISO 45001:2018, p. 2.

OH&S management system."[45] Key here is providing time, training, and resources, as well as removing barriers for effective participation. Clause 5.4.d requires "emphasiz[ing] the consultation of non-managerial workers" in a range of activities, including: OH&S policy (d.2); assigning organizational roles, responsibilities and authorities as applicable (d.3); planning, establishing, implementing, and maintaining an audit programme (d.8); and others. In like fashion, clause 5.4.e requires "emphasiz[ing] the participation of non-managerial workers" in a range of activities, including: identifying hazards and assessing risks and opportunities (e.2); determining actions to eliminate hazards and reduce OH&S risks (e.3); investigating incidents and nonconformities and determining corrective actions (e.7); and others.[46] These requirements can, in some circumstances, create a challenge where the relationship between workers and employers is regulated by law. That said, the thinking behind emphasizing workers who are often closest to the hazard, is to leverage their direct knowledge and experience in the ways of work, designed versus actual, that impact health and safety risks.

3.4.2.4.3 OH&S Policy The OH&S policy provides context and trajectory for an OHSMS. ISO 45001:2018 defines policy (3.14) generically as the "intentions and direction of an organization, as formally expressed by its top management,"[47] and OH&S policy (3.15) specifically as the "policy to prevent work-related injury and ill health to workers and to provide safe and healthy workplaces."[48] This section (5.2) provides guidance and defines the requirement for establishing, implementing, and maintaining an OH&S policy that:

> ...includes a commitment to provide safe and healthy working conditions for the prevention of work- related injury and ill health and is appropriate to the purpose, size and context of the organization and to the specific nature of its OH&S risks and OH&S opportunities; provides a framework for setting the OH&S objectives; includes a commitment to fulfill legal requirements and other requirements; includes a commitment to eliminate hazards and reduce OH&S risks; includes a commitment to continual improvement of the OH&S management system; and, includes a commitment to consultation and participation of workers, and, where they exist, workers' representatives.[49]

---

45 ISO 45001:2018, p. 10.
46 ISO 45001:2018, pp. 10–11.
47 ISO 45001:2018, p. 4.
48 ISO 45001:2018, p. 4.
49 ISO 45001:2018, p. 9.

In addition, ISO 45001:2018 requires that the OH&S policy is "available as documented information; communicated within the organization; available to interested parties as, appropriate; and relevant and appropriate."[50]

3.4.2.4.4 *Organizational Roles, Responsibilities, and Authorities* Requirements to define OH&S roles, responsibilities and authorities are contained in ISO 45001:2018 as well as with earlier OHSMS approaches. This section (5.3) requires that "Top management shall ensure that the responsibilities and authorities for relevant roles within the OH&S management system are assigned and communicated at all levels within the organization... [and that] workers at each level of the organization shall assume responsibility for those aspects of the [OHSMS] over which they have control."[51] Requirements for assigning responsibility and authority include OHSMS conformance with ISO 45001, and reporting on the OHSMS's performance to top management.

3.4.2.5 Planning (6)

ISO 45001:2018's planning section (6) contains requirements that are familiar to EHSS professionals and have been bedrock in EHSS management for decades. These include hazard identification and assessment of risks and opportunities (6.1.2); determination of legal requirements and other requirements (6.1.3); and, generating OH&S objectives and planning to achieve them (6.2).

The standard requires that (6.1.1):

> ...when planning for the OH&S management system, the organization shall consider the issues referred to in 4.1 (context), the requirements referred to in 4.2 (interested parties) and 4.3 (the scope of its OH&S management system) and determine the risks and opportunities that need to be addressed to: give assurance that the OH&S management system can achieve its intended outcome(s); prevent, or reduce, undesired effects; and achieve continual improvement.[52]

It further points out the need for proactive management of change during the planning process, where it states:

> ...the organization, in its planning process(es), shall determine and assess the risks and opportunities that are relevant to the intended outcomes of the OH&S management system associated with changes in the

---

50 ISO 45001:2018, p. 10.
51 ISO 45001:2018, p. 10.
52 ISO 45001:2018, p. 11.

organization, its processes or the OH&S management system. In the case of planned changes, permanent or temporary, this assessment shall be undertaken before the change is implemented.[53]

3.4.2.5.1 Actions to Address Risks and Opportunities While all of this section's (6) requirements are important, perhaps most significant are the requirements related to "actions to address risks and opportunities." Central to the EHSS management is the identification of hazards, and assessing and prioritizing associated risks. ISO 45001:2018, as have other OHSMSs, provides robust guidance on this process that establishes a foundation for a number of actions, such as setting objectives and determining controls. As already mentioned, the inclusion of the term "opportunities" is relatively new in the historical development of OHSMS approaches. An example of when an "opportunity" might occur is when an organization is updating a process. The organization can choose the one with the greatest OH&S improvement even though it may be more difficult or costly to implement.

Robust requirements are included related to hazard identification (6.1.2.1), "The organization shall establish, implement, and maintain a process(es) for hazard identification that is ongoing and proactive."[54] From an audit perspective, consideration needs to be given to what constitutes "ongoing" and how to demonstrate this. In well-functioning OHSMSs, this issue points to establishing feedback channels for hazard-identification data that arise from any number of activities, such as audits, accident reports, or worker complaints. The requirement to be proactive, while not absent in intent in early OHSMS approaches, is more clearly stated here.

Of particular interest related to hazards is the requirement that the process for hazard identification (6.1.2.1.a) also take into account "how work is organized, social factors (including workload, work hours, victimization, harassment, and bullying), leadership and the culture in the organization."[55] Language in this section makes clear that hazard identification extends to "locations not under the direct control of the organization" that have an impact on the organization's workers and workplaces." (6.1.2.1.e.3). This includes multi-employer work locations and "situations not controlled by the organization and occurring in the vicinity of the workplace that can cause injury and ill health to persons in the workplace." (6.1.2.1.f.3)[56]

---

53 ISO 45001:2018, p. 18.
54 ISO 45001:2018, p. 12.
55 ISO 45001:2018, p. 12.
56 ISO 45001:2018, p. 12.

A distinction is made between OH&S risks and "other risks to the OH&S management system" (6.1.2.2). OH&S risks refer to what could be considered traditional risks, such as chemical exposure, slips, trips, falls, etc. Risks to the OH&S management system refer to things that can affect OH&S performance, such as day-to-day operations and decision-making; regulatory changes; the organizational culture; and changes in resources; among others. A methodology for assessing OH&S risks is required that needs to "be defined with respect to their scope, nature, and timing to ensure they are proactive rather than reactive."[57]

3.4.2.5.2 Legal and Other Requirements Determination of legal and other requirements is a common element of OHSMS approaches. These requirements include, for example, governmental regulations, applicable nongovernmental consensus standards, and internal company standards. 45001's Annex (A.6.1.3) suggests "requirements" to consider. From an audit perspective, it is valuable to demonstrate what "requirements" have been considered, and then, which ones are selected for inclusion in the OHSMS's planning. It is also valuable to define how knowledge/requirements will be updated, and how often.

3.4.2.5.3 Planning Action While common, and somewhat *pro forma*, the standard requires that actions are planned to "address these risks and opportunities (6.1.2.2 and 6.1.2.3), legal requirements and other requirements (6.1.3), [and] prepare for and respond to emergency situations (8.2)" (6.1.4.a). It is also required that the organization plan "how to integrate and implement the actions into its OH&S management system processes or other business processes, [and] evaluate the effectiveness of these actions."[58] The explicit requirement of integration with business processes is a significant evolution from earlier OHSMS approaches and provides a valuable concept and tool for EHSS professionals.

3.4.2.5.4 Objectives Requirements for OH&S objectives have been integral to OHSMSs since the earliest formulations. In ISO 45001:2018, the term objective is defined as a "result to be achieved."[59] Several "notes" associated with this definition indicate that objectives can be strategic, tactical, or operational and can relate to different levels or parts of an organization. The term "OH&S" objective is defined separately as an "objective (3.16) set by the

---

57 ISO 45001:2018, p. 12.
58 ISO 45001:2018, p. 14.
59 ISO 45001:2018, p. 4.

organization (3.1) to achieve specific results consistent with the OH&S policy (3.15)."[60] ISO 45001:2018 (6.2.1) requires that an organization "establish OH&S objectives at relevant functions and levels in order to maintain and continually improve the OH&S management system and OH&S performance."[61] The standard provides a number of requirements associated with "planning how to achieve [the] OH&S objectives" (6.2.2). These include: "...what will be done; what resources will be required; who will be responsible; when will it be completed; how will the results be evaluated, including indicators for monitoring; and how the actions to achieve OH&S objectives will be integrated into the organization's business processes."[62] As indicated earlier, the explicit mention of integration "into the organization's business processes" is a significant evolution with 45001:2018. While developing procedures associated with OH&S objectives and plans is not required, the standard does require that "documented information" associated with them be created and retained.

3.4.2.6 Support (7)
Elements that support an OHSMS's operation are found in ISO 45001 Section 7. These include:

- Resources (7.1) "needed for the establishment, implementation, maintenance, and continual improvement" of the OHSMS,
- Competent (7.2) personnel (workers),
- Awareness (7.3) by workers of aspects of the OHSMS, consequences of actions, and hazards/risks,
- Communication (7.4) process(es), and
- Documented information (7.5).

Aspects of each of these elements are found in OHSMSs since their inception.

3.4.2.6.1 Resources This standard does not include requirements for specific resources, rather these need to be determined by the organization. Annex A (A.7.1) provides some examples of common resources including: "human, natural, infrastructure, technology, and financial."[63]

---

60 ISO 45001:2018, p. 4.
61 ISO 45001:2018, p. 14.
62 ISO 45001:2018, p. 14.
63 ISO 45001:2018, p. 32.

3.4.2.6.2 Competence

Concepts related to worker competence have been part of standard EHSS management practices for decades, primarily, in reference to required education and training. ISO 45001:2018 goes deeper where competence is defined as (3.23) the "ability to apply knowledge and skills to achieve intended results."[64] As with earlier OHSMSs, ISO 45001 provides guidance on this, where it requires (7.2) that the organization:

> ...determine the necessary competence of workers that affects or can affect its OH&S performance; ensure that workers are competent (including the ability to identify hazards) on the basis of appropriate education, training or experience; where applicable, take actions to acquire and maintain the necessary competence, and evaluate the effectiveness of the actions taken; and, retain appropriate documented information as evidence of competence.[65]

An important point here is "maintain[ing] the necessary competence, and evaluat[ion] of the effectiveness of the actions taken."

3.4.2.6.3 Awareness

Closely related to competence is awareness of the items, issues, etc. associated with competency. Awareness leads to recognition where competence relates to the application of knowledge and skills in managing risks. Understanding and knowledge of workplace hazards and risks is the historical focus here. ISO 45001:2018, as do earlier approaches, provides additional requirements regarding understanding and knowledge (awareness) of the OH&S policy and its objectives; the worker's impact on OHSMS effectiveness; and, the consequences of not conforming with OHSMS requirements. The standard also requires that workers are aware of "the ability to remove themselves from work situations that they consider present an imminent and serious danger to their life or health, as well as the arrangements for protecting them from undue consequences for doing so."[66]

3.4.2.6.4 Communication

The presence of a robust communication system, within an OHSMS is essential for a well-functioning OHSMS and a defining feature that differentiates it from pre-OHSMS approaches of OH&S management. Communication recommendations and requirements are found in all OHSMS approaches. ISO 45001:2018 requires that the organization "...establish, implement

---

64 ISO 45001:2018, p. 5.
65 ISO 45001:2018, p. 14.
66 ISO 45001:2018, p. 15.

and maintain the process(es) needed for the internal and external communications relevant to the OH&S management system, including determining: on what it will communicate; when to communicate; with whom to communicate; how to communicate." The "with whom" to communicate includes internally among the various levels and functions of the organization; among contractors and visitors to the workplace; and, among other interested parties. When determining communication needs associated with OH&S, "diversity aspects" need to be considered, including gender, language, culture, literacy, and disability. And, as part of its communication system, the organization needs to "respond to relevant communications" related to the OHSMS.

3.4.2.6.5 Documented Information Documentation and "controlled" document guidance and requirements are found in all OHSMS approaches. In ISO 45001:2018, documented information (3.24) is defined as "information required to be controlled and maintained by an organization and the medium on which it is contained."[67] Prior OHSMS standards referred to documents and records, both of which are covered under the term, documented information. Document management is a historical aspect of OH&S management and is familiar to OH&S professionals. ISO 45001:2018 contains criteria for "creating and updating" (7.5.2) and for the "control of documented information" (7.5.3). Actions in these two sections are aimed at ensuring that documented information is: up-to-date; reviewed and approved for suitability and adequacy; and, adequately protected. A key aspect of this section's requirements is "preventing unintended use of obsolete documented information" (A.7.5).

3.4.2.7 Operation (8)

All OHSMS approaches address aspects of the system relative to the organization's operations. In ISO 45001:2018, the "operation" (8) section contains two subsections – operational planning and control (8.1), and emergency preparedness and response (8.2). Concepts and ideas contained in these subsections are familiar to historic OH&S program management. Subsection 8.1 contains important topics, that include: elimination of hazards and reducing OH&S risks (8.1.2); management of change (8.1.3); and, procurement (8.1.4).

A new concept in ISO 45001, and all ISO management system standards, is the requirement for maintaining and controlling "processes" and not just the hazard itself. An example would be how well the process of planning works to make improvements in the management of risks. This is an important function when

---

67 ISO 45001:2018, p. 5.

conducting internal audits of the management system. Traditional EHSS that focuses on the direct control of hazards based on the hierarchy of controls is also a part of operational requirements.

3.4.2.7.1 Eliminating Hazards and Reducing OH&S Risks ISO 45001:2018 contains requirements for the elimination of hazards and reduction of OH&S risks. The standard states (8.1.2) that:

> ...the organization shall establish, implement and maintain a process(es) for the elimination of hazards and reduction of OH&S risks using the following hierarchy of controls: eliminate the hazard; substitute with less hazardous processes, operations, materials or equipment; use engineering controls and reorganization of work; use administrative controls, including training; and, use adequate personal protective equipment.[68]

Examples are provided in the standard's Annex A (A.8.1.2) that "illustrates measures that can be implemented at each level" of the stated hierarchy of controls.

3.4.2.7.2 Management of Change An increasingly important function in OH&S management is "management of change" (MOC). This concept is contained in all OHSMS approaches and represents a significant advancement in OH&S management in recent decades. MOC addresses activities that require examining OH&S risks associated with new projects, processes or substances, or when existing operations or processes are modified. Changes to an organization can disrupt established protections and controls which require evaluation and active management not only of the end result but also the actions taken to get there. MOC also includes considerations related to human resources as well as organizational strategies. EHSS professionals should consider MOC to be an important element of a comprehensive OH&S management system. The extent and complexity of a MOC process should be consistent with the potential risk of catastrophic failure of the new or modified process.

ISO 45001:2018 states (8.1.3) that:

> ...the organization shall establish a process(es) for the implementation and control of planned temporary and permanent changes that impact OH&S performance, including: new products, services and processes, or

---

68 ISO 45001:2018, p. 18.

changes to existing products, services and processes (including – workplace locations and surroundings, work organization, working conditions; equipment, and work force); changes to legal requirements and other requirements; changes in knowledge or information about hazards and OH&S risks; developments in knowledge and technology.[69]

The standard further requires that a review is done of "...the consequences of unintended changes, [and] taking action to mitigate any adverse effects, as necessary."[70]

3.4.2.7.3 Procurement Procurement-related requirements address the procurement of products and services "to ensure their conformity to" the OHSMS. Establishing, implementing, and maintaining process(es) to do this is required in ISO 45001:2018. This includes how the organization procures materials, such as chemicals and equipment, as well as people who perform activities that may, in some way, impact health and safety.

Related to contractors, ISO 45001:2018 (8.1.4.2) states that:

...the organization shall coordinate its procurement process(es) with its contractors, in order to identify hazards and to assess and control the OH&S risks arising from: the contractors' activities and operations that impact the organization; the organization's activities and operations that impact the contractors' workers; and, the contractors' activities and operations that impact other interested parties in the workplace.[71]

The standard requires that contractors conform with the organization's OHSMS requirements and that the "procurement process(es) define and apply occupational health and safety criteria for the selection of contractors." Annex A (8.1.4.2) offers provides helpful guidance on contractor and OHSMS coordination.

In addition, the standard contains requirements for outsourcing activities and defines "outsource" (verb) as to "make an arrangement where an external organization (3.1) performs part of an organization's function or process."[72] A note to this definition states that the "external organization is outside the scope

---

69 ISO 45001:2018, p. 32.
70 ISO 45001:2018, p. 32.
71 ISO 45001:2018, p. 34.
72 ISO 45001:2018, p. 6.

of the management system (3.10), although the outsourced function or process is within the scope."[73] Relating to outsourcing, the standard states that (8.1.4.3):

> ...the organization shall ensure that outsourced functions and processes are controlled, [and that] the organization shall ensure that its outsourcing arrangements are consistent with legal requirements and other requirements and with achieving the intended outcomes of the OH&S management system. The type and degree of control to be applied to these functions and processes shall be defined within the OH&S management system.[74]

In practice, this means evaluation of the capabilities of the company that will perform the outsourced work that can impact the intended outcomes of your organization's OHSMS. The evaluation should include an assessment of whether the outsourced workers and their company can perform the required work in a legal and safe manner.

3.4.2.7.4 Emergency Preparedness and Response Being prepared for, and responding to emergencies is an historic component of OH&S management. All OHSMS approaches address this and contain guidance and requirements related to it. ISO 45001:2018 requires (8.2) that "The organization shall establish, implement and maintain a process(es) needed to prepare for and respond to potential emergency situations, as identified in 6.1.2.1 [hazard identification]."[75] This includes:

a) "establishing a planned response to emergency situations, including the provision of first aid,
b) providing training for the planned response,
c) periodically testing and exercising the planned response capability,
d) evaluating performance and, as necessary, revising the planned response, including after testing and, in particular, after the occurrence of emergency situations,
e) communicating and providing relevant information to all workers on their duties and responsibilities,
f) communicating relevant information to contractors, visitors, emergency response services, government authorities and, as appropriate, the local community,

---

73 ISO 45001:2018, p. 6.
74 ISO 45001:2018, p. 19.
75 ISO 45001:2018, p. 19.

g) taking into account the needs and capabilities of all relevant interested parties and ensuring their involvement, as appropriate, in the development of the planned response."[76]

The standard requires that "...documented information on the process(es) and on the plans for responding to potential emergency situations"[77] be maintained and retained.

3.4.2.8 Performance Evaluation (9)

Many requirements in this ISO 45001:2018 section (9) are familiar to EHSS professionals and are aligned with earlier-generation OHSMSs. In addition to addressing general monitoring, measurement, analysis, and performance evaluation (9.1) issues, this section contains requirements related to the evaluation of compliance (9.1.2); internal auditing (9.2); and, management review (9.3).

3.4.2.8.1 Monitoring, Measurement, Analysis, and Performance Evaluation ISO 45001:2018 outlines a general requirement that (9.1.1) "The organization shall establish, implement, and maintain a process(es) for monitoring, measurement, analysis and performance evaluation."[78] To do this, the organization needs to determine what needs to be monitored and measured in order to evaluate OH&S and OHSMS performance, and the extent to which this is impacted, or dictated by legal and other requirements; activities and operations related to identified hazards, risks, and opportunities; progress toward achievement of the organization's OH&S objectives; and, effectiveness of operational and other controls.[79]

Included in these requirements is ensuring "that monitoring and measuring equipment is calibrated or verified as applicable, and is used and maintained as appropriate."[80]

3.4.2.8.2 Evaluation of Compliance Evaluation of compliance with governmental regulation is a common activity in IH/OH&S management. ISO 45001:2018 provides guidance and related requirements. The standard states (9.1.2) that "the organization shall establish, implement and maintain a process(es) for evaluating

---

76 ISO 45001:2018, p. 19.
77 ISO 45001:2018, p. 19.
78 ISO 45001:2018, p. 19.
79 ISO 45001:2018, p. 19.
80 ISO 45001:2018, p. 20.

compliance with legal requirements and other requirements.[81]" And that it "...shall determine the frequency and method(s) for the evaluation of compliance; evaluate compliance and take action if needed; maintain knowledge and understanding of its compliance status with legal requirements and other requirements; and, retain documented information of the compliance evaluation result(s)."[82]

3.4.2.8.3 Internal Audit OH&S auditing is a common activity in organizations historically focused on determining compliance with governmental regulatory standards, rules, and laws as well as its own voluntary requirements. However, the focus of the internal audit prescribed in ISO 45001:2018 is whether the OHSMS is effectively implemented, including compliance considerations, in conformance with the standards as well as the organization's own OHSMS requirements. It states, "the organization shall conduct internal audits at planned intervals to provide information on whether the OH&S management system conforms to the organization's own requirements for its OH&S management system, including the OH&S policy and OH&S objectives; and, the requirements of [45001:2018]."[83] Further, internal OHSMS audits need to address whether the OHSMS "is effectively implemented and maintained."

Criteria and requirements are included related to the internal audit program's structure and activities (9.2.2). Included are issues related to audit planning, methods, frequency, scope, auditor qualification/character, and reporting. Internal audit criteria need to be established, along with ensuring auditors selected to do audits are objective and impartial. Results need to be reported to "relevant managers," and "...relevant audit results need to be reported to workers, workers' representatives (where they exist), and relevant interested parties."[84] The organization is required to "...take action to address nonconformities and continually improve its OH&S performance; and, retain documented information as evidence of the implementation of the audit programme and the audit results."[85]

For more information on auditing and the competence of auditors, users of ISO 45001:2018 are directed to ISO 19011:2018, *Guidelines for auditing management systems*.[86]

---

81 ISO 45001:2018, p. 20.
82 ISO 45001:2018, p. 20.
83 ISO 45001:2018, p. 20.
84 ISO 45001:2018, p. 21.
85 ISO 45001:2018, p. 21.
86 International Organization for Standardization (2018). *Guidelines for Auditing Management Systems, ISO 19011*, Geneva, Switzerland.

3.4.2.8.4 Management Review A legacy element in the OHSMS approach is management review (9.3). ISO 45001:2018 contains management review guidance and requirements. The standard states that "top management shall review the organization's OH&S management system, at planned intervals, to ensure its continuing suitability, adequacy and effectiveness."[87] The key here is evaluation of the OHSMS's suitability, adequacy, and effectiveness. Top management needs to determine whether maintaining the OHSMS is aligned with strategic objectives, and, if not, should maintaining conformance to it be continued; and, if it is, ensure that proper resources and support are being given to it. Criteria and requirements for conducting management reviews include the types of information and data that need to be considered.

ISO 45001:2018 requires that:

> the outputs of the management review shall include decisions related to: the continuing suitability, adequacy and effectiveness of the OH&S management system in achieving its intended outcomes; continual improvement opportunities; any need for changes to the OH&S management system; resources needed; actions, if needed; opportunities to improve integration of the OH&S management system with other business processes; and, any implications for the strategic direction of the organization.[88]

Top management is required to "communicate the relevant outputs of management reviews workers, and, where they exist, workers' representatives."[89]

3.4.2.9 Improvement (10)

The term "improvement" finds its way into ISO 45001:2018 from ISO's MSS requirements. A number of improvement-related activities and requirements are bundled here: incident, nonconformity, and corrective action responses (10.2); and, continual improvement (10.3).

3.4.2.9.1 Incident, Nonconformity, and Corrective Action ISO 45001:2018 (10.2) states that "The organization shall establish, implement, and maintain, a process(es) including reporting, investigating and taking action to determine and manage incidents and nonconformities."[90] Specific requirements include: timely response; conducting root cause analysis, with worker involvement; assessing

---

87 ISO 45001:2018, p. 21.
88 ISO 45001:2018, p. 22.
89 ISO 45001:2018, p. 22.
90 ISO 45001:2018, p. 22.

potential historical trends; ensuring that findings are fed back into the planning process. The standard defines nonconformity, incident, and corrective action as follows:

- *Nonconformity (3.34)*: the "non-fulfillment of a requirement."[91] This relates to the requirements of ISO 45001, as well as requirements that the organization establishes for itself.
- *Incident (3.35)*: is an "occurrence arising out of, or in the course of, work that could or does result in injury and ill health."[92] An incident where no injury and ill health occurs, but has the potential to do so, may be referred to as a "near-miss," "near-hit," or "close call." Although there can be one or more nonconformities related to an incident, an incident can also occur where there is no nonconformity.
- *Corrective action (3.36)*: is an "action to eliminate the cause(s) of a nonconformity or an incident and to prevent recurrence."[93] This term is one of the common terms and core definitions in ISO MSSs.

3.4.2.9.2 Continual Improvement A hallmark of the OHSMS approach is continual improvement. 45001:2018 continues this trajectory, stating in 10.3 that:

> ...the organization shall continually improve the suitability, adequacy and effectiveness of the OH&S management system: by enhancing OH&S performance; promoting a culture that supports an OH&S management system; promoting the participation of workers in implementing actions for the continual improvement of the system; and, communicating the relevant results of continual improvement to workers, and where they exist, workers' representatives.[94]

The standard defines continual improvement as a "recurring activity to enhance performance,"[95] and reinforces the idea that continual does not mean continuous, that is, the activity does not need to take place all of the time without interruption. The addition of a separate clause on improvement also emphasizes that the clauses in ISO 45001 do not work independently but rather together. Improvements identified here result in part from monitoring and assessments, including management reviews which consider possible new objectives.

---

91 ISO 45001:2018, p. 7.
92 ISO 45001:2018, p. 7.
93 ISO 45001:2018, p. 7.
94 ISO 45001:2018, p. 23.
95 ISO 45001:2018, p. 8.

3.5 COSO Enterprise Risk Management Framework

Enterprise risk management (ERM) is a term used for approaches, methods, and frameworks that look at organizational risk from the perspective of the entire organization, including not just financial, but also, for example operational, strategic, and reputational risks. ERM ideas have been advanced by the Committee of the Sponsoring Organizations of the Treadway Commission (COSO) which was formed in 1985 as a joint initiative of five private sector organizations.[96] Although EHSS is not explicitly addressed in ERM terminology, there is value for EHSS/ORM professionals in understanding the COSO ERM framework, learning the language of business embedded in it, and gaining fluency in the ideas of ERM. By using the language of ERM which is familiar to senior management and executives, the EHSS professional can play a larger role in organizational leadership and deepen the integration of EHSS practices in the organization.

3.5.1 Evolution

In 2004, COSO published *Enterprise Risk Management – Integrated Framework* which presented an ORM framework that has been a driver in defining ERM. The ERM framework was updated in 2017 with *Enterprise Risk Management: Integrating Strategy and Performance*. From the 2004 to the 2017 revision, there was a shift from viewing risk management as an isolated process to one that supports an organization in achieving objectives, creating value, and increasing attention to strategic considerations, which address:[97]

- alignment of strategy and business objectives with mission, vision, and values,
- implications and consequences from the strategic choices (aka decision-making), and
- risks and uncertainties in executing strategies.

---

96 These five private sector organizations are: The American Accounting Association (AAA), the American Institute of Certified Public Accountants (AICPA), Financial Executives International (FEI), the Institute of Internal Auditors (IIA), and the Institute of Management Accountants (IMA).
97 Committee of Sponsoring Organizations of the Treadway Commission (2017). *Enterprise Risk Management – Integrating with Strategy and Performance, Executive Summary*, 3. Association of International Certified Professional Accountants.

This shift in definition of ERM from process to value is seen in the 2014 and 2017 versions of the framework (Table 3.2).[98,99]

Table 3.2 COSO ERM 2004 – ERM as process versus 2017 – ERM as value.

2004 – ERM as process[98]	2017 – ERM as value[99]
"Enterprise risk management is a process, effected by an entity's board of directors, management and other personnel, applied in strategy setting and across the enterprise, designed to identify events that may affect the entity, and manage risk to be within its risk appetite, to provide reasonable assurance regarding the achievement of entity objectives."	"Enterprise risk management: The culture, capabilities, and practices, integrated with strategy-setting and its performance, that organizations rely on to manage risk in creating, preserving, and realizing value."

In the 2017 version, it is emphasized that ERM "...does not refer to a functional group, or department within an entity."[100] Rather, it is deeply embedded throughout the entire organization.

The 2004 framework consists of four "objective categories," eight "interrelated components," and four "entity units" as depicted in Figure 3.5. The objective categories are strategic, operations, reporting, and compliance. The interrelated components are internal environment, objective-setting, event identification, risk assessment, risk response, control activities, information and communication, and monitoring. The entity units are defined as subsidiary, business unit, division, and entity level.

Internal environment "...encompasses the tone of an organization, and sets the basis for how risk is viewed and addressed by an entity's people, including risk management philosophy and risk appetite, integrity and ethical values, and the environment in which they operate."[101]

3.5.2 Overview – 2017 Version

The 2017 ERM framework places emphasis on "...managing risk through: recognizing culture; developing capabilities; applying practices; integrating with strategy-setting and performance; managing risk to strategy and business

---

98 Committee of Sponsoring Organizations of the Treadway Commission (COSO) (2004). *Enterprise Risk Management – Integrated Framework, Executive Summary* Framework, p. 16.
99 Committee of Sponsoring Organizations of the Treadway Commission (COSO) (2017). *Enterprise Risk Management – Integrating with Strategy and Performance*, p. 10.
100 COSO (2017), p. 4.
101 Committee of Sponsoring Organizations of the Treadway Commission (COSO) (2004). *Enterprise Risk Management – Integrated Framework, Executive Summary* Framework, p. 5.

3.5 COSO Enterprise Risk Management Framework

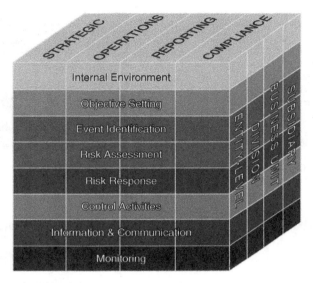

Figure 3.5 2004 COSO ERM framework.

objectives; and linking to value."[102] A framework is presented that focuses on "...strategy in the context of mission, vision, and core values, and as driver of an entity's overall direction."[103] These relationships are shown in Figure 3.6[104], "strategy in context." Included are the relationships of business objectives and performance; possibility of strategy not aligning, implications from the strategy chosen; and, risk to strategy and performance. Later chapters explore these dynamics and how they play out in integrated thinking, decision-making, and, action.

The 2017 framework presents a set of 20 risk management principles[105] that are organized into five interrelated components. The components are Governance and Culture; Strategy and Objective-Setting; Performance; Review and Revision; and, Information, Communication, and Reporting. Figure 3.7[106] shows these in relation to the "strategy in context" dynamics depicted in Figure 3.6.

A description of the 20 principles can be found in Appendix 3.B at the end of this chapter. EHSS/ORM professionals will find value in understanding and embracing all twenty principles. However, there is particular value in focusing on principles 3, 4, 6, 7, 10, and 14 that can assist with navigating the Risk Matrix (Chapter 8).

---

102 Committee of Sponsoring Organizations of the Treadway Commission (COSO) (2017). *Enterprise Risk Management – Integrating with Strategy and Performance*, p. 10.
103 COSO 2017, p. 13.
104 COSO 2017, p. 13
105 These are called components and principles.
106 COSO 2017, p. 21.

Figure 3.6 COSO ERM depiction of strategy in context. *Source:* Adapted from COSO 2017.

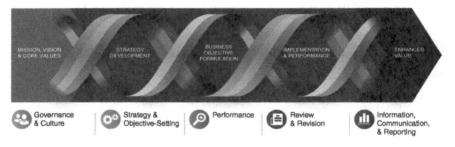

Figure 3.7 COSO ERM, risk management components. *Source:* Adapted from COSO 2017.

Principle #3 – defines desired culture: The organization defines the desired behaviors that characterize the entity's desired culture.

Principle #4 – demonstrates commitment to core values: The organization demonstrates a commitment to the entity's core values.

Principle #6 – analyzes business context: The organization considers potential effects of business context on risk profile.

Principle #7 – defines risk appetite: The organization defines risk appetite in the context of creating, preserving, and realizing value.

Principle # 10 – identifies risk: The organization identifies risk that impacts the performance of strategy and business objectives.

Principle # 14 – develops portfolio view: The organization develops and evaluates a portfolio view of risk.

The value and impact that these principles – along with the other 14 – have on integrated thinking, decision-making, and action is addressed in later chapters, with particular emphasis on the Risk Matrix's Social-Human elements.

While "risk aware decision making" is not explicitly stated in the 2017 framework documents, COSO asserts that the ERM framework does focus on risk awareness.[107] The importance of increasing risk awareness is addressed in Chapter 5 and continues through the remaining chapters.

3.6 Environmental, Social, and Governance

ESG has significant organizational risk ramifications. There are numerous aspects to this that range from determining what external requirement are relevant to an organization, and then, what internal structures are needed to meet them. A brief overview of terminology and primary frameworks is provided here.

3.6.1 Overview and Terminology

The terms ESG, sustainability, and corporate social responsibility (CSR) are often used interchangeably. Here, ESG is used to capture all three. In operating within a company, it is important to understand how these three terms are defined and used at the board- and C-Suite levels, as well as in company operations.

3.6.1.1 Sustainability

Sustainability is the broadest concept of these three. Its classic definition – which originates from the United Nations World Commission on Environmental Development (Brundtland Commission) in 1987, is – "Development that meets the needs of the present without compromising the ability of future generations to meet their own needs."[108] The predominant topic associated with sustainability is climate change. In 2015, the United Nations formulated 17 Sustainable Development Goals (SDGs) that were "…designed to serve as a shared blueprint for peace and prosperity for people and the planet now and into the future."[109]

3.6.1.2 Environmental, Social, and Governance

While ESG has roots in numerous areas, such as social responsibility, the triple bottom line (TBL), and sustainability in general, its formalization is pegged to a 2004 United Nations led initiative to "…better integrate environmental, social, and

---

107 COSO (2017). Power point presentation. https://www.coso.org/Documents/COSO-ERM-Presentation-September-2017.pdf (accessed 23 September 2019).
108 Our Common Future (1987).
109 https://en.wikipedia.org/wiki/Sustainable_Development_Goals (accessed 14 February 2023).

corporate governance issues in asset management, securities brokerage services and associated research functions."[110]

ESG's historical focus has been on financial asset decision-making; specifically, where investments could be placed that were aligned with sustainability and TBL goals. While its finance orientation is still dominant, there has been increased pressure to impact company governance through ESG; the term "stakeholder capitalism" is associated with this. There has been increasing emphasis on social and human capital within this context. There is a wide range of NGOs, financial advisories, and third-party rating agencies which have prepared HC/ESG frameworks and reporting schemes.

It is important to gain awareness and understanding of the tensions that have evolved in the ESG space from the increased emphasis on human and social capital.

It is often asked, "what is the difference between 'sustainability' and 'ESG'?" In some ways, it depends on one's perspective. There is a general consensus that "sustainability," usually thought of in terms of climate change, encompasses ESG as a subset of it.

3.6.1.3 Corporate Social Responsibility

Of the three terms, ESG, sustainability, and CSR, CSR is the oldest. CSR tends to focus externally as opposed to ESG's internal focus (e.g. governance). Corporate philanthropy and support of social causes date back to the late 1800s. This term "corporate social responsibility" was coined in 1953 by Howard Bowen who is often cited as the "father of CSR."[111] The idea of a "social contract" was put forth by the Committee for Economic Development in 1971. Core to this was the idea of business obligation to serve the needs of society which is embedded in the notion of "license to operate."

3.6.1.4 Materiality

This is possibly the most important concept in the sustainability/ESG space. For the layperson, materiality can be thought of "that which is important to creating and maintaining value." Financial materiality is a key distinction in sustainability/ESG reporting; this historically focuses on events (e.g. climate change)

---

110 The Global Compact (2004). *Who Cares Wins: Connecting Financial Markets to a Changing World*. Recommendations by the financial industry to better integrated environmental, social and governance issues in analysis, assessment management and securities brokerage. Swiss Federal Department of Foreign Affairs (Bern) and the United Nations (New York). https://www.unepfi.org/fileadmin/events/2004/stocks/who_cares_wins_global_compact_2004.pdf.
111 Bowen, Howard R. (1953). *Social Responsibilities of the Businessman*. University of Iowa Press.

relevant to the reporting entity, to inform investors of issues affecting enterprise valuation. Several definitions of materiality follow:

Harvard's Law School Forum on Corporate Governance:

> Materiality is an accounting principle which states that all items that are reasonably likely to impact investors' decision-making must be recorded or reported in detail in a business's financial statements using GAAP standards. Essentially, materiality is related to the significance of information within a company's financial statements. If a transaction or business decision is significant enough to warrant reporting to investors or other users of the financial statements, that information is "material" to the business and cannot be omitted.[112]

The International Sustainability Standards Board (ISSB)[113]:

> In the context of sustainability-related financial disclosures, information is material if omitting, misstating or obscuring that information could reasonably be expected to influence decisions that primary users of general purpose financial reports make on the basis of those reports, which include financial statements and sustainability-related financial disclosures and which provide information about a specific *reporting entity*.[114]

The International Financial Reporting Standards (IFRS) Foundation:

> Information is material if omitting, misstating or obscuring it could reasonably be expected to influence decisions that the primary users of general-purpose financial statements make on the basis of those financial statements which provide financial information about a specific reporting entity. – IAS 1 *Presentation of Financial Statements*.[115]

---

112 Harvard Business School: https://online.hbs.edu/blog/post/what-is-materiality (accessed 16 February 2023).
113 ISSB is introduced later in this chapter (§3.6.5).
114 International Sustainability Standards Board (2023): IFRS, General Requirement for Disclosure of Sustainability-related Financial Information. Page 8. https://www.ifrs.org/content/dam/ifrs/publications/pdf-standards-issb/english/2023/issued/part-a/issb-2023-a-ifrs-s1-general-requirements-for-disclosure-of-sustainability-related-financial-information.pdf?bypass=on. (accessed 12 July 2023).
115 IFRS Foundation (2020). Consultation paper on sustainability reporting. September 2020. London. https://www.ifrs.org/content/dam/ifrs/project/sustainability-reporting/consultation-paper-on-sustainability-reporting.pdf (accessed 18 February 2023).

Frances Schwartzkopff with Bloomberg:

> What is materiality? At the basic level it is an accounting principle, referring to something that may have an impact on – be material to – how a company performs. A material risk can threaten targets or goals – something of keen interest to investors. In the context of ESG, this is known as single materiality and means mainly ESG factors that may pose a threat or opportunity to a business and its bottom line, such as extreme weather. It doesn't tell you anything about how "green" a company's business practices are, but rather how vulnerable its earnings may be to ESG risks.[116]

When considering materiality, one should ask, "value to whom?" Historically the focus has been from an investment and shareholder perspective. As sustainability/ESG has evolved, so have concepts of materiality – evolving from a strictly shareholder focus to a broader distinction of stakeholder. This dynamic, shareholder versus stakeholder, is at the core of what is referred to as "stakeholder capitalism," and the source of ESG-related political tensions.

3.6.1.5 Materiality Beyond Financial Reporting – Double/Impact Materiality

Concepts of double and impact materiality have evolved due to increased attention on meeting sustainability/ESG reporting needs beyond the impact to enterprise valuation issues, such as impacts that an enterprise can or does have on the climate, environment, society, and people.

The difference between financial and impact materiality is suggested by SASB:

> Financial materiality in the context of sustainability information represents the sustainability factors that are material to short, medium, and long-term enterprise value. Impact materiality captures the significant impacts an organization has on the economy, environment, and people that are not captured by enterprise value. There are a variety of users with a range of objectives who want to understand an organization's positive and negative contributions to sustainable development.[117]

---

116 Schwartzkopff, F (2022). What New ESG Approach 'Double Materiality' Means – and Why JPMorgan Is a Fan. Bloomberg News. https://www.bloomberg.com/news/articles/2022-09-21/what-double-materiality-means-for-esg-and-jpmorgan-quicktake (accessed 17 February 2023).
117 SASB. https://sasb.wpengine.com/about/sasb-and-other-esg-frameworks/ (accessed 16 February 2023).

Frances Schwartzkopff with Bloomberg offers:

> What is double materiality? That's where greenness comes in. 'Double materiality' adds the risks a company's activities pose to the environment and society to those that it potentially faces internally. How such things should be accounted for in corporate reports remains the subject of intense debate. For now, reports vary wildly, making it hard for investors to compare one company or fund with another and make informed decisions.

European Commission guidelines and regulations introduced a double materiality perspective beginning in 2017 with the Commission's Nonbinding Guidelines on Nonfinancial Reporting, which became binding in 2019. In addition to assessing "financial materiality in the broad sense of affecting the value of a company," impacts of the company's activities beyond the fence line which affect environmental and social materiality needed to be included.[118]

Figure 3.8 graphically depicts the double materiality perspective presented in the EU's Nonfinancial Reporting Directive.[119]

Double materiality is an important concept to understand, as many of the sustainability/ESG reporting tensions originate from it. In the shareholder/stakeholder dynamic, there has been increased expectations regarding what is in financial statements, what is reported, and how materiality is addressed. Mattias Tager at the London School of Economics observes: "Accounting standards are not neutral, but they systematically affect capital allocation and market dynamics. Decades of global harmonisation have veiled the fact that accounting practices are simply social conventions and not exact or objective measures."[120]

3.6.2 Human Capital

Ideas about human capital have been around at least since the writings of the 18th-century economist Adam Smith. The element of human capital, has been included in corporate sustainability reports since the 1990s. In recent years, there has been increased attention given to human and social capital within ESG reporting.

---

118 European Commission (2019). *Guidelines on Reporting Climate-related Information*, 6. Brussel, Belgium.
119 European Commission (2019). *Guidelines on Reporting Climate-related Information*. Brussel, Belgium. https://ec.europa.eu/finance/docs/policy/190618-climate-related-information-reporting-guidelines_en.pdf (accessed 24 February 2023). p. 7.
120 Tager, M. (2021). Double materiality: what is it and why does it matter? London School of Economics and Political Science. https://www.lse.ac.uk/granthaminstitute/news/double-materiality-what-is-it-and-why-does-it-matter/ (accessed 26 October 2021).

122 | 3 Frameworks

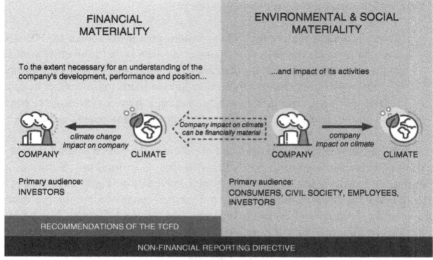

* *Financial materiality is used here in the broad sense of affecting the value of the company, not just in the sense of affecting financial measures recognised in the financial statements.*

Figure 3.8 Double materiality example.

The application of "capital thinking" in ESG is relatively new. The term "natural capital" was first used in the 1970s and adopted into environmental economics in the late 1980s. The application of natural capital thinking in the ESG space led to the formation of The Economics of Ecosystems and Biodiversity (TEEB) for the Business Coalition in 2012, which became the Natural Capital Coalition in 2014. In 2018, the Social and Human Capital Coalition was founded, and in 2020, the two coalitions joined to form the Capitals Coalition. The new entity characterizes itself as a "...global collaboration transforming the way decisions are made by including value provided by nature, people, and society." *The Capitals Coalition presents a common concept of capital as natural, social, human, and produced capital.* The coalition defines human capital as "...the knowledge, skills, competencies and attributes embodied in individuals that facilitate the creation of personal, social, and economic well-being."[121]

The advent of "One Report" and the formation of the International Integrated Reporting Committee (IIRC) in 2010 introduced and began to formalize ideas of integrated thinking within the ESG space. "Capitals thinking" is reflected in the IIRC framework, which identifies six capitals as financial, manufactured, intellectual, human, social and relationship, and natural. The IIRC defines human capital as "the knowledge, skills, competencies and other attributes embodied in

---

121 Capitals Coalition (2019), Social and Human Protocol, p. 11.

individuals or groups of individuals acquired during their life and used to produce goods, services, or ideas in market circumstances".[122]

In August 2020, the U.S. SEC published a rule that, beginning in November 2020, required public companies to provide:

> [a] description of the registrant's human capital resources, including the number of persons employed by the registrant, and any human capital measures or objectives that the registrant focuses on in managing the business (such as, depending on the nature of the registrant's business and workforce, measures or objectives that address the development, attraction, and retention of personnel).[123]

Studies indicate wide variation in what companies report pertaining to human capital. One study that looked at 100 company reports identified six different broad categories of human capital[124]:

1) Employment and Labor Type
2) Job Stability
3) Wages, Compensation, and Benefits
4) Workforce Diversity, Equity, and Inclusion
5) Occupational Health and Safety
6) Training and Education

Two metrics identified in the study for occupational health and safety were absenteeism rate and total recordable incidence rate (TRIR). Note that these are lagging indicators, as opposed to leading ones.[125]

3.6.3 Reporting and Performance Criteria

ESG, sustainability, CSR, and human capital performance and reporting criteria are developed and established by both government and NGOs. These are in the form of nonmandatory standards companies can use, and criteria developed by ESG rating agencies that are used to assess company performance.

In the early 2020s there was significant consolidation of NGO standards-developers. During this time, the International Sustainability Standards Board

---

122 Note, the IIRC is now part of the International Sustainability Standards Board (ISSB).
123 SEC (August 26, 2020), "Modernization of Regulation S-K Items 101, 103, and 105." https://www.sec.gov/rules/final/2020/33-10825.pdf. 101(c)(2)ii. Accessed December 21, 2021.
124 Omens, A. (2021): "The Current State of Human Capital Disclosure." Blog post, October, 31, 2021. Harvard Law School Forum on Corporate Governance. https://corpgov.law.harvard.edu/2021/10/31/the-current-state-of-human-capital-disclosure/. Accessed February 28, 2022.
125 Example leading indicators include an entity's management system (e.g., ISO 45001) and elements within the system, such as leadership, participation, policies, exposure assessment, and training.

(ISSB) evolved as the dominant standards-defining entity. Entities that have merged or been consolidated with ISSB include:

- *Sustainability Accounting Standards Board (SASB)*: Formed in 2011 in the context of U.S. securities disclosures. Merged with ISSB in August 2022.
- *Climate Disclosure Standards Board (CDSB)*: Formed in 2007 as part of the Carbon Disclosure Project that started in 2002. Merged with ISSB, in January 2022.
- *International Integrated Reporting Committee (IIRC)*: Formed in 2010 and merged with SASB in 2021 to form the *Value Reporting Foundation* which merged with ISSB in August 2021.

3.6.4 Global Reporting Initiative (GRI)

The Coalition for Environmentally Responsible Economies (CERES) was formed in Boston, in 1989. This NGO pushed for companies to provide public reporting on their sustainability efforts and progress. This was followed by the first United Nations conference on environment and development in Rio de Janeiro in 1992. This conference officially promoted social and economic development conducted in a sustainable manner.

The Global Reporting Initiative (GRI) was formed in 1997 by CERES and the Tellus Institute, with support from the United Nations Environmental Programme (UNEP). GRI and ISSB executed an agreement (MoU) in March 2022 "seeking to coordinate work programmes and standard-setting activities."[126]

GRI has become the gold standard for sustainability reporting relative to occupational health and safety (GRI 403: Occupational Health And Safety 2018). The GRI framework has undergone several changes in organization and has expanded its scope. The second UN conference called RIO+10 was held in Johannesburg, South Africa in 2002. In this conference, GRI published a recommended set of metrics for companies to include in their sustainability reporting. There were additional conferences held by the UN on sustainability which included 2012 again in Rio de Janeiro (RIO+20) and most recently in New York City in 2015. At the New York conference, they published their goals for 2030 which were adopted by 193 member nations. The major emphasis for their goals is in the social and economic arenas in addition to environmental concerns such as protecting world habitats in the face of climate change.

---

126 ISSB press release, 23 June 2022. https://www.ifrs.org/news-and-events/news/2022/06/issb-and-gri-provide-update-on-ongoing-collaboration/ (accessed 18 February 2023).

On a parallel path, Paul Hawken published the "Ecology of Commerce" in 1993 which provided a restorative path for business. Industrial ecology developed concurrently with the UN Environmental Program, GRI and footprint methodologies.[127] Life-cycle analysis grew out of the ISO 14001 standard, an international agreement on environmental management first published in the mid-1990s. The global footprint network, founded in 2003, relied on input-output tables to characterize environmental impacts. Input-output tables have been successfully used to measure the effects of industrial activity on the environment.[128] They have been suggested as a way to measure industrial activity on worker health and well-being.[129] The Global Resource Footprint of Nations[130] describes how water, land, and materials are embodied in trade and final consumption. The development of these tools has inspired, invigorated and informed consumers to act on behalf of both the environment and workers. Fairtrade groups and consumer advocates have lobbied to improve environmental and occupational health.

GRI standards are grouped within two categories: universal standards, and three topic-specific standards. Topic-specific standards include economic (series 200), environmental (series 300), and social (series 400). Within the social series, occupational health and safety is addressed in GRI 403, which has 10 disclosure categories.[131]

3.6.5 International Sustainability Standards Board

The International Sustainability Standards Board (ISSB) was established under the International Financial Reporting Standards (IFRS) Foundation in 2021, and the board's first meeting was held in July 2022. Their initial focus has been on the development and promulgation of two standards. The term "Exposure Draft" refers to the draft documents that lead up to an actual standard. The two issues (standards) under discussion since its inception are:

1) General Requirements for Disclosure of Sustainability-related Financial Information (S1), and
2) Climate-related Disclosures (S2).

---

127 (Graedel and Allenby, 2010; Boons and Howard-Grenville, 2009; Batty and Hallberg, 2010).
128 (Hendrickson, Lave and Matthews, 2006; Murray and Wood, 2010).
129 (Scanlon, Gray et al., 2013; Kijko et al., 2015).
130 (Tukker, A., Bulavskaya, T., Giljum, S., de Koning, A., Lutter, S., Simas, M., Stadler, K., Wood, R. 2014).
131 https://www.globalreporting.org/media/zripcgu2/gri-403-occupational-health-and-safety-2018-presentation.pdf. Accessed, April 12, 2019.

At its February 16, 2023 meeting, ISSB discussed the Exposure Drafts for these two items, set a goal to ballot them by the end of Q2 2023, and to promulgate them on 1 January 2024. The Board tentatively decided to introduce a requirement to permit, but not require, preparers to consider the GRI Standards and the European Sustainability Reporting Standards (ESRS) in identifying disclosures about sustainability-related risks and opportunities.

3.6.6 Value Reporting Foundation, IIRC, SASB[132]

SASB was founded in 2011; its primary aim was to develop standards for use in corporate filings to the U.S. Securities and Exchange Commission (SEC). The intention was to provide investors with comparable, nonfinancial information about the companies whose stocks they or their investment funds owned and to allow investors and financial analysts to compare performance on critical ESG issues within an industry.

SASB states that its standards focus on "financially material issues because our mission is to help businesses around the world report on the sustainability topics that matter most to their investors." This focus has been recognized by firms such as BlackRock, as referenced in Larry Fink's 2020 letter to CEOs, which said, "…BlackRock believes that the Sustainability Accounting Standards Board (SASB) provides a clear set of standards for reporting sustainability information across a wide range of issues, from labor practices to data privacy to business ethics."[133]

SASB's emphasis on financial materiality sets it apart from other sustainability reporting standards, such as those of the Global Reporting Initiative (GRI), by focusing on "…a company's impacts on the broader economy, environment and society to determine its material issues."[134]

There are 77 industry-specific SASB standards that are organized within 11 sectors. The sectors are Consumer Goods; Extractive and Materials Processing; Financials; Food and Beverage; Health Care; Infrastructure; Renewable Resources and Alternative Energy; Resource Transformation; Services; Technology and Communications; and Transportation.

SASB identifies 26 material areas as follows.

---

132 These entities merged with the International Sustainability Standards Board in 2021. They are called out separately here for historical purposes.
133 https://www.blackrock.com/americas-offshore/en/larry-fink-ceo-letter. Accessed 4 June 2021.
134 https://en.wikipedia.org/wiki/Sustainability_Accounting_Standards_Board (accessed 19 February 2023).

Environment:

- GHG Emissions
- Air Quality
- Energy Management
- Water and Wastewater Management
- Waste and Hazardous Materials Management
- Ecological Impacts

Social capital:

- Human Rights and Community Relations
- Customer Privacy
- Data Security
- Access and Affordability
- Product Quality and Safety
- Customer Welfare
- Selling Practices and Product Labeling

Human capital:

- Labor Practices
- Employee Health and Safety
- Employee Engagement, Diversity, and Inclusion

Business model and innovation:

- Product Design and Lifecycle Management
- Business Model Resilience
- Supply Chain Management
- Materials Sourcing and Efficiency
- Physical Impacts of Climate Change

Leadership and governance:

- Business Ethics
- Competitive Behavior
- Management of the Legal and Regulatory Environment
- Critical Incident Risk Management
- Systemic Risk Management

3.7 Transcending Paradigms

> The ancient Egyptians built pyramids because they believed in an afterlife. We build skyscrapers because we believe that space in downtown cities is enormously valuable. (Except for blighted spaces, often near the skyscrapers, which we believe are worthless.) Whether it was Copernicus and Kepler showing that the earth is not the center of the universe, or Einstein hypothesizing that matter and energy are interchangeable, or Adam Smith postulating that the selfish actions of individual players in markets wonderfully accumulate to the common good, people who have managed to intervene in systems at the level of paradigm have hit a leverage point that totally transforms systems.[135]

In circling back to Donnella Meadows's "Leverage Points: Places to Intervene in a System" that was introduced in Chapter 1 (Table 1.1), I focus on her two highest leverage points:

1) the power to transcend paradigms.
2) the mindset or paradigm out of which the system – its goals, structures, rules, delays, parameters – arise.

When thinking about these two leverage points it's valuable to consider the mindsets and paradigms from which the content of this chapter has risen. In Chapter 2, I offered insights into how norms, laws, regulations, and ways of responding to these have evolved. In Section 2.1.1, I suggested that a forward-looking place is the end point of five evolution sequences, which are impact, field, capitals, stakeholders and, value, and purpose. I built on these in developing the Risk Matrix presented in Chapter 8. Three initiatives that also helped inform the development, and I would suggest represent Meadows's notion of transcending paradigms, are 1) NIOSH's Total Worker Health initiative; 2) the Robert Wood Johnson Foundation (RWJF) and Global Reporting Initiative's (GRI) Culture of Health for Business initiative; and 3) the capitals-related ideas and protocols developed by the Capitals Coalition. I suggest that these initiatives demonstrate Meadows's idea of transcending paradigms. Each of these initiatives is briefly introduced here and discussed further in later chapters.

---

135 Meadows, D. (1999). *Leverage Points: Places to Intervene in a System*, 18. Hartland, VT: The Sustainability Institute.

3.7.1 NIOSH Total Worker Health

The National Institute for Occupational Safety and Health (NIOSH) is the federal institute responsible in the United States for conducting research and making recommendations for the prevention of work-related injury and illness. It was established by the Occupational Safety and Health Act of 1970. Following a long history of innovation, the Total Worker Health (TWH) approach was formalized in 2011 as the cumulation of a series of workshops and agency research that began in 2003 with the *Healthier U.S. Workplace Initiative*. A *Total Worker Health* approach is defined as, "policies, programs, and practices that integrate protection from work-related safety and health hazards with promotion of injury and illness–prevention efforts to advance worker well-being."[136]

Five defining elements of TWH are:

Defining element of TWH 1: Demonstrate leadership commitment to worker safety and health at all levels of the organization.

Defining element of TWH 2: Design work to eliminate or reduce safety and health hazards and promote worker well-being.

Defining element of TWH 3: Promote and support worker engagement throughout program design and implementation.

Defining element of TWH 4: Ensure confidentiality and privacy of workers.

Defining element of TWH 5: Integrate relevant systems to advance worker well-being.[137]

Aspects of these defining elements are expanded (transcending paradigms) in the Risk Matrix by considering the social-human dimension of organizational risk management. TWH Element 5 in particular is a platform for the integrated thinking, decision-making, and action promoted and supported by the Risk Matrix.

NIOSH offers a comprehensive list of "issues relevant to advancing worker well-being through Total Worker Health." These are depicted in Figure 3.9[138].

---

136 NIOSH (2016). Fundamentals of total worker health approaches: essential elements for advancing worker safety, health, and well-being. By Lee MP, Hudson H, Richards R, Chang CC, Chosewood LC, Schill AL, on behalf of the NIOSH Office for Total Worker Health. Cincinnati, OH: U.S. Department of Health and Human Services, Centers for Disease Control and Prevention, National Institute for Occupational Safety and Health. DHHS (NIOSH) Publication No. 2017-112. p. 1.
137 NIOSH (2016), p. 3.
138 NIOSH (2016), p. 17.

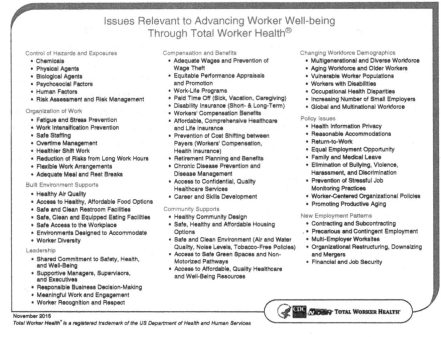

Figure 3.9 Issues relevant to advancing worker well-being through total worker health. *Source:* Adapted from NIOSH (2016).

3.7.2 Culture Health for Business (COH4B)

The RWJF has been funding research on the Culture of Health (COH) for a number of years. In 2013, they engaged the Rand Corporation to develop a COH framework, and in 2017 they began working with GRI to conduct the research that led to the COH4B "framework" that contains 16 "practices" (principles) that are arranged in four categories (strategic, policy and benefits, workforce and operations, and community). The COH4B framework is presented in "A Culture of Health for Business: Guiding Principles to Establish a Culture of Health for Business." This was published in April 2019.

The Culture of Health for Business (COH4B) Framework, which resulted from the RWJF/GRI collaboration, "is a pioneering, holistic framework on the role of business in impacting the health and well-being of its stakeholders, linked to a curated set of principles and business practices. It is a multi-stakeholder-developed, evidence-based public tool which companies can use to take comprehensive action on the health and well-being of employees, families, and communities."[139] The

---

139 GRI (2020). Fact sheet, using the GRI standards with the culture of health for business framework. https://www.globalreporting.org/public-policy-partnerships/strategic-partners-programs/culture-of-health-for-business (accessed 18 March 2022).

3.7 Transcending Paradigms | 131

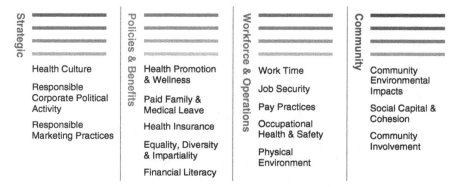

Figure 3.10 COH4B practices. *Source:* GRI (2020)/with permission of Global Reporting Initiative.

framework identifies 16 practices that are organized into 4 categories. These are depicted in Figure 3.10.[140]

The framework's research was robust. Through my engagement with the Center for Safety and Health Sustainability, I supported the effort's advisory committee in document reviews and workshops. In the final report, the 16 practices were analyzed against 5 groupings of health determinants (individual behavior, social environment, biology and genetics, and physical environment), and 2 groupings of positive impact profile outcomes (business outcomes, and health outcomes).[141]

Aspects of the COH4B framework and research findings are woven into the Risk Matrix (Chapter 8) and its applications (Chapters 9 and 10).

3.7.3 Capitals Coalition

The Capitals Coalition is mentioned at numerous junctures in this book. This entity is at the forefront of "transforming the way decisions are made by including the value provided by nature, people, and society." At the core of their work is the development of tools, frameworks, and initiatives based on a capitals approach, which "identifies, measures and values the flows and benefits that come from natural, social, human, and/or produced/financial stocks. This integrated

---

140 GRI (2020). Fact sheet, using the GRI standards with the culture of health for business framework. https://www.globalreporting.org/public-policy-partnerships/strategic-partners-programs/culture-of-health-for-business (accessed 18 March 2022).
141 RWJF/GRI (2019). *Culture of Health for Business, Guiding Principles to Establish a Culture of Health for Business.* p. ix. https://www.globalreporting.org/public-policy-partnerships/strategic-partners-programs/culture-of-health-for-business/ (accessed 12 October 2019).

132 | *3 Frameworks*

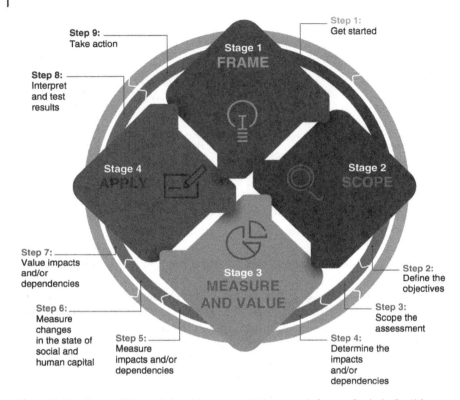

Figure 3.11 Steps of the social and human capital protocol. *Source:* Capitals Coalition (2019)/with permission of Capitals Coalition.

thinking can be used to carry out assessments or produce integrated accounts. By providing robust and actionable information that includes all the capitals we can better inform decisions."[142]

The Social and Human Capital Protocol (SHCP) along with the Natural Capital Protocol (NCP) provide foundational principles for measuring and valuing social, human and natural capitals. These protocols are presented as four-stage frameworks that each have 9 steps; the SHCP framework is shown in Figure 3.11.[143] I go into these steps in Chapter 7 (Decision-making), and Chapter 9 (Matrix in Action). Of particular interest in developing integrated thinking,

---

142 Capitals Coalition (2020). Our shared ambition. p. 2.
143 Capitals Coalition (2019). Social and Human Capital Protocol, p. 7.

decision-making, and action skills is the determination of impact and dependency pathways. These pathways are immensely valuable in beginning to see interconnections in a risk field. This is discussed further in Chapter 9 (Matrix in Action).

Appendix 3.A   ISO 3100:2018 Principles

This standard contains eight risk management principles. These are depicted in Figure 3.3.[144]

a) Integrated

Risk management is an integral part of all organizational activities.

b) Structured and comprehensive

A structured and comprehensive approach to risk management contributes to consistent and comparable results.

c) Customized

The risk management framework and process are customized and proportionate to the organization's external and internal context related to its objectives.

d) Inclusive

Appropriate and timely involvement of stakeholders enables their knowledge, views, and perceptions to be considered. This results in improved awareness and informed risk management.

e) Dynamic

Risks can emerge, change or disappear as an organization's external and internal context changes. Risk management anticipates, detects, acknowledges, and responds to those changes and events in an appropriate and timely manner.

f) Best available information

The inputs to risk management are based on historical and current information, as well as on future expectations. Risk management explicitly takes into account any limitations and uncertainties associated with such information

---

144 International Organization for Standardization (2018). *Risk Management – Guidelines, ISO 31000;*. 3–4. Geneva, Switzerland.

and expectations. Information should be timely, clear and available to relevant stakeholders.

g) Human and cultural factors

Human behavior and culture significantly influence all aspects of risk management at each level and stage.

h) Continual improvement

Risk management is continually improved through learning and experience.

Appendix 3.B COSO ERM (2017) Principles

The COSO ERM framework (2017) contains twenty principles that are organized within five interrelated components.[145]:

1) *Governance and culture*: Governance sets the organization's tone, reinforcing the importance of, and establishing oversight responsibilities for, enterprise risk management. Culture pertains to ethical values, desired behaviors, and understanding of risk in the entity.
   1) *Exercises board risk oversight*: The board of directors provides oversight of the strategy and carries out governance responsibilities to support management in achieving strategy and business objectives.
   2) *Establishes operating structures*: The organization establishes operating structures in the pursuit of strategy and business objectives.
   3) *Defines desired* culture: The organization defines the desired behaviors that characterize the entity's desired culture.
   4) *Demonstrates commitment to core* values: The organization demonstrates a commitment to the entity's core values.
   5) *Attracts, develops, and retains capable* individuals: The organization is committed to building human capital in alignment with the strategy and business objectives.
2) *Strategy and objective-setting*: Enterprise risk management, strategy, and objective-setting work together in the strategic-planning process. A risk appetite is established and aligned with strategy; business objectives put strategy into practice while serving as a basis for identifying, assessing, and responding to risk.
   6) *Analyzes business* context: The organization considers potential effects of business context on risk profile.

---

145 Committee of Sponsoring Organizations of the Treadway Commission (2017). *Enterprise Risk Management – Integrated Framework, Executive Summary* Framework, p. 6, and 10.

7) *Defines risk appetite*: The organization defines risk appetite in the context of creating, preserving, and realizing value.
8) *Evaluates alternative* strategies: The organization evaluates alternative strategies and their potential impact on risk profile.
9) *Formulates business* objectives: The organization considers risk while establishing the business objectives at various levels that align and support strategy.

3) *Performance*: Risks that may impact the achievement of strategy and business objectives need to be identified and assessed. Risks are prioritized by severity in the context of risk appetite. The organization then selects risk responses and takes a portfolio view of the amount of risk it has assumed. The results of this process are reported to key risk stakeholders.
   10) *Identifies risk*: The organization identifies risk that impacts the performance of strategy and business objectives.
   11) Assesses severity of risk: The organization assesses the severity of risk.
   12) *Prioritizes risks: The organization prioritizes risks as a basis for selecting responses* to risks.
   13) *Implements risk responses*: The organization identifies and selects risk responses.
   14) *Develops portfolio view*: The organization develops and evaluates a portfolio view of risk.

4) *Review and revision*: By reviewing entity performance, an organization can consider how well the enterprise risk management components are functioning over time and in light of substantial changes, and what revisions are needed.
   15) *Assesses substantial change*: The organization identifies and assesses changes that may substantially affect strategy and business objectives.
   16) Reviews risk and performance: The organization reviews entity performance and considers risk.
   17) Pursues improvement in enterprise risk management: The organization pursues improvement in enterprise risk management.

5) *Information, communication, and reporting*: Enterprise risk management requires a continual process of obtaining and sharing necessary information, from both internal and external sources, which flows up, down, and across the organization.
   18) *Leverages information systems*: The organization leverages the entity's information and technology systems to support enterprise risk management.
   19) *Communicates risk information*: The organization uses communication channels to support enterprise risk management.
   20) *Reports on risk, culture, and performance*: The organization reports on risk, culture, and performance at multiple levels and across the entity.

Suggested Reading

Capitals Coalition (2019). *Social and human capital protocol.* https://capitalscoalition.org/capitals-ap- proach/social-human-capital-protocol/ (accessed 14 August 2021).

Committee of Sponsoring Organizations of the Treadway Commission (2017). *Enterprise Risk Management – Integrating with Strategy and Performance.* Association of International Certified Professional Accountants.

International Organization for Standardization (2002). *Risk Management – Vocabulary, Guide 73.* Geneva: International Organization for Standardization.

International Organization for Standardization (2015). *Environmental Management Systems – Requirements with guidance for Use, ISO 14001.* Geneva, Switzerland: International Organization for Standardization.

International Organization for Standardization (2018). *Occupational Health and Safety Management Systems – Requirements with Guidance for use, ISO 45001.* Geneva, Switzerland: International Organization for Standardization.

International Organization for Standardization (2018). *Risk Management - Guidelines, ISO 31000.* Geneva, Switzerland: International Organization for Standardization.

International Organization for Standardization (2021). *Governance of organizations – Guidance, ISO 37000.* Geneva, Switzerland: International Organization for Standardization.

National Research Council (1983). *Risk Assessment in the Federal Government: Managing the Progress.* Washington, DC: National Academy Press.

National Research Council (2009). *Science and Decisions: Advancing Risk Assessment.* Washington, DC: National Academy Press.

NIOSH (2016). *Fundamentals of total worker health approaches: essential elements for advancing worker safety, health, and well-being.* By Lee MP, Hudson H, Richards R, Chang CC, Chosewood LC, Schill AL, on behalf of the NIOSH Office for Total Worker Health. Cincinnati, OH: U.S. Department of Health and Human Services, Centers for Disease Control and Prevention, National Institute for Occupational Safety and Health. DHHS (NIOSH) Publication No. 2017-112. https://www.cdc.gov/niosh/docs/2017-112/pdfs/2017_112.pdf?id=10.26616/NIOSHPUB2017112 (accessed 21 February 2017).

RWJF/GRI (2019). *Culture of health for business, guiding principles to establish a culture of health for business.* https://www.globalreporting.org/public-policy-partnerships/strategic-partners-programs/culture-of-health-for-business/ (accessed 17 July 2019).

4

Conformity Assessment and Measurement

CONTENTS

4.1 Frameworks and Guidelines, 139
    4.1.1 National Research Council, 139
    4.1.2 ISO Committee on Conformity Assessment (CASCO), 140
    4.1.3 Inference Guidelines and Decision-Making Currency, 140
4.2 Measurement, 141
4.3 Auditing, 143
    4.3.1 Historical Background, 143
    4.3.2 Types of Audits, 144
Suggested Reading, 146

All evaluation approaches constitute a pedagogy of some kind.[1]

Michael Quinn Patton

Conformity assessment was briefly introduced in Chapter 2 (Risk Logics) and is revisited in later chapters. Conformity assessment is a much-discussed topic in numerous circles, including regulatory and standards development, product certifications, and EHSS management system certifications. At the core of conformity assessment are the questions, "who is the judge, who can be a/the judge, and are there unknown judges?" As EHSS management developed over the years and as the risk field depiction offered in Chapter 2 (Figures 2.2 and 2.7) demonstrates, numerous judges can be identified. Conformity-assessment became more familiar in EHSS with the introduction of management systems into organizational

---

1 Patton, Michael Quinn (2017). *Editor's Notes, Pedagogy of Evaluation.* New Directions for Evaluation, Wiley. No. 155, Fall 2017, p. 9.

Organizational Risk Management: An Integrated Framework for Environmental, Health, Safety, and Sustainability Professionals, and their C-Suites, First Edition. Charles F. Redinger.
© 2025 John Wiley & Sons, Inc. Published 2025 by John Wiley & Sons, Inc.

practices and their certification by third-party entities in the 1990s. In early years, the term "registered" was used to refer to an organization's management system when it was found to be in conformance with a standard by a third-party registrar. Over time, this shifted from "registered" to "certified." Third-party certification of ISO management systems is commonly used in conformity assessment activity. Such external certifications are not required by ISO. Rather, these are driven by market or stakeholder requirements.

Occasionally, the distinction between a management system standard (MSS) and system certification can become blurred. That is, organizations can and do use MSSs to develop internal systems. This does not necessarily mean that the system needs third-party verification (e.g. certification). The reasons for external verification are related to the question, of who will be the judge. The judge refers to whoever is making the assessment and why, or who is making decisions based on the assessment. The assessment involves determining how, and the extent to which, an organization is conforming to a given standard, norm, regulation, and so on. In the context of a management system, the judge is the third-party registrar or certification entity. With regulatory compliance, the regulating agency is the judge. The entities (individual, team/department, enterprise/company, community) included in the Risk Matrix's (Chapter 8) actors/motive force dimension can also be considered judges.

A central component of assessments are audits, within which measurement methods are used to establish a foundation. Decision-making is a central activity in EHSS/ORM. Within decision-making are other important considerations that include operational distinctions, data collection forms, and types of variables.

Pedagogy refers to methods and practices of teaching. As we move forward with the conformity-assessment topics in this chapter, I point out connections with organizational learning, particularly with double-loop learning (§5.6.2.2). An important feature borrowed from the program evaluation field distinguishes between summative and formative evaluations. Traditional EHSS/ORM audits tend to be summative; that is, they present results (outcomes) but devote little attention to explaining why the results were what they were. Thus, if the results are disappointing, there is little information about how to improve performance. Over time, there have been improvements that attempt to explain the results so that organizations can learn and make any necessary mid-course corrections; this is characterized as formative evaluation. The generative field construct is introduced in Chapter 10 (§10.2) where I revisit the topics of formative assessments, evaluations, audits, etc., and frame them as pedagogies (tools for learning) that can escalate impact and are integral in evolving generative EHSS/ORM fields.

This chapter is influenced by the conformity-assessment frameworks and practices that have evolved in, and are associated with, ISO. This evolution is characterized and used to frame related concepts and practices as well as the evolving social–human era, namely in terms of ESG, human capital, and DE&I.

Conformity assessment is an important concept that relates to auditing and determining whether a consensus standard, such as ISO 31000, or a framework, such as COSO's ERM, has been implemented true to the standard or framework. Two definitions of conformity assessment are:

- "The determination of whether a product or process conforms to particular standards or specifications. Activities associated with conformity assessment include testing, certification, and quality assurance."[2]
- "The process of conformity assessment demonstrates whether a product, service, process, claim, system, or person meets the relevant requirements. Such requirements are stated in standards, regulations, contracts, programs, or other normative documents."[3]

4.1 Frameworks and Guidelines

Over the years, various conformity assessment frameworks have been developed by public and private organizations including the National Research Council and the ISO.

4.1.1 National Research Council

In 1995, the NRC's Science, Technology, and Economic Policy Board published a report titled "Standards, Conformity Assessment, and Trade." The report was issued pursuant to Public Law 102-245, which was enacted to gain a better understanding of the interrelationships between standards, conformity assessment, and international trade.

The NRC report presents a generic conformity-assessment model that can be applied to manufacturing, service delivery, or quality management systems. The model presents a conformity-assessment hierarchy with three levels that include assessment (primary), accreditation (secondary), and recognition (tertiary).

> The primary level represents measurement activities, including auditing. Workplace air sampling or wastewater sampling are examples of assessment activities, as are management system audits. The secondary level addresses the formal qualifications of the entities performing primary level activities and the bodies that provide confirmation of those qualifications. Examples are the Certified Safety Professionals (CSP) and

---

2 National Research Council (1995). *Standards, Conformity Assessment, and Trade: Into the 21st Century*. Washington, DC: National Academy Press.
3 ISO. https://www.iso.org/conformity-assessment.html (accessed 13 November 2023).

Certified Industrial Hygienists (CIH) who perform workplace assessments. The CSP and CIH designations are given respectively by the Board of Certified Safety Professionals (BCSP) and the Board for Global EHS Credentialing (BGC). The certification functions performed by the BCSP and BGC represent secondary level activities.[4]

The tertiary level addresses the standards, regulations, and people who recognize accreditations.

4.1.2 ISO Committee on Conformity Assessment (CASCO)

The development of conformity-assessment policies and procedures in ISO is the responsibility of ISO's Committee on Conformity Assessment (CASCO). CASCO, in its current form, was created in 1985 as a successor to the former CERTICO – the ISO committee on certification – which was initially established in 1970. The original focus of CERTICO was on the principles and practices of product certification. The change in name from CERTICO to CASCO explicitly reflected the extension of the field of certification to all conformity-assessment activities. In particular, the extension included quality systems, thereby responding to market demand and to the publication of the ISO 9000 series standards. CASCO's primary task is to develop guidance documents for use by ISO Technical Committees (TC) in the development of specific ISO standards.

A trend in CASCO is to prepare generic guidelines whereby TCs can develop conformity-assessment schemes unique to the TC's specific industry/service sectors while maintaining some level of uniformity between the conformity-assessment work of different TCs (e.g. accreditation of conformity-assessment bodies: laboratories, inspection bodies, certification bodies, etc.).

4.1.3 Inference Guidelines and Decision-Making Currency

When considering risk management decision-making, it is valuable to understand the frameworks within which one is operating, the principles and logics underpinning it, and the criteria used to judge performance. A term introduced in the *Red Book*[5] is "inference guideline," which is defined as "...the principles followed by risk assessors in interpreting and reaching judgments based on scientific data." Such guidelines establish performance criteria and are integral in making

---

4 Redinger, C.F. and Levine, S.P. (1998): "Analysis of Third-Party Certification Approaches using The Michigan Occupational Health and Safety Conformity Assessment Model." *Am. Ind. Hyg. Assoc. J.* 59:802.
5 This is the colloquial term used for the National Research Council publication (1983). *Risk Assessment in the Federal Government: Managing the Progress.* Washington, DC: National Academy Press.

conformity assessment decisions. Both inference guidelines and conformity assessment frameworks measure performance objectives, either in quantitative terms (interval) such as occupational exposure limits (OELs); in qualitative terms such as exposure banding levels (ordinal); or with programmatic or systems assessments, such as management systems (nominal and ordinal).

4.2 Measurement

Few topics are more central to organizational life than measurement. It is important to understand its technical aspects and its role in conformity assessment and decision-making. "What gets measured, gets done" is a common phrase we hear in our organizational lives. Many EHSS/ORM professionals are fixated on metrics, indicators, KPIs, dashboards, analytics, and so on. I think it is fair to say that there has been a dramatic increase in the focus on data, specifically data mining,[6] and the use of analytics.

It is valuable to consider "how do we know what we know?" This is a question traditionally addressed in philosophy, specifically epistemology, and, more recently, in cognitive science. However, for this book, the focus is less academic and more on the practical day-to-day aspects of organizational risk decision-making. Beyond our formal training, life experiences, and inputs from numerous sources (e.g. the news, professional journals), "what we know" about EHSS/ORM risk decision-making largely comes from organization-generated outputs such as audits and assessments.

A hierarchical relationship among the topics in this chapter can be depicted as:

Variable → Indicator → Measurement → Assessment → Conformity Assessment.

For this discussion, the focus is on measurement, which many of us in EHSS/ORM often think of as mostly synonymous with assessment. What follows is intended to be a brief introductions to a handful of measurement concepts.

Measurement in EHSS/ORM is technical, involving such disciplines as engineering, chemistry, economics, statistics, and organizational theory and behavior. It is beyond the scope of this book to provide skills to develop EHSS/ORM measurement tools. What I do provide here are some basic concepts, a sense of the complexities involved, and definitions of certain key terms that are important to understand because they appear in later chapters.

Measurement activities focus on the key components of EHSS/ORM that have already been mentioned: (1) implementation, (2) outputs, and (3) outcomes.

---

6 Society for Risk Analysis (SRA) annual conference 2019, plenary session (12/10/19), suggestion that "mining" large data sets, and analytics could replace traditional scientific methods/practices.

Note that outcomes can be short-term, mid-term or long-term (long-term outcomes are often referred to as impacts in ESG contexts). Every variable that you identify as useful for decision-making will have to be operationalized, which means that an indicator for each variable must be well defined. For example, if you're measuring the variable *temperature*, you would need to specify the measuring instrument, the scale (centigrade) and the level of accuracy required (e.g., +/−0.1 °C). Or, if you were measuring a variable about whether you produced quarterly reports to management about decreasing a particular emission, the indicator would simply be 1 = "Yes" and 2 = "No."[7] Everyone involved in collecting performance data needs to use these same indicators and analytical methods consistently. In addition, the analytical methods and reporting requirements should be specified for each variable.

In measuring the performance of a risk management system (RMS), you will develop variables with indicators at the nominal, ordinal, and interval levels:

1) **Nominal Variables**: Variables whose attributes have only one characteristic of exhaustiveness and mutual exclusiveness. Examples include sex, religious affiliation, political party affiliation, and birthplace. In the RMS, an example is the presence of an ORM policy statement, or an audit program procedure.
2) **Ordinal Variables**: Variables with attributes that we can logically rank order. The different attributes of ordinal variables represent relatively more or less of the variable. Variables of this type include social class, conservatism, and alienation. In the RMS, there might be a scale to rank order the quality of the ORM policy, or the audit program procedure.
3) **Interval Variables**: For the attributes composing some variables, the actual distance separating these attributes does have meaning. For these, the logical distance between attributes can be expressed in meaningful standard intervals. For example, in the Fahrenheit temperature scale, the difference or distance between 80° and 90° is the same as that between 40° and 50°.
4) **Ratio Variables**: The attributes composing a ratio variable, besides having all the structural characteristics mentioned previously, are based on a true zero point. The Kelvin temperature scale is one such measurement.[8] In the RMS, and EHSS/ORM, ratio variables are not common.

---

7 Poister, Theodore H. (2003). *Measuring Performance in Public and Nonprofit Organizations*. San Francisco, CA: Jossey-Bass: A Wiley Imprint.
8 Babbie, Earl. (2011). *The Basics of Social Science Research*. Belmont, CA: Wadsworth, Cengage Learning.

Understanding the level of measurement that you are using is important because, depending on the type of indicator for a given variable, different kinds of statistical analyses can be performed. For example, calculating the average of a nominal variable (e.g. where urban = 1, suburban = 2, and rural = 3) would be meaningless. Rather, you could show the counts and the percentage of households in each category.

The three-dimensional Risk Matrix presented in Chapter 8 contains 168 cells and, within each cell, data regarding various risks could potentially be collected. However, staff time and budget constraints will force you to focus on a limited number of cells and variables within those cells. Even with a reduced number of variables, it can still be challenging to make sure that each variable and its indicator(s) are measuring what it's intended to measure (validity) and repeated measurements of the same variable using the same indicator will produce the same results (reliability). In Chapter 9, I will provide more insight into this process and a few examples.

When we get to Chapter 8 and the discussion of the Risk Matrix's social-human element, these definitions and an understanding of the complexities of performance measurement will become particularly important.

Several references are provided at the end of this chapter that can help guide you through the measurement concepts introduced here.

4.3 Auditing

4.3.1 Historical Background

The term "audit" is historically associated with regulatory compliance determination(s). More recently, in the MSS arena, it is associated with determining conformance to standards such as ISO 14001, ISO 50001, and ISO 45001. The term has also been associated with performance assessment, monitoring, and measurement. It is common for EHSS/ORM standards to have internal auditing requirements. Internal auditing is a familiar activity in organizations. In larger organizations, it is common to have a dedicated internal audit function. EHSS/ORM professionals are also familiar with auditing, primarily from a regulatory compliance perspective.

It is important that EHSS/ORM professionals performing management system audits are well trained in this type of auditing because it requires an expanded skillset beyond traditional assessments of compliance. For example, there is greater emphasis on conducting interviews and interpreting documents; requiring the auditor to make assessments and judgments that go beyond simply following a checklist or assessing regulatory compliance. Decision-making in MSS auditing is based more on nominal and ordinal data than interval or ratio.

Audit guidelines unique to ISO 9001:1987 were first published in 1990 and 1991 and focused on auditing quality systems. ISO published guidelines for auditing environmental management systems in 1996, when it also published ISO 14001. In 2002, ISO consolidated its quality and environmental audit documents into a single auditing standard (ISO 19011). This standard was subsequently revised in 2011 and 2018. EHSS/ORM professionals now use ISO 19011:2018[9] to support risk management system audits.

4.3.2 Types of Audits

Audits are commonly characterized as first-, second-, or third-party audits.

4.3.2.1 First Party – Internal Audits

First-party audits refer to audits that are conducted within an organization by members of the organization. Many organizations have comprehensive internal audit programs that assess both regulatory compliance and conformance with nongovernmental standards or management systems.

Internal audits should be conducted by personnel who are technically competent, have audit training, and are capable of making unbiased and independent assessments. In order to maintain their audit independence, internal auditors should *not* have direct responsibility for activities at the site being audited. An internal audit will generally assemble several individuals who are experienced within their respective disciplines and who may even know the facility and its operations. The audit team should be sufficiently large and broad to address all the anticipated issues. However, when working with teams drawn from internal sources only, this can sometimes be difficult to achieve. Every audit also requires deft team leadership and communication skills that may be difficult for internal personnel to have acquired or keep current if they do not perform audits on a regular basis. Lead auditors should have training in managing audits. Governmental bodies, professional associations, and for-profit organizations are sources for training in compliance and management system auditing.

There are distinct advantages to the organization of using an internal audit team. The costs associated with the use of an internal audit team can be lower than those associated with an external audit team. Another advantage is that the time commitment internal auditors must make can have a significant impact on the organization. For example, robust audits allow for the development and sharing of best practices within the organization. Participating in audits also provides auditors and those being audited with an excellent way to upgrade their

---

9 International Organization for Standardization (2018). *Guidelines for auditing management systems, ISO 19011*. Geneva, Switzerland.

regulatory knowledge and technical and leadership skills. Once internal auditors, who do not work at a particular satellite site within the larger company, have completed their work, they can return with knowledge they can use at the site where they work. On the other hand, there are advantages to external audits that are discussed in the next section.

Internal auditing and internal audit departments are often unrecognized pedagogical and organizational learning resources. It is not uncommon for auditors to be received by auditees much like "cops on the beat," as clients have referred to them. I return to this idea in Chapter 10 (§10.3.4) in terms of the pedagogy of evaluation and suggest that the internal audit department can provide leverage in creating generative fields.

4.3.2.2 Second- and Third-Party External Audits

Second and third-party audits are considered external audits. They are conducted by personnel who are not members of the organization that is being audited. In the case of second-party audits, within a supply chain, a customer performs an audit of a supplier. Second-party audits are common in quality management circles (e.g. ISO 9001), but over time they have increasingly included EHSS.

Third-party audits refer to audits performed by people independent of the organization being audited. These are typically performed by consultants, and, in the management system arena, by what are called third-party registrars. Third-party audits are common in the management system arena.

There are several advantages to using external auditors. First, it should provide access to highly qualified individuals best suited to evaluate the site and its unique operations. Second, there should be fewer time and resource constraint issues that plague internal audit programs because auditors are not being pulled off an already full work schedule. Finally, a report from an unbiased outside entity has the advantage of being perceived as more objective because it avoids the potential for intraorganizational biases.

The concerns of organized labor about external auditors, if any, need to be addressed. A concern that can surface is the potential bias that external auditors might not report bad findings for fear of losing future work with the organization. This concern can be addressed by using credible third-party auditors who hold themselves to a high level of ethical conduct.

4.3.2.3 Hybrid Approaches

Audit programs in some organizations use hybrid audit teams that include both internal company representatives and external consultants. This approach yields benefits with deep organizational understanding from the internal team members and external expertise from the consultant. An increasing trend combines compliance and management system audit functions, thus auditing both at the same time. In fact, compliance with internal and regulatory compliance is part of what is covered during management system audits at the process level. Caution should be taken to ensure that neither is diluted when combined.

Suggested Reading

Gwet, K.L. (2012). *Handbook of Inter-Rater Reliability*. Gaithersburg, MD: Advanced Analytics, LLC.

Heckard, U. (2015). *Mind On Statistics*. Stamford, CT: CENGAGE Learning.

International Organization for Standardization (2018). *Guidelines for Auditing Management Systems, ISO 19011*. Geneva, Switzerland: International Organization for Standardization.

Montgomery, D.C. (2013). *Introduction to Statistical Quality Control*. Hoboken, New Jersey: John Wiley & Sons.

National Research Council (1983): *Risk Assessment in the Federal Government: Managing the Progress*. Washington, DC: National Academy Press.

National Research Council (1995). *Standards, Conformity Assessment, and Trade: Into the 21st Century*. Washington, DC: National Academy Press.

Podems, D. (2019). *Being an Evaluator: Your Practical Guide to Evaluation*. New York: Guilford Press.

Rossi, P. and Freeman, H. (1993). *Evaluation: A Systemic Approach*. Newbury Park: Sage Publications.

Salkind, N.J. *Statistics for People Who (Think They) Hate Statistics*. Thousand Oaks, CA: SAGE Publications, Inc.

Taylor, J.R. *An Introduction to Error Analysis: The Study of Uncertainties In Physical Measurement*. Sausalito, CA: University Science Books.

Tobias, P.A. and Trindade, D.C. (1995). *Applied Reliability*. New York: Chapman & Hall/CRC.

Weiss, C. (1972). *Evaluation: Method for Studying Programs and Policies*. New Jersey: Prentice Hall.

Wholey, J., Hatry, H., and Newcomer, K. (ed.) (1994). *Handbook of Practical Program Evaluation*. San Francisco: Josey-Bass.

Part II

Leverage

5

Awareness in Risk Management

CONTENTS
5.1 Origins and Development, 152
5.1.1 Genesis and Fourth-Generation Risk Management, 152
5.1.2 Early Years, 1999–2018, 152
5.1.3 Current Iteration Risk Field → Risk Matrix (ABRM v.2), 155
5.2 Defining Awareness, 156
5.2.1 Standards and Frameworks, 157
5.2.2 Paying Attention, 159
5.3 Orientations, Perspectives, and Mental Models, 160
5.3.1 Decision-Making Prequel – Bias and Heuristics, 161
5.3.2 Being-Doing, 162
5.4 Leverage and Seven Risk Awareness Elements, 162
5.4.1 Awareness, 163
5.4.2 Internal State, 163
5.4.3 Risk, 164
5.4.4 Purpose, 164
5.4.5 Value Creation and Preservation, 164
5.4.6 Decision-Making Processes, 165
5.4.7 Generative Field, 165
5.5 Language as Currency, 165
5.5.1 Future-Based Language, 167
5.5.2 Carriers of Meaning, 168
5.6 Shifting Mindset and Paradigms, 169
5.6.1 Revisiting A → B, 170
5.6.2 A Learning Context, 171
5.6.3 Capitals Coalition's Four Shifts Model, 175
5.6.4 Anatomy and Physiology of Shifts, 176
Suggested Reading, 176

Organizational Risk Management: An Integrated Framework for Environmental, Health, Safety, and Sustainability Professionals, and their C-Suites, First Edition. Charles F. Redinger.
© 2025 John Wiley & Sons, Inc. Published 2025 by John Wiley & Sons, Inc.

5 Awareness in Risk Management

> *The internal state of people and teams, and their level of awareness, has a direct impact on the success of an intervention and quality of results produced.[1]*
>
> Otto Scharmer

> *Until people can start to see their habitual ways of interpreting a situation, they can't really step into a new awareness.[2]*
>
> Francisco Varela

My consideration of how awareness can impact organizational risk management (ORM) has evolved over decades of work in the development of management system models and standards, work in organizations, and through the training received at MIT's Sloan School of Management and the Society for Organizational Learning. While I had paid attention to nonfinancial risk management (NFRM) frame works since the 1990s, it was not until 20 years later that I began to think about the development of a scalable NFRM framework for organizations. Three key factors that influenced that development were awareness, health, and a 360 degree Perspective.

This chapter builds on the observation in Chapter 2 that ORM outcomes are a function of decision-making processes that happen within systems and frameworks that are driven by people and teams in an organization. This observation lays a foundation for the core logic model depicted in Figure 2.1, and the Risk Matrix presented in Chapter 8.

While it seems obvious that the human element of ORM is important, historically, it has not been given as much attention as regulatory compliance and technical aspects. The dynamics of awareness, cognition, our orientations, perspectives, and mental models, and their role in impacting risk outcomes are highlighted here. Otto Scharmer points to this in the earlier quote. In addition, this chapter offers the basic underpinnings of a transformational shift in our ORM policies and behaviors. Awareness of our orientations, perspectives, and mental models is fundamental to integrated thinking, decision-making, and action necessary to navigate in the social–human era. Francisco Valera points to the importance of becoming aware of our "habitual ways of interpreting situations," and Donnella Meadows points to the notion of transforming systems, and transcending paradigms.

The work of Scharmer, Valera, and Meadows has ultimately led to the need for escalating impact and creating generative fields, both of which are addressed in Chapters 9 and 10. Up to this point, I have covered risk logics, frameworks, and conformity assessment. These three areas serve as the necessary foundation for considering new ways of thinking and approaches to ORM.

---

1 Scharmer, Otto: *Leading from the Emerging Future.* p. 18–19.
2 Valera et al. (2002); *On Becoming Aware.* John Benjamins Publishing Company. Philadelphia.

Notions of leverage were introduced in Chapter 1 and are developed further in this chapter. The focus here is on awareness, which sets the stage for Chapter 6 where the focus is on *motive force* (e.g. leadership). These chapters in turn set the stage for decision-making in Chapter 7.

Awareness of the physical aspects of operations, processes, and frameworks is considered a given for EHSS/ORM professionals. Traditionally, when we look to improve things or institute new initiatives, we take actions with respect to these physical aspects. However, change or transformation that persists requires more than just putting in place a new system or framework or upgrading the existing one. The physical aspects are not the whole picture; beyond the physical are people and culture. Both are needed. The Risk Matrix presented in Chapter 8 offers an example of new ways of operating in organizations.

There is already awareness about a lot of things in an organization – awareness of systems and programs, for example. As already suggested, in most instances this awareness is filtered through a compliance lens. The Risk Matrix is an integrated approach – a framework – through which a company's risk management activities are defined, managed, and assessed.

Increasing awareness is one of the highest, goal that leaders have for their workforce. I have seen this through dozens of engagements, as well as when teaching at Tulane University and while teaching many professional development courses. At Tulane, we were approached by BP to develop an internal course on "anticipation" (i.e. awareness) following the 2005 explosion at their Texas City facility; BP concluded that anticipation was not well developed in their personnel. I have worked for numerous companies, where the engagements boiled down to the specific need to increase awareness. Work done with the Singapore Ministry of Manpower focused on increasing the awareness of their inspectors. As a result, the companies and the Singapore Ministry reported improvement in key outcomes which they attributed to increased awareness.

In an engagement with a Fortune 200 company to develop and implement an integrated risk management system (RMS), it quickly became apparent one afternoon that we were going in circles. To Varela's point on awareness, the breakthrough happened when we stepped back from the immediate task, and identified and evaluated their assumptions – mostly negative – about developing and implementing new initiatives in the company; themes discovered in this exercise were failure, disappointment, and cynicism related to past events. As we teased out these themes by distinguishing them and getting them out in the open, the team's awareness shifted, and there was a notable increase in energy, vitality, and creative thinking. Several unpredictable things followed, the most significant of which was the idea of approaching the CEO directly for implementation support. The CEO's response was, "I've been waiting for you to come to me." Through this shift and implementation of the RMS, not only did EHS performance improve but there was also an improvement in the turnover rates across the company. The work done to reveal the assumptions and interpretations of key past events opened up the possibility of a shift-in-thinking, out of which there was a shift-in-context. This set the stage for a significant upgrade of the RMS.

In our EHSS/ORM practices, many of our activities relate to increasing the awareness of hazards and risks through training in anticipation, recognition, evaluation, and control. We measure how effective we have been using a number of metrics and indicators, the data for which were obtained through audits. In the regulatory/technical field, we have a high level of awareness. In the organizational and social–human field, however, there is more work to be done.

5.1 Origins and Development

5.1.1 Genesis and Fourth-Generation Risk Management

Four generations of ORM are presented in §2.1.1. I actively practiced in the second-generation, focusing on regulatory compliance, and worked to evolve the third-generation, focusing on "beyond compliance" consensus standards. Through helping organizations design, build, implement, and operate management systems (MS), it became more and more apparent that there were limitations with how far these could go in impacting performance. In those early days, I characterized these generally as "culture" issues. As time went on, I began a quest to learn more about what then appeared to be areas where impact could be improved. As a result of the discoveries made in that endeavor I have come to characterize the fourth-generation in terms of value and purpose.

5.1.2 Early Years, 1999–2018

Awareness-Based Risk Management (ABRM)™ evolved out of research, organizational engagements, and interviews between 2006 and 2013. During that time, I served on the US Technical Advisory Group to ISO 31000, and participated in programs both at MIT's Sloan School of Management and the Society for Organizational Learning. In these travels, ideas of increasing perspective and awareness kept surfacing. In asking people about the four generations, and, in particular, the fourth, I discerned general agreement on their meanings and value. From this, I presented a poster at the Society for Risk Analysis's annual conference in 2017 that focused on ABRM; it was titled "Awareness-based risk management: Seeing, transforming, and unleashing organizational capacity." It won the award for best poster and became a catalyst for the production of this book.

5.1.2.1 Integrated Model

During 2012–2013 work began with looking at various models and bodies of work.[3] These were:

- Deming's quality work (Plan-Do-Check-Act)
- ISO's first-generation MSS framework (Policy-Planning-Implementation and Operation-Checking-Management Review)
- Systems-thinking/dynamics (Purpose-Vision-Mental Models-Structures-Patterns-Events)
- Organizational learning (Seeing Systems-Collaborating Across Boundaries-Creating Desired Futures)
- Industrial health's ARECC model (Anticipation-Recognition-Evaluation-Control-Confirm).

The output from this work is depicted in Table 5.1, and Figures 5.1 and 5.2.

Unique to these was integrating ISO MS frameworks, along with organizational learning, and living systems concepts. This included building in the second-order change and double loop learning concepts. These are depicted as awareness, context, and reflection.

5.1.2.2 Second-Order Change

From the earliest days, I have viewed MSs from an organizational learning perspective and not simply from one of registration or certification. Of course, there are benefits (outputs and outcomes) derived from implementing and

Table 5.1 Awareness-based risk management framework v.1.

1) Awareness
2) Context
3) Commitment
4) Planning
5) Execution
6) Evaluation
7) Reflection

---

3 Notes: (1) ISO's second-generation MSS framework had been evolving, but its first uses were not promulgated until 2015. (2) Industrial health's ARECC model evolved over time. The second "C" for confirm was beginning to be used in 2012–2013. By 2023 it was entrenched. (3) Parallel to ABRM's development, I and other colleagues began to include a second "A" to depict awareness, thus it became AARECC.

5 Awareness in Risk Management

Deming & 1st Generation ISO MS	2nd Generation ISO MS	Systems Thinking	Organizational Learning	Industrial Health		ABRM Framework
		Purpose Vision	Creating Desired Futures			AWARENESS
	Context of the Organization	Mental Models		Anticipation	Business Continuity / Organizational Health	CONTEXT
Policy	Leadership		Collaboration Across Boundaries			COMMITMENT
Plan Planning	Planning Support	Structures & Systems		Recognition		PLANNING
Do Imp/Op	Operation			Control		EXECUTION
Check Checking	Performance Evaluation			Evaluation Confirm		EVALUATION
Act Mgmt Rev	Improvement		Seeing Systems			REFLECTION

Business Continuity – maintaining sustainability and resilience
Organizational Health – ability to flourish

Figure 5.1 Awareness-Based Risk Management™ framework development (v.1).

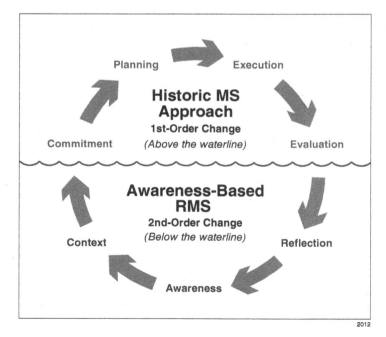

Figure 5.2 ABRM framework v.1 (2012) and distinguishing first- and second-order change.

operating a MS. This can be referred to as first order change. However, over time there are deeper changes in orientations, perspectives, and mental models (internal state). These are second-order changes, as reflected in the double loop learning concept. Looking at these through the ABRM framework (2012) in Figures 5.1 and 5.2 yielded:

First-order: Commitment, Planning, Execution, Evaluation; and,
Second-order: Awareness, Context, Reflection.

5.1.2.3 360 Perspective

A 360 degree risk management perspective is an approach that promotes a thorough examination of stakeholder domains and their interrelationships with the organization. It stresses exploration of organizational and employee "blind spots," an inquiry into understanding what might not be known;[4] breaking down and acting outside organizational siloes; and ensuring that there is wide input to the risk management process, and that "all voices are heard."

5.1.2.4 Stakeholder Domains

Identifying stakeholders in organizational governance is important, particularly so with risk management. An evolution from a traditional shareholder and agency theory perspective to a wider cohort was presented in Chapter 2. As I have worked with organizations, the list depicted in Table 5.2 has evolved. It is offered here to provide examples of how you can begin to identify stakeholders.

5.1.3 Current Iteration Risk Field → Risk Matrix (ABRM v.2)

ABRM has continued to evolve framing ORM as occurring at the intersection of three fields as described in Chapter 2 (§2.6); regulatory/technical, organizational, and social–human. These fields are reflected in the three-dimensional Risk Matrix (Chapter 8). Key distinctions in the evolved ABRM construct are:

- Integrated thinking, decision-making, and action
- Generative fields, with a sequence of Social → Organizational → Risk
- Value creation and preservation
- A capitals orientation
- Purpose

---

4 "Blind-spot" is a central distinction in Scharmer's *Theory U* body-of-work.

Table 5.2 Stakeholder domains.

	EHSS department
	Executives and Leadership Team Function Area Managers Technicians and Coordinators EHSS Related Contractors and Vendors
1 2 3	Strategic Partners – Internal (HR, Engineering, Legal) Operations/Business Function (Employees, Supervisors, etc.) C-Suite, Corporate, and Business Units
4 5 6	Contractors Strategic Partners – External Supply/Value Chain Partners and Vendors
7 8 9	Neighbors/Immediate Community Customers/consumers Board of Directors
10 11 12 13 14 15 16	Governmental Agencies NGOs/ Standards-Developers/Conformity Assessment Bodies Broader Society/Media Shareholders Competitors Academia, Research Bodies Legal Community and Judiciary

5.2 Defining Awareness

Awareness is defined in different ways depending on the discipline. Famously, in psychology, Freud's model of personality structure presents levels of awareness. Francisco Varela's cognitive psychology body of work explores different aspects of awareness and consciousness, a key observation of which is indicated in this chapter's opening quotation. The New Oxford American Dictionary defines awareness as, *Knowledge or perception of a situation or fact.*[5] Synonyms offered for "aware" are cognizant, conscious, sensible, awake, alert, watchful, and vigilant.

---

5 New Oxford American Dictionary, Apple OS 13.5.2.

5.2.1 Standards and Frameworks

Awareness entered into ISO MS's frameworks with the development of the organization's generic framework, Annex SL, included in MSS Section 7 – Support. Examples of the ISO MSS language follow.

Annex SL 2013[6]
7.3 Awareness
Persons doing work under the organization's control shall be aware of:

- the XXX policy.
- their contribution to the effectiveness of the XXX management system, including the benefits of improved XXX performance.
- the implications of not conforming with the XXX management system requirements.

14001:2015[7]
7.3 Awareness
The organization shall ensure that persons doing work under the organization's control are aware of:

a) the environmental policy.
b) the significant environmental aspects and related actual or potential environmental impacts associated with their work.
c) their contribution to the effectiveness of the environmental management system, including the benefits of enhanced environmental performance.
d) the implications of not conforming with the environmental management system requirements, including not fulfilling the organization's compliance obligations.

A.7.3 Awareness
"Awareness of the environmental policy should not be taken to mean that the commitments need to be memorized or that persons doing work under the organization's control have a copy of the documented environmental policy. Rather, these persons should be aware of its existence, its purpose, and their role in achieving the commitments, including how their work can affect the organization's ability to fulfil its compliance obligations."

---

6 ISO/IEC Directives, Part 1, Consolidated ISO Supplement, 2013, p. 143. Note: the "xxx" is what is stated in Annex SL. It reflects a holding place for specific uses.
7 ISO 14001:2015, p. 11, pp. 26–27.

50001:2018[8]
7.3 Awareness
"Persons doing work under the organization's control shall be aware of:

a) the energy policy (see 5.2).
b) their contribution to the effectiveness of the EnMS, including achievement of objectives and energy targets (see 6.2), and the benefits of improved energy performance.
c) the impact of their activities or behavior with respect to energy performance.
d) the implications of not conforming with the EnMS requirements."

45001:2018[9]
7.3 Awareness
Workers shall be made aware of:

a) the OH&S policy and OH&S objectives.
b) their contribution to the effectiveness of the OH&S management system, including the benefits of improved OH&S performance.
c) the implications and potential consequences of not conforming to the OH&S management system requirements.
d) incidents and the outcomes of investigations that are relevant to them.
e) hazards, OH&S risks and actions determined that are relevant to them.
f) the ability to remove themselves from work situations that they consider present an imminent and serious danger to their life or health, as well as the arrangements for protecting them from undue consequences for doing so.

A.7.3 Awareness
"In addition to workers (especially temporary workers), contractors, visitors, and any other parties should be aware of the OH&S risks to which they are exposed."

ISO 31000: 2018 does not explicitly address awareness as a stand-alone element. It does point to awareness in its principles, specifically in "d) Inclusive," where it states, "Appropriate and timely involvement of stakeholders enables their knowledge, views, and perceptions to be considered. This results in improved *awareness* and informed risk management."[10] It is also identified in relation to stakeholders

---

8 ISO 50001:2018, p. 13.
9 ISO 45001:2018, p. 15.
10 ISO 31000:2018, p. 3.

and framework implementation, it states, "Successful implementation of the framework requires the engagement and _awareness_ of stakeholders."[11] And finally, in terms of communication and consultation:

> The purpose of communication and consultation is to assist relevant stakeholders in understanding risk, the basis on which decisions are made, and the reasons why particular actions are required. Communication seeks to promote awareness and understanding of risk, whereas consultation involves obtaining feedback and information to support decision-making.[12]

5.2.2 Paying Attention

> Pay attention is an economic term. Attention can be thought of as a currency and a scarce resource: give it to one thing, and by definition you can't give it to something else. The state of being aware, alert, and cognitive (System 2) requires paying attention.[13]
>
> Paul Dolan, London School of Economics

"Paying attention" is related to awareness. As simple as it sounds, paying attention is actually a practice, if not a skill, that is highlighted in numerous disciplines, including system dynamics, systems thinking, behavioral economics, and decision science. It seems too basic even to say that in EHSS/ORM paying attention is central in production and operations settings and is central in auditing. It is a bit of a chicken-and-egg inquiry to ask which comes first – awareness of something, or paying attention to it– or vice versa – or are they synonymous?

For our purposes, it is not critical to tease out the sequence depicted in Figure 5.3. But it is essential to reflect on this bundle as we move forward toward Chapter 7 where we will look at decision-making processes. As I have asked people over the years for their view on the order of this sequence, the only constant has been that decision-making should be at the end.

Paying attention → Observe → Awareness → Sense making → Decision-making

Figure 5.3 Decision-making precursors.

---

11 ISO 31000:2018, p. 8.
12 ISO 31000:2018, p. 9.
13 Keynote address by Professor Paul Dolan at the 2015 IOHA/BOHS Conference in London England. He is a Professor of Behavioural Science in the Department of Social Policy at the London School of Economics and Political Science.

5.3 Orientations, Perspectives, and Mental Models

In this chapter's opening quotation Otto Scharmer refers to "the internal state of people and teams…" within which orientations, perspectives, and mental models are embedded. Box 5.1 defines each of these three, which are always involved with decision-making and subsequent actions and play a foundational role in being[14] and identity.[15] Becoming aware of these at the individual, team/department, and enterprise level is a necessary first step in developing integrated thinking, decision-making, and action skills.

Levine's Lever (Chapter 1, Figures 1.2 and 1.3) offers a simple depiction of cause and effect. Awareness of these internal state elements (Box 5.1) is necessary; the further up the lever one goes, the greater the need for transformative shifts in systems/structures. More likely than not, developing a Risk Matrix (Chapter 8) with the incorporation of the social–human element, and navigating an expanded decision-making process, will necessitate shifts in this lever.

Box 5.1   Definitions – Orientation, Orient, Perspective, and Mental Model

Orientation (noun): "The determination of the relative position of something or someone (especially oneself). A person's basic attitude, belief, or feelings in relation to a particular subject or issue."[16]

Orient (verb): "Align or position (something) relative to the points of a compass or other specified position. Find one's position in relation to new and strange surroundings."[17]

Perspective (noun): "A particular attitude toward or way of regarding something; a point of view."[18]

Mental model: "The semipermanent tacit 'maps' of the world which people hold in their long-term memory, and the short-term perceptions which people build up as part of their everyday reasoning process."[19]

---

14 *Def* "Existence. The nature or essence of a person." New Oxford American Dictionary, Apple OS 13.5.2.
15 *Def* "The fact of being who or what a person or thing is. The characteristics determining who or what a person or thing is." New Oxford American Dictionary (Apple OS bundle).
16 New Oxford American Dictionary (in Apple OS 10.15.7).
17 New Oxford American Dictionary (in Apple OS 10.15.7).
18 New Oxford American Dictionary (in Apple OS 10.15.7).
19 Senge, P. et al. (1994). *The Fifth Discipline Field Book*. Doubleday, p. 237. Quote attributed to Art Kleiner.

5.3.1 Decision-Making Prequel – Bias and Heuristics

Historically, decision-making models posit that people will attempt to maximize their expected utility.[20] Initial work in this area suggested that decision-making frameworks and approaches overcame human biases were logical, rational, and measurable, and led to outcomes that were predictable. Over time, evidence mounted that suggested this is not necessarily the case, and pointed to the probabilistic nature of decision analysis and its accompanying uncertainty. At the core of risk decision-making is efficacy, namely, whether the means by which the risk is to be managed can plausibly produce the desired result while at the same time minimizing unintended consequences.

Breakthroughs in cognitive psychology have shown that people treat positive uncertainty (gain) quite differently from negative uncertainty (loss). In short, people have the tendency to value more that which they currently have and potentially could lose, than they value that which they currently do not have and potentially could gain. The decisions people make when faced with uncertainty relate to how the mind processes information. Cognitive biases are different than biases rooted in beliefs or preferences; cognitive biases relate to actual and systematic judgments that people make where reliability wanes. Heuristics (rules of thumb for the mind) have been identified that underpin irrational judgments and systematic biases, for example, the availability heuristic, representativeness heuristic, and anchoring/adjustment heuristic.[21] Heuristics and biases underpin the workings of the human mind and intuitive judgments made by people. They (we) also make decisions based on stories, framing, and context rather than actual data and calculation.

---

20 Expected utility in decision theory is the expected value of an action to an agent, calculated by multiplying the value to the agent of each possible outcome of the action by the probability of that outcome occurring and then summing those numbers. The concept of expected utility is used to elucidate decisions made under conditions of risk. According to standard decision theory, when comparing alternative courses of action, one should choose the action that has the greatest expected utility. www.britannica.com/topic/expected-utility (accessed 22 January 2019).
21 Tversky, A. and Kahneman, D. (1974). "Judgment under uncertainty: heuristics and biases – biases in judgment reveal some heuristics of thinking under uncertainty." *Science* 185, 1124–1131. This is also the focus of their book, along with Paul Slovic, *Judgment under Uncertainty: Heuristics and Biases* (1982).

5.3.2 Being-Doing

> Being alters action; context shapes thinking and perception. When you fundamentally alter the context, the foundation on which people construct their understanding of the world, actions are altered accordingly. Context sets the stage; being pertains to whether the actor lives the part or merely goes through the motions.[22]
>
> Tracy Goss et al.

The system dynamics iceberg presented in Figure 2.6 offers a powerful way to begin to see connections between internal state dynamics and outcomes. Another way to frame this is in terms of "being" and "doing." Where being is a function of internal state and doing is the actions related to the internal state.

In the iceberg model, outcomes are commonly thought of as only the events observed above the waterline. I suggest that these should also include the structures and patterns of behavior. I will expand on this dynamic in Chapter 9 (Matrix in Action) and Chapter 10 (Escalate Impact).

5.4 Leverage and Seven Risk Awareness Elements

There is a multitude of things we need to be, should be, or are aware of in ORM endeavors. Seven "awareness topics" are presented in this section (Box 5.2) I refer to these *as risk awareness elements*. They contain areas, topics, and items that have risen to the top in my organizational engagements and teaching. As you proceed in reading this book, and as time goes on after you set it down, consider tailoring them to meet your needs. The context for these seven is awareness itself, the awareness of awareness – bringing attention and intention to your awareness.

In the introduction, I framed this book in terms of "leverage to create" new clearings and capacities that increase the ability to generate and preserve value and increase resilience. I introduced the term leverage and offered Professor Levine's simple leverage model along with Donnella Meadows's list of places to intervene in a system. I suggest that these *seven risk awareness elements* provide higher leverage points on Levine's Lever, and access to Meadows's highest

---

22 Goss, T. et al. (1993). The reinvention roller coaster: risking the present for a powerful future. *Harvard Business Review*. November-December 1993. p. 101.

> **Box 5.2   Seven Risk Awareness Elements**
>
> 1) Awareness
> 2) Internal state
> 3) Risk
> 4) Purpose
> 5) Value creation and preservation
> 6) Decision-making processes
> 7) Generative field

intervention points, which are, "the mindset or paradigm out of which the system – its goals, structure, rules, delays, parameters – arise," and "the power to transcend paradigms."[23]

5.4.1   Awareness

> Addressed in this chapter's introduction and above in Section 5.2.

While it may seem redundant to include awareness here, the point is to highlight the value and leverage provided by bringing attention and intention to your awareness, i.e. that which you are aware of, and that to which you pay attention.

Suggestions on expanding perspective and awareness are offered in Section 5.6. The Risk Matrix can also be framed as a risk awareness framework that promotes integrated thinking, decision-making, and action.

5.4.2   Internal State

> Addressed in this chapter, Section 5.3, and in Chapter 7, Section 7.1.2.

Awareness of orientations, perspectives, and mental models perhaps provides the greatest leverage in creating new clearings and capacities, and in transcending paradigms. Awareness of these is critical when considering risk decision-making heuristics and biases.

---
23 Meadows, D. (1999). *Leverage Points: Places to Intervene in a System.* Hartland, VT: The Sustainability Institute, p. 3.

5.4.3 Risk

> Addressed in: Chapter 2, Section 2.2; Chapter 7, Section 7.5; and Chapter 8, Section 8.5.1.1.

Becoming aware of how risk is perceived and defined in an organization, by individuals, teams/departments, the organization itself, communities, and stakeholders is critical from a leverage perspective. The relevance of Paul Slovic's observation in §2.2 cannot be over-emphasized. Whether you agree with it or not, it is smart to consider how risk perceptions and orientations play out in an organization, especially when the social–human dimension of risk is addressed.

5.4.4 Purpose

> Addressed in: Chapter 1, Section 1.1.1; Chapter 2, Section 2.1.2; Chapter 8, Section 8.5.2.1; and Chapter 10, Section 10.3.

Awareness (§5.4.1) of organizational purpose and risk (§5.4.3), sets the stage for risk decision-making. It provides a filter through which risk decision-making inputs and processes are considered (§7.2.2).

As mentioned in earlier chapters, organizational purpose has evolved as a critical input to organizational governance (§2.1.2). It is also observed in the evolution of ESG-related frameworks and thinking.

5.4.5 Value Creation and Preservation

> Addressed in: Chapter 1, Section 1.1.1; Chapter 6, Section 6.3; Chapter 8, Section 8.5.2.2; and, Chapter 8, Section 8.5.2.3.

Value considerations are at the fore of ORM. Given its importance, I repeat here one of the opening quotations from Chapter 2.

> The purpose of risk management is the creation and protection of value. It improves performance, encourages innovation, and supports the achievement of objectives.[24]

A foundation for value and valuation issues is started in Chapter 1 (§1.1.1), and it is developed further in Chapter 6 (§6.3). Risk Matrix elements Nos. 7 and 8, Social-human engagement (§8.5.2.2), and Leadership (§8.5.2.3) provide guidance for frameworks and standards.

---

24 ISO 31000:2018, p. 2.

5.4.6 Decision-Making Processes

> Addressed in: Chapter 2, Section 2.1; Chapter 7, Sections 7.2 and 7.3; and Chapter 8, Section 8.5.2.4.

Decision-making is at the core of ORM. As suggested in Chapter 1 and Chapter 2, as ORM has become more complex, so has risk decision-making. Chapter 7 is devoted to this topic. Given its central nature, it is critical, to understand not only the organization's RDM process(es) but also how these processes affect risk-related outcomes. A dynamic here is being aware of the roles of the risk field actors (individual, team/department, enterprise/company, and community), and how these roles affect RDM.

It is critical to become aware of the impact and dependency pathways, and stocks and flows that travel through these pathways. In some circles, the notion of "seeing systems" is used to distinguish this phenomenon. Impact-dependency pathways are a central piece in capitals-thinking and the flow of value between capitals, namely social, human, and natural capital.

5.4.7 Generative Field

> Addressed in: Chapter 2, Section 2.6; Chapter 10, Section 10.2.

The field construct is introduced in Chapter 1 and is found on the frameworks' evolution sequence in Chapter 2, Section 2.1.1.2. At numerous junctures, it is suggested that there is value in viewing a progression from a systems perspective to a field perspective. This progression has been framed in terms of Donnella Meadows's highest leverage point – transcending paradigms.

A distinction between field and generative field is made in Chapter 10. Characteristics of a generative field include interiority, expansion of engagement, and a health orientation. That is the health of the organization itself, and people and communities associated with it. A generative field can also be framed in terms of capitals and value, especially social, human, and natural capitals.

5.5 Language as Currency

> An organization is a factory that manufactures judgments and decisions... My aim for water cooler conversations is to improve the ability to identify and understand errors of judgment and choice, in others and eventually in ourselves, by providing a richer and more precise language to discuss them... Learning medicine consists in part of learning the language of

medicine. A deeper understanding of judgments and choices also requires a richer vocabulary than is available in everyday language.[25]

Daniel Kahneman, Princeton University

Language and words matter. They can be viewed as a currency that helps access new clearings and capacities that increase ability to generate and preserve value and increase resilience. They impact personal and collective thinking, which in turn impact systems, structures, and outcomes. Considering new language – new words – is key in the evolution to fourth-generation risk management and transcending paradigms.

Box 5.3   Definitions: Language and Currency[26]

Language (n): the principal method of human communication, consisting of words used in a structured and conventional way and conveyed by speech, writing, or gesture. A nonverbal method of expression or communication. A system of communication used by a particular country or community.
Currency (n): a system of money in general use in a particular country. The fact or quality of being generally accepted or in use. The time during which something is in use or operation.

Building on Kahneman's observation, in an interview with *Harvard Business Review* for their 75th anniversary issue, Charles Handy, a leading business futurist said: "We are the unconscious prisoners of our language. While most of the time this constraint matters little, at times of momentous change in culture or society, our use of old words to describe new things can hide the emerging future from our eyes."[27]

That words communicate meaning is taught in basic communication training. It is reinforced by people who critique our writing and speaking. Their importance is not reinforced as much in our technical and scientific endeavors. The regulatory/technical field tends to be driven by data (numbers) that are interval or ratio in nature; its decision-making processes and standards tend to be well-defined and concrete. In organizational and social fields, the role of words takes on increased importance in decision-making and communication . It is when we enter the domain of possibility – creating it, seeing it, standing in it, committing to it – that the role of words and awareness of them, along with vocabulary and language are critical.

I pay close attention to defining terms, and making distinctions throughout this book. To help with this, definitions for these are provided in Boxes 5.3 and 5.4.

---

25 Kahneman, D. (2011). *Thinking Fast and Slow*, 3–4, 418. New York: Farrar, Straus and Giroux.
26 New Oxford American Dictionary, Apple OS 13.5.2.
27 HBR Editors (1997). Looking ahead: implications of the present. *Harvard Business Review*, September-October, 1997. Reprint 97503, p. 7.

Box 5.4 Definitions: Definition, Distinguish, Distinguishing, and Distinction[28]

Definition (n): a statement of the exact meaning of a word, especially in a dictionary. An exact statement or description of the nature, scope, or meaning of something. The action or process of defining something. The degree of distinctness in outline of an object, image, or sound, especially of an image in a photograph or on a screen. The capacity of an instrument or device for making images distinct in outline.

Distinguish (v): recognize or treat (someone or something) as different. Be an identifying or characteristic mark or property of.

Distinguishing (adj): characteristic of one thing or person, so serving to identify it; distinctive.

Distinction (n): a difference or contrast between similar things or people. The separation of things or people into different groups according to their attributes or characteristics.

A dynamic in transcending paradigms as suggested by Donnella Meadows is the use of words and language, possibly in ways they have not been used previously. Examples in this work are risk field, Risk Matrix, and generative field. While the terms value, capital, and purpose are familiar, I suggest that the contexts and ways in which they are offered in these chapters are different, and that they are tools that can be used in shifting mindsets and transcending paradigms.

5.5.1 Future-Based Language

Steve Zaffron and Dave Logan have contributed a valuable body of work in their book, *The Three Laws of Performance: Rewriting the Future of Your Organization and Your Life*.[29] The three laws of performance they offer are: (1) how people perform correlates to how situations occur to them; (2) how a situation occurs arises in language; and (3) future-based language transforms how situations occur to people. Of interest here is how they point to the role of language, and in particular "future-based" language in impacting performance. They also call future-based language, "generative language," and juxtapose it with descriptive language. Future-based and generative language, they assert, "… has the power to create new futures, to craft vision, and to eliminate the blinders that are preventing people from seeing possibility."[30] They offer a three-step process of "blanking the canvas" to create an environment where future-based/generative language can arise. These

---

28 New Oxford American Dictionary, Apple OS 13.5.2.
29 Zaffron, S. and Logan, D. (2009). *The Three Laws of Performance, Rewriting the Future of your Organization and your Life*. Jossey-Bass, San Francisco (3LOP).
30 3LOP, p. 69.

are: "(1) seeing that what binds and constrains us is not the facts, its language – in particular, descriptive language; (2) articulating the default future, and asking, 'do we really want this as our future?'; and (3) completing issues from the past."[31]

I suggest that new risk language is coming into focus in the emerging social-human era, and evolutions in ORM (§2.1.1), some of which can be characterized as future-based and generative. Building on Kahneman's and Handy's observations, Zaffron and Logan point to identifying language that provides "access to fulfilling what is possible." This is an important concept as we look at creating generative fields in Chapter 10, and what it takes to transcend paradigms as offered by Donnella Meadows.

The new language introduced in the risk field/matrix construct will play a major role in solving ORM challenges with greater speed and efficiency. Addressing ORM through the lens of the field/matrix construct has the potential to transform fundamental assumptions about organizational performance, organizational risk, and well-being, both inside and outside an organization's fence line.

5.5.2 Carriers of Meaning

Carriers of meaning are "fundamental mechanisms that allow us to account for how ideas move through space and time, who or what is transporting them, and how they may be transformed by their journey."[32] They are embedded in communications – verbal and nonverbal – and are an integral aspect of organizational performance.

Numerous types of carriers are identified in the institutional theory literature. These include symbolic carriers (rules, norms, language, myths, stories, standards, and schemes), rational carriers (networks and role systems), activities as carriers (decision-making, routines, and monitoring), and artifacts as carriers.

As we move forward together and consider generative fields in Chapter 10, I put forward for your consideration a number of "carriers" in ORM generally, and risk decision-making specifically. "An emerging theme recognized [in the institutional theory literature] is that carriers are never neutral modes of transmission but affect the nature of the message and the ways in which it is received."[33] This observation points to the tensions and paradoxes that EHSS/ORM professionals navigate in their risk decision-making endeavors.

It is important to be aware of carriers of meaning that are embedded in the words and language used in communications, as well as in actions.

---

31 3LOP, pp. 73–76.
32 Richard, S.W. (2014): *Institutions and Organizations, Ideas, Interests and Identities*, 4e, 95. Thousand Oaks, CA: Sage Publications, Inc.
33 Scott (2014), p. 96.

5.6 Shifting Mindset and Paradigms

> Paradigms are the source of systems. From them, from shared social agreements about the nature of reality, come system goals, and information flows, feedbacks, stocks, flows, and everything else about systems.[34]
>
> Donnella Meadows

Change, shift, transformation – regardless of which term is used, challenges lurk, oftentimes in insurmountable ways. Anyone with experience in implementing something new in an organization, or even merely improving a seemingly simple program or process, has experienced this. Early in my career of working with organizations in developing and implementing risk MSs, I quickly learned that there was much more at play than simply the "nuts and bolts" of what was being implemented.

It is common to frame these challenges in terms of culture. Peter Drucker famously wrote, "Culture eats strategy for breakfast." Volumes are written on implementation and shifting paradigms. In her *Leverage Points* paper, Meadows references Thomas Kuhn's seminal book[35] on paradigms and sums up his view on shifting paradigms as,

> You keep pointing at the anomalies and failures in the old paradigm, you keep speaking louder and with assurance from the new one, you insert people with the new paradigm in places of public visibility and power. You don't waste time with the reactionaries; rather you work with active change agents and with the vast middle ground or people who are open-minded.[36]

We grow, evolve, and advance through a complex cycle of learning and acting on what we have learned. Einstein's scientific breakthroughs are well known. He was obviously a curious person who thought deeply about many things – he famously observed, "We cannot solve our problems with the same level of thinking that created them." Embedded in this observation are notions of awareness, paying attention to what is happening, considering our orientation to problems, and thinking about solutions. These are not new ideas to you, and I bet you can quickly come up with examples where you have had shifts in your thinking both personally and professionally. I highlight this observation here to reinforce the idea that there are numerous shifts in both orientation and thinking that are wise, and necessary, to meet ORM challenges.

---

34 Meadows, D. (1999). *Leverage Points: Places to Intervene in a* System, 18. Hartland, VT: The Sustainability Institute.
35 Kuhn, T. (1962). *The Structure of Scientific Revolution*. Chicago University Press.
36 Meadows, D. (1999). *Leverage Points: Places to Intervene in a System*, 18. Hartland, VT: The Sustainability Institute.

I have introduced various reasons for shifting the way we have historically approached ORM. I've argued that the historical ways we frame and practice ORM are archaic and have diminished ability to meet post-2020 needs, it is important to view ORM as occurring within a larger risk field, and we should highlight the social-human dimension in a/the risk field.

Embracing the social-human dimension of ORM requires shifts in orientation and thinking for many of us. I have suggested that ORM is an enterprise full of paradoxes and tension that are generated by a stew of personal, organizational, and social values, perspectives, and orientations. Keeping this notion of shifts in thinking in your toolbox will be valuable as we move forward. As we consider risk decision-making as well as when we begin to consider what it takes to bring a field into existence, this toolbox will provide a way of testing whether our ideas are consistent with the kind of shifts we are seeking. A question to ponder here is, "what leads to shifts in orientation and thinking?"

There is already "awareness" about a lot of things in an organization – awareness of systems, programs, and culture. As suggested earlier, in many instances, ORM-related awareness is filtered through a compliance orientation. Key to this point of departure from the *status quo*, compliance-based orientation, is generating a shift-in-thinking, to one organized around what matters, is aligned with organizational purpose, and increases resilience and fulfillment.

Various ways to bring about this shift and unleash organizational capacity are addressed in later chapters where I argue that the best way to do this is to embrace and commit to improving human, community, environmental, and global health. A key component of this is to understand the range of facets of the social–human era and to begin to integrate them into your ORM approach/field.

5.6.1 Revisiting A → B

Change, transformation, and shifts are all terms that can be used to describe the simple causal model of A → B, the phenomenon of getting from point A to B. Change management is a prominent topic in business as is transformation in personal development – the word shift is used to depict the range of these change, transformation, etc. phenomena. As touched on earlier, the symbol "→" is used to depict the "shift" vehicle that has been described in organizational science as the intervention, plan, program, and process. In Chapter 8, I offer the Risk Matrix as a shift vehicle to address the ORM imperatives identified in Chapter 1.

> "The mindset or paradigm out of which the system – its goals, structures, rules, delays, parameters – arise,"[37] provides a means for impacting said system.

---

37 Meadows, D. (1999). *Leverage Points: Places to Intervene in a System*. Hartland, VT: The Sustainability Institute, p. 3. This is the second top leverage point identified by Meadows. Her entire scale of intervention points is discussed in Chapter 1 (§1.2.3).

5.6 Shifting Mindset and Paradigms

A foundation for this concept is depicted in Figure 2.1 which introduces two general concepts: (1) systems/frameworks, and (2) social and human capital, i.e. people. Of course, there is a very wide range of inputs, outputs, and outcomes, with a wide range of complexity that can be presented in the A → B causal model. Examples include sourcing a new plant, cleaning up legacy chemical waste sites, or implementing an EHS MS. With the suggested social–human era and the increased attention to the "social" of ESG, the focus becomes *what to shift* in the "outcome space" depicted in Figure 2.1.

5.6.2 A Learning Context

Learning is requisite to growth and development. Here we examine two aspects of organizational learning, double loop learning, and transformational learning, both of which are components of systems thinking.

Box 5.5   Definition: Learn

Learn (v): Gain or acquire knowledge of or skill in (something) by study, experience, or being taught. Commit to memory. Become aware of (something) by information or from observation.[38]

5.6.2.1 Organizational Learning

> The field of organizational learning explores ways to design organizations so that they fulfill their function effectively, encourage people to reach their full potential, and, at the same time, help the world to be a better place.[39]

Organizational learning and the learning organization are relatively new concepts. Seminal work in this area was done at Harvard University and MIT in the 1970s principally through the scholarship of Chris Argyris, Donald Schon, Donnella Meadows, and Peter Senge. Peter Senge's groundbreaking book, *The Fifth Discipline: The Art & Practice of the Learning Organization*, presents five disciplines: personal mastery, mental models, shared vision, team learning, and systems thinking. Senge writes:

> I call systems thinking the fifth discipline because it is the conceptual cornerstone that underlies all of the five learning disciplines of this book. All are concerned with a shift of mind from seeing parts to seeing wholes, from seeing people as helpless reactors to seeing them as active participants in shaping their reality, from reacting to the present to creating the future. Without systems-thinking, there is neither the incentive nor the

---
38 New Oxford American Dictionary, Apple OS 13.5.2.
39 Pegasus Communications, Inc. http://209.237.164.151/aboutol.html#systems%20 (accessed 14 June 2023).

means to integrate the learning disciplines once they have come into practice. As the fifth discipline, systems thinking is the cornerstone of how learning organizations think about their world.[40]

The concepts embodied in the organizational learning and learning organizations' bodies of work provide a foundation for, "creating new clearings and capacities that increase ability to generate and preserve value, and increase resilience," as stated in Chapter 1. It is from this foundation that the generative field construct is evolved in Chapter 10 (§10.2), including its ability within a risk field context to provide leverage for impacting organizational performance in the social–human era.

5.6.2.2 Double Loop Learning

A powerful concept in the organizational learning and systems thinking literature is double loop learning. While systems thinking basics are presented in Chapter 7 (§7.2), it is worthwhile here to provide a brief preview of a key system component – feedback.

The heating system example discussed later in §7.2.4 highlights the signals between a thermostat and heating source. This can be characterized as a single loop learning phenomenon whereby the thermostat responds to room temperature by sending an on or off signal to the heating source. This would be characterized as a double loop system (or learner), "if the thermostat asked itself such questions as why it was set at 68 degrees, or why it was programmed as it was?"[41]

EHSS MSs provide another example with ISO 14001 or ISO 45001. For the most part, and as commonly conceived and used, these MSs represent a single loop learning process. There is feedback on the planning process, and as needed, changes are made to EHS procedures and processes. As new risks are identified, or things learned in audits or investigations, the system can be or is modified. Historically, the primary reason these systems are implemented is in response to markets and customers. Certification of a system is typically the driver.

Over time, as these systems mature, they can morph into double loop learning tools, whereby they have an impact on all organization performance. This is reflected in the *evolution sequences* presented in Chapter 2 (§2.1.1) and Figure 2.2 where there is increased attention to the action strategies and techniques that produce impacts. When organizational learning, or a culture of learning context is present, the MS management review element is used as a double loop learning tool. This happens when these high-level reviews expands beyond simply looking at MS outputs and outcomes, to consider impacts beyond MS scope.

Figure 5.4 illustrates single and double loop learning.[42]

---

40 Senge, P. (2006): *The Fifth Discipline: The Art & Practice of the Learning Organization*. Doubleday. p. 69.
41 Argyris, C. (1992). *On Organizational Learning*, 8. Cambridge MA: Blackwell Publishers.
42 Figure no 5.3 provided by Andrew Bryant, Self-Leadership International P/L. https://www.selfleadership.com/blog/double-loop-learning-survive-thrive (accessed 24 April 2022).

5.6 Shifting Mindset and Paradigms | 173

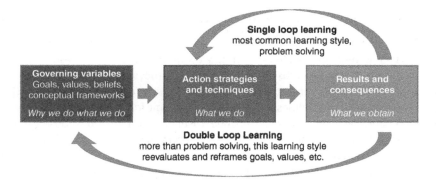

Figure 5.4 Double loop learning.

5.6.2.3 Transformational Learning

Transformational learning is an exploration of how deep-seated values, beliefs, and assumptions shape the ways in which we frame and react to situations.[43]

In the organizational learning and systems thinking literature, learning at the individual level is framed in terms of transformational learning; some, such as Jack Mezirow, refer to this as transformative learning.[44] Characteristics of transformational and transformative learning include participants learning "to surface their hidden assumptions, opinions, and emotions, and build new, [including] shared views of the business issues." [45]

The Leading Learning Communities (LLC) program originated at MIT in 1992 in collaboration with Electronics Data Systems (EDS). The program's goal was to develop a learning infrastructure that adopted "skills and disciplines of a learning organization."[46] The program included four week-long sessions that

---

43 Kofman F. Transformational learning: a blueprint for organizational change. The Systems Thinker. https://thesystemsthinker.com/transformational-learning-a-blueprint-for-organizational-change/ (accessed 18 November 2022).
44 Mezirow was on the faculty at Teachers College, Columbia University for many years and is credited with significant advances in education. To learn more about his work, see: Mezirow J. (1978). "Perspective Transformation". *Adult Education Quarterly* 28/2 (1978) 100-110. Mezirow J. (1991). *Transformative Dimensions of Adult Learning.* San Francisco 1991. Mezirow J. (1997). "Transformative Learning: Theory to Practice". Eds. S. Imel, J. M. Ross-Gordon & J. E. Coryell. *New Directions for Adult and Continuing Education* (1997) 5-12.
45 Kofman F. Transformational learning: a blueprint for organizational change. The Systems Thinker. https://thesystemsthinker.com/transformational-learning-a-blueprint-for-organizational-change/ (accessed 18 November 2022).
46 Kofman F. Transformational learning: a blueprint for organizational change. The Systems Thinker. https://thesystemsthinker.com/transformational-learning-a-blueprint-for-organizational-change/ (accessed 18 November 2022).

Box 5.6 Transformational Learning Tools[47]

1) *Beginner's mind*: Approach every situation with the assumption that there is something you can learn from it. View disagreements as treasures that can open windows into different reasoning processes.
2) *Fluid framing*: Recognize that the way we interpret an event or situation is only one of many possibilities. Explore alternative logic for diverse points of view.
3) *Observation/assessment differentiation*: Distinguish observations (observable data) from assessments (subjective interpretations); do not treat opinions as indisputable facts.
4) *Advocacy/inquiry*: Share your data and reasoning so that others understand your logic. Inquire into others' data and reasoning so that you understand their logic.
5) *Commitment conversations*: Create and sustain a context of trust and internal commitment. Coordinate actions through effective requests, offers, and promises. Deal constructively with breakdowns through complaints and apologies that preserve relationships and correct mistakes.
6) *Shared context*: Use dialogue to explore diverse perspectives, create common understanding, and negotiate parameters for future conversations.
7) *Verbal aikido*: Respond to challenges and negative assessments by blending with and redirecting their energy with harmony.
8) *Check-in and check-out*: Take time at the beginning and end of meetings to say a few words about what is on your mind and become present. Bring concerns and issues into the open.
9) *Role-play*: Adopt another person's position to understand his or her view or put yourself in an imaginary situation to speculate concretely on what you might say or do.
10) *System mapping*: Use causal loop diagrams to map the systemic consequences of alternative policies.
11) *The ladder of inference*: Reverse the unconscious process through which we select a subset of observations, add meanings, and draw conclusions to explore how this is conditioned by the beliefs, assumptions, and values of our mental model.
12) *Congruence between private and public conversations*: Process your automatic assessments in a way that they can be shared without damaging a relationship. Make dilemmas discussable.

(*Source*: Kofman (2002)/Leverage Networks, Inc/ Public Domain.)

---

47 Kofman F. Transformational learning: a blueprint for organizational change. The Systems Thinker. https://thesystemsthinker.com/transformational-learning-a-blueprint-for-organizational-change/ (accessed 18 November 2022).

focused on transformational learning, systems-thinking and strategic analysis, personal and interpersonal structures that prevent learning, and leadership." Box 5.6 shows a number of tools used in LLC's program.

Fred Kofman writes that "the 'tools' of transformational learning are a set of practices that may appear simple, yet are not easy to implement. They demand a significant change in mental models, self-image, and consciousness."[48] This points to dynamics associated with organizational energy and the gravitational pull of systems, and the inertia impeding shifts.

The LLC program and its tools are mentioned here to reflect a deeper level of organizational learning that addresses individual learning and shifts (e.g. transformation). The shifts point to the right side of Figure 2.1 in Chapter 2. Other aspects of the LLC program will be discussed in Chapters 9 and 10.

5.6.3 Capitals Coalition's Four Shifts Model

The Capitals Coalition (CapCo) offers a transformative change model to impact organizational decision-making. At the model's core is shifting (changing) the math, conversation, rules, and system. These are defined as:

> *Change the math*: Provide decision-makers with holistic data that reshapes their decision-making calculus, leading to outcomes that deliver benefits across the system.
> *Change the conversation*: Accessible and influential communications, and best practices promoted and championed in our community are cultivated.
> *Change the rules*: Transform the incentives offered by investors, governments, shareholders, regulators, and rating agencies to reward those who adopt a capitals approach.
> *Change the system*: Ensure that the value of nature, people, and society informs decision-making at all levels and delivers holistic value across the capitals.[49]

Capitals and the coalition's work are touched on throughout this book. The shifts CapCo identifies are included here to support a concept that will be introduced in Chapter 8 (§8.2.2), "templating," which is offered as a powerful tool in bridging the three elements (regulatory/technical, organization, social–human)

---

48 Kofman F. Transformational learning: a blueprint for organizational change. The Systems Thinker. https://thesystemsthinker.com/transformational-learning-a-blueprint-for-organizational-change/ (accessed 18 November 2022).
49 Capitals Coalition (2020). Our value report. p. 13. https://capitalscoalition.org/wp-content/uploads/2021/08/Our-Value-Report_2020-21-.pdf (accessed 18 February 2022).

of the Risk Matrix's Contexts/Drivers dimension (§8.3). This dynamic is addressed further in Chapter 9 (§9.3.2). CapCo's four shifts (math, conversation, rules, and system) provide a template that can be used in shifting the ORM context.

5.6.4 Anatomy and Physiology of Shifts

Use of the terms "anatomy" and "physiology" to frame shifts reflects the "living system" nature of organizations and activities within them, including shifts.[50] It reflects a systems-based orientation, and the presence of structure(s), the flow of energy, and internal and external feedback mechanisms. Addressing the functions and mechanisms of organizations in this way is well established in the business and organizational science literature.

Similar to the way an artist uses a brush and paint, and a carpenter uses a hammer, level, and saw, we can identify the competencies needed, and processes and tools used to produce organizational shifts. A handful that I have observed and used are offered here and addressed throughout this book.

1) Creating a foundation for a shift (§10.3)
2) Awareness of current conditions (§2, Figure 2.1)
3) Clarity on the "B" in the A → B model. Embedded here is vision and purpose (§10.3)
4) Competency with the *seven risk awareness elements* (§5.4)
5) Understanding of language as currency (§5.5)
6) Motive force (§6.1)
7) Structures that provide an environment for "→" (§8, §9 and §10)

Suggested Reading

Argyris, C. (1992). *On Organizational Learning*, 8–12. Cambridge, MA: Blackwell Publishing, Inc.

Cartwright, S. (2002). Double-loop learning: a concept for leadership educators. *Journal of Leadership Education* 1 (1): 68–71.

Kim, D. (1993). The link between individual and organizational learning. *MIT Sloan Management Review* 35 (1): https://doi.org/10.1016/B978-0-7506-9850-4.50006-3.

---

50 The concept of "living company" is the subject of a seminal book written by Arie deGeus. Of interest here are characteristics that DeGeus associates with a "living company", including, they learn, have coherence, are resilient, build relationships, grow, and develop. *The Living Company: Habits for Survival in a Turbulent Business Environment* by Arie deGeus. Harvard Business School Press, Boston. 2002.

deGeus, A. (2002). *The Living Company: Habits for Survival in a Turbulent Business Environment*. Boston: Harvard Business School Press.

Goss, T. and Pascale, R. (1993). The reinvention roller coaster: risking the present for a powerful future. *Harvard Business Review* 71(6): 97–108

Handy, C. (1997). Looking ahead: implications of the present; citizen corporation. *Harvard Business Review*. Harvard Press 75: 18.

Kehneman, D. (2011). *Thinking, Fast and Slow*. New York: Farrer, Straus and Giroux.

Kahneman, D., Slovic, P., and Tversky, A. (ed.) (1982). *Judgment under Uncertainty: Heuristics and Biases*. Cambridge, U.K.; New York: Cambridge University Press.

Meadows, D. (2008). *Thinking in Systems*. White River Junction, VT: Chelsea Green Publishing Company.

Scharmer, O. and Kaufer, K. (2013). *Leading from the Emerging Future*. San Francisco: Berrett-Koehler Publishers, Inc.

Senge, P. (2006). *The Fifth Discipline: The Art and Practice of the Learning Organization, updated*. New York: Doubleday.

Scott, W.R. (2014). *Institutions and Organizations, Ideas, Interests and Identities*, 4e. Thousand Oaks, California: Sage Publications, Inc.

Siegel, D. (2018). *Aware – The Science and Practice of Presence*. TarcherPerigee.

Varela, F. (2003). *On Becoming Aware*. Amsterdam: John Benjamins Publishing Company.

Vogel, B. and Bruch, H. (2012). Organizational energy. In: *Oxford Handbook of Positive Organizational Scholarship* (ed. K. Cameron and G. Spreitzer), 691–702. New York, NY: Oxford University Press.

Zaffron, S. and Logan, D. (2009). *The Three Laws of Performance: Rewriting the Future of your Organization and your Life*. San Francisco: Jossey-Bass.

6

Field Leadership – Motive Force

CONTENTS
6.1 Motive Force, 181
6.1.1 Organizational Energy, 181
6.1.2 Culture of Health, 183
6.2 Field "Actors" – Individuals, Teams/Departments, Enterprise, Community, 184
6.2.1 Interiority, Accountability, 185
6.2.2 The Hats You Wear – Designer, Builder, Operator, Participant, 185
6.3 Creating Value, 186
6.3.1 Why This Is Important, 186
6.3.2 ISO 31000:2018 and COSO's ERM Framework, 186
6.3.3 ISO 37000:2021, Section 6.2 – Value Generation, 187
6.3.4 Capitals, 188
6.4 Leadership and Participation in Frameworks, 191
6.4.1 ISO 37000:2021, 191
6.4.2 COSO's Enterprise Risk Management, 193
6.4.3 ISO 31000:2018, 194
6.4.4 ISO MSS Examples – ISO 14001:2015 and ISO 45001:2018, 195
6.5 Emerging Leadership Paradigms, 196
6.5.1 System Leadership – Senge, Hamilton, Kania, 196
6.5.2 Responsible Leadership – Accenture, World Economic Forum, 197
Suggested Reading, 198

Organizational Risk Management: An Integrated Framework for Environmental, Health, Safety, and Sustainability Professionals, and their C-Suites, First Edition. Charles F. Redinger.
© 2025 John Wiley & Sons, Inc. Published 2025 by John Wiley & Sons, Inc.

Awareness alone is not sufficient for change.[1]

Aquilino, Michael, et al.
Duke University

I am convinced that great leadership in the 21st century is a matter of endowing groups of individuals with a satisfying sense of us and channeling their collective energy productively toward noble ends.[2]

George Halvorson
CEO Kaiser Permanente, 2002–2014

There is no shortage of organizational leadership theories, ideas, and practices. My purpose here is not to repeat them but rather to offer an idea of field leadership that builds on them. Leadership might not even be the best term, because the phenomenon I highlight is that of generating context, motive force, and engagement that drives the organization toward fulfillment of purpose from an "all-hands on deck" perspective. A more precise term is field leadership, is:

> *the practice of integrating context and motive force that generates engagement toward fulfilling organizational purpose, and value creation.*[3]

Fulfillment of organizational purpose and the creation of value are not stand-alone activities. They necessitate the alignment of organizational functions and engagement at all levels. I suggest that the risk management function provides a portal through which the organization's values and culture can more clearly be seen and leveraged to impact the fulfillment of organizational purpose.

Field leaders have a wide perspective. While productivity and profits predominate, they also have a multi-capital perspective that drives integrated thinking, decision-making, and action. Embracing health (human, social, and organizational) as a core value is a fundamental orientation. One key here is an orientation toward seeing and having an awareness of "the whole" as a field within which its systems, people, stakeholders, artifacts, etc., operate. And with this awareness, one can better understand the interconnections and health of the field's components.

---

1 Aquilino, M. et al. (2017). *The Power of Awareness and Choice in Effective Leadership*, 9. Durham, NC: Leadership Program in Integrative Healthcare at Duke University.
2 Halvorson, G. (2014). Getting to 'Us'. *Harvard Business Review*, September 2014, p. 38.
3 Refreshing from Chapter 2, Section 2.1.2, ISO 37000:2021 defines *organizational purpose* as, an "organization's meaningful reason to exist." And that "The [organization's] governing body should ensure that the organization's reason for existence is clearly defined as an organizational purpose. This organizational purpose should define the organization's intentions toward the natural environment, society and the organization's stakeholders. The governing body should also ensure that an associated set of organizational values is clearly defined."

I invite you to consider your and your team's leadership perspectives, legacy, and Buckminster Fuller's profound, poetic, invocation – "Call me Trim Tab."

6.1 Motive Force

Motive (n): "A reason for doing something, especially one that is hidden or not obvious. (adj) Producing physical or mechanical motion. Causing or being the reason for something."
Force (n): "Strength or energy as an attribute of physical action or movement."[4]

Field leadership is not a hierarchical phenomenon centered around single entities with organizational authority over others. A completely different style of leadership occurs in a field (social, organizational, risk). Yes, there are of course still hierarchical leaders as reflected in organizational charts, but with an "all hands on deck" orientation, leadership in a traditional sense is less of a factor. This is replaced by the motive force of teams driven by purpose, value generation, and value preservation.

Key components related to generating motive force are organizational energy, culture of health, culture of learning, and the social–human context/driver. Stocks, flows, and impact-dependency pathways are introduced in Chapter 7 (Decision-Making) and applied in Chapter 9 (Matrix in Action). Motive force is one of the dynamics that is characterized by these components.

6.1.1 Organizational Energy

Consistency in behavior *builds trust* in a relationship and a context in which energy can be created. Energy is part of everyday talk and experience in organizational life. It is associated with people's motivation and willingness to exert effort. Analyzing the energy in *social networks* can allow for the identification of broad patterns. Once these patterns are revealed, network participants can take action to create, or at least not destroy, energy and enthusiasm.[5]

Robert Cross and Andrew Parker
"Charged Up: Creating Energy in Organizations"

Energy is ubiquitous. We feel it, observe it, and harness it. For many of us, energy was inherent in the physical or biological sciences we studied. It is a topic in daily

---
4 Apple OS dictionary, New Oxford American Dictionary.
5 Journal of Organizational Excellence, Autumn 2004, Wiley Periodicals, Inc. 3.

news cycles and stories on climate change and the costs of oil, electricity, gas, etc. Considering "energy" in organizations is relatively new in organizational and business science literature. References began to appear in the 1960s and, notably in the seminal work of social psychology by Katz and Kahn (1966), where energy is referred to as a resource that is generated and used within organizations.[6] This notion of organizational energy has intrigued me since my public administration graduate work in the mid-1990s, and continued to capture my interest throughout my doctoral work at the University of Michigan.

Over the past several decades, there has been increasing attention devoted to organizational energy. Cross, Parker, and others in Positive Organizational Science have developed definitions, constructs, and approaches in this area. Much of this work focuses on the social–human aspects of organizations in addressing things such as building trust, communication, and characterizing social networks.

Organizational energy is a function of "the extent to which and organization, division, or team has mobilized its emotional, cognitive, and behavioral potential to pursue its goals. Simply put, it is the force with which a company (or division or team) works."[7] Attributes of organizational energy are:

1) "Organizational energy comprises the organizations' activated emotional, cognitive, and behavioral potential;
2) Organizational energy is a collective attribute – it comprises the shared human potential of a company (unit or team); and,
3) Organizational energy is malleable."

Cross and colleagues continue:

> Energy is created in conversations that balance several dimensions of an interaction. Hitting the midpoint, or sweet spot of these five dimensions, rather than the extremes, is the challenge in generating energy. The five dimensions identified are: compelling vision, ability to meaningfully contribute, full engagement, sense of progress, hope is present.[8]

This creation of energy comes through the integration of the social–human into the organization's risk management efforts/approaches, or, as I am suggesting, into the organization's risk field. The Risk Matrix provides a structure and environment for this. The three elements of it are people, structures, and contexts/

---

6 Katz, D. and Kahn, R. (1966). *The Social Psychology of Organizations*. New York: Wiley.
7 Bruch, H. and Vogel, B. (2011). *Fully Charged: How Great Leaders Boost Their Organization's Energy and Ignite High Performance*, 5. Boston: Harvard Business Review Press.
8 Cross, R. et al. (2003). What Creates Energy in Organizations? *MIT Sloan Management Review*, Summer 2003, vol 44, no. 4. Reprint 4445. pp. 53–55.

drivers. People are at the center of generating organizational energy. Structures and frameworks provide support, and context provides drivers.

6.1.1.1 People

The term "actors" is used in the Risk Matrix. This refers to individuals, teams, and departments, the enterprise (company), and communities. Of course, all of these are composed of people. The Matrix's z-axis depicts the organizational entities that design, build, manage and operate the risk field, along with those who are impacted by it.

6.1.1.2 Structures

Structures include the risk management frameworks such as ISO-based management systems and COSO's enterprise risk management frameworks. They also include organizational design, which is reflected in organizational charts and reporting hierarchy. Included here is how organizational performance measurement is done as expressed in audits, metrics, etc. The framework's evolution sequence in Chapter 2 (§2.1.1.2) – process, program, system, and field – represents a network of structures that provide for the creation of organizational energy.

6.1.1.3 Contexts/Drivers

Drivers is a commonly used term in business and refers to, "a factor which causes a particular phenomenon to happen or develop."[9] It is closely linked with context, which, along with "driver", is one of the Risk Matrix's three dimensions. The contexts/drivers elements identified in this dimension are regulatory/technical, organizational, and social–human. Regulatory compliance and associated technical components are historical drivers that establish the context for much of ORM. While compliance cannot be ignored, over time, there has been increased attention given to organizational performance and non-regulatory drivers. From things such as corporate social responsibility (CSR), and ESG, there has been increased attention to external impacts. That is, the focus has shifted beyond those of only the shareholder to a broader group of stakeholders; the concept double-materiality captures this dynamic.

6.1.2 Culture of Health

Culture of health for business (COH4B) was introduced in Chapters 1 (§1.4) and 2 (§2.1.5) and is revisited again in Chapter 10 (§10.3) in the context of creating generative fields. Attention is given to COH4B by the GRI/Robert Wood Johnson

---

9 New Oxford American Dictionary, Apple OS 13.5.2.

Foundation (GRI/RWJF) because of its role in impacting organizations. As an advisor in the development of the GRI/RWJF framework, I reviewed research, heard testimonials, and learned about organizations that had developed strong cultures of health. While COH4B was not framed in terms of motive force, it was evident that attention to a culture of health generates motive force that impacts organizational performance.

6.2 Field "Actors" – Individuals, Teams/Departments, Enterprise, Community

> The problem with leadership today is that most people think of it as being made up of individuals, with one person at the top. But if we see leadership as the capacity of a system to co-sense and co- shape the future, then we realize that all leadership is distributed – it needs to include everyone. To develop collective capacity, everyone must act as a steward for the larger eco-system. To do that in a more reliable, distributed, and intentional way, we need: A social grammar: a language; a social technology: methods and tools; and a new narrative of societal evolution and change.[10]
>
> Otto Scharmer
Gathering in Generative Social Fields
Garrison Institute, 2018

In a field, whether a social, organizational, or risk field, leadership and the generation of motive force are distributed. There are any number of ways to address the notion of distributed. In this book, distributed leadership and the generation of motive force represented in the Risk Matrix's Actors/Motive Force dimension (§8.4). The field literature identifies actors as including "...individuals, associations of individuals, populations of individuals, organizations, associations of organizations, and populations of organizations."[11]

The Risk Matrix's Actors/Motive Force dimension contains four elements: (1) individual, (2) team/department, (3) enterprise/company, and (4) community. This Matrix dimension plays a significant role in generating motive forces and organizational energy that produce ORM outcomes. Its four elements are defined in greater detail in Chapter 8 (§8.4) and Chapter 9 (§9.2.2 and §9.4.3).

---

10 Scharmer, O. (2018). How to Cultivate the Social Field. Garrison Institute conference, 1–3 October 2018. Conference introduction document, p. 11. Excerpt from *The Essentials of Theory U*, Berrett-Koehler.
11 Scott, W.R. (2014). *Institutions and Organizations*. Sage Publications, Inc., 228.

6.2.1 Interiority, Accountability

In the context of this discussion, interiority is one of the key way to distinguish between systems and fields. Interiority refers to being inside the field not viewing it from the outside as a system. It means thinking inside the field, engaging other actors/entities within the field, and your degree of accountability within the field. Scharmer makes this distinction in discussing Social Field Theory in the following terms:

> What differentiates social fields from social systems is their degree of interiority. Social systems are social fields seen from the outside (the third-person view). At the moment we cross the boundary between them and step inside a social system – that is, at the moment we begin to inquire into its interiority by turning the camera around (from the third-person to the first-person view) – we switch the perspective from the social system to the social field.[12]

I revisit this distinction in Chapter 10 in the context of generative fields. At this point in our journey, we need to consider the shifts that happen when we step inside a system and how this impacts things like accountability and engagement. I suggest that a key attribute of social field leadership is the generation of motive force and organizational energy.

As with the internal state perspective discussed in Chapter 5, this relates to a situation in which a person's orientation and perspective is from inside the system, engaged as a member of a field, as opposed to someone who is not engaged as such and is outside the system. In both instances, the person is a member/actor. The key issues are the level of engagement, and the internal state dynamics which are addressed in Chapter 5 (Scharmer's quote, §5.3 and §5.4.2). Interiority is a foundational concept, as you will see when we get to generative fields in Chapter 10 (§10.2 and §10.3).

6.2.2 The Hats You Wear – Designer, Builder, Operator, Participant

It is valuable to think about and identify the multiple roles and accountabilities people have in organizational life. In this chapter, we address field leadership, motive force, organizational energy, etc. Chapter 5 addressed awareness and issues related to internal states. Later chapters address decision-making (§7), Risk

---

12 Senge, Peter, Scharmer, Otto, Boell, Mette (2015). Towards a Lexicon for Investigating Generative Social Fields. Academy for Contemplative and Ethical Leadership. Reprinted in Garrison Institute event (2018) document, p. 15.

Matrix design (§8), and its operation (§9). The point here is to draw attention to the range of roles and accountabilities you, your team, and those impacted by your actions can have. Thinking of these might not seem relevant. However, I suggest these considerations are critical as we begin to navigate the multidimensional Risk Matrix. This is especially true when considering roles and accountabilities in the actor/motive force dimension (individual, team/department, enterprise/company, community), while navigating the Contexts/Drivers dimension (regulatory/technical, organizational, social-human).

6.3 Creating Value

Value is presented in various ways in the risk management, ESG, and organizational purpose arenas. Terms used when discussing values include generation, creation, protection, preservation, and delivery. I am using the term "creating value" here to reflect both creating and protecting/preserving value. Definitions related to this section are provided in Box 6.1.[13]

6.3.1 Why This Is Important

This topic is included here because it is central to organizational risk management and it is a driver in generative fields (social, organizational, risk, etc.). This thread is picked up in Chapter 10 (§10.2). Foundations for each of these are offered here.

A subtle but important dynamic is to distinguish between the terms "value" and "values." Both are important, but their meanings are often confused. The focus here is on the value of something as it relates to its worth (often thought of in terms of financial metrics), as opposed to values that relate to principles or standards of behavior as indicated in the above definitions. For instance, both terms are found throughout COSO's ERM framework in parallel but in different contexts.

6.3.2 ISO 31000:2018 and COSO's ERM Framework

ISO 31000:2018 addresses value in terms of creation and protection of value. The standard states, "The purpose of risk management is the creation and protection of value."[14] A range of value aspects are addressed throughout the COSO ERM

---

13 New Oxford American Dictionary. Apple OS 13.5.2.
14 ISO 31000:2018, p. 2.

6.3 Creating Value | **187**

> **Box 6.1   Definitions – Create, Value, Worth, and Values**
>
> *Create* (v): "Bringing (something) into existence. Cause (something) to happen as a result of one's actions."
> *Value* (n): "The regard that something is held to deserve; the importance, worth, or usefulness of something. The material or monetary worth of something. The worth of something compared to the price paid or asked for it."
> *Worth* (n): "The value equivalent to that of someone or something under consideration; the level at which someone or something deserves to be valued or rated. An amount of a commodity equivalent to a specified sum or money."
> *Values* (n): "A person's principles or standards of behavior; one's judgement of what is important in life."

framework. It states, "A discussion of enterprise risk management begins with this underlying premise: every entity – whether for-profit, not-for-profit, or governmental – exists to provide value for its stakeholders. This publication is built on a related premise: all entities face risk in the pursuit of value."[15] It continues by connecting value to decision-making, stating, "The value of an entity is largely determined by the decisions that management makes – from overall strategy decisions through to day-to-day decisions. Those decisions can determine whether value is created, preserved, eroded, or realized."[16]

6.3.3   ISO 37000:2021, Section 6.2 – Value Generation

ISO's organizational governance standard has been touched on in previous chapters. In Chapter 2 (§2.1.2), it was introduced generally in terms of its overall risk governance context. Figure 2.4 illustrates its structure. It was revisited in Chapter 3 (§3.3.4), where the focus was on its "risk governance" principle, 6.9. The standard offers 11 governance principles that are divided into three categories: primary, foundational, and enabling. There are 4 foundational principles, the first of which is "value generation." The standard states, "The governing body should define the organization's value generating objectives such that they fulfill the organizational purpose in accordance with the organizational values and the natural environment, social, and economic context within which it operates."[17] Notice here the use of "value" and "values."

---

15   COSO, p. 36.
16   COSO, p. 38.
17   ISO 37000:2021, p. 11.

6 Field Leadership – Motive Force

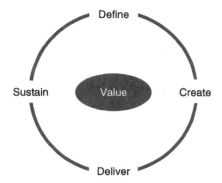

Figure 6.1 ISO 37000's organizational value-generation model. *Source:* Adapted from ISO 37000:2021.

The standard offers a simple value-generation model in Figure 6.1[18] as well as recommending criteria for defining, creating, delivering, and sustaining value. A key aspect of this is to understand "stakeholder expectations, regulatory frameworks, technological change, and the present and potential future natural environment, social and economic issues."[19]

6.3.4 Capitals

A capital orientation provides a structure and language for considering value generation, preservation, and risk decision-making. This goes beyond the traditional use of externalities in environmental economics when considering the implications of single and double materiality in an ESG context.[20] In Chapter 7, I expand on this in terms of risk decision-making, integrated thinking, and action in using the Risk Matrix. The tools and methods discussed are impact and dependency pathways.

6.3.4.1 Defining Capitals

Capital is a common element in business and finance. It includes things such as money, property, and people. The New Oxford American Diction defines it as:

> Wealth in the form of money or other assets owned by a person or organization or available or contributed for a particular purpose such as starting a company or investing. A valuable resource of a particular kind.

---

18 ISO 37000:2021, p. 16.
19 ISO 37000:2021, p. 15.
20 See Chapter 3 (§3.6.1.4 and 3.6.1.5) to refresh on single and double materiality.

Organizations in the ESG and integrated reporting communities have identified numerous capitals, including financial, manufactured, natural, social, human, intellectual, and produced.

6.3.4.1.1 *International Integrated Reporting Council (IIRC)* In their integrated reporting framework, the International Integrated Reporting Council (IIRC) identifies six capitals: financial, manufactured, intellectual, human, social and relationship, and natural. Definitions for these are:[21]

- *Financial capital*: "The pool of funds that is: available to an organization for use in the production of goods or the provision of services; and, obtained through financing, such as debt, equity or grants, or generated through operations or investments."
- *Manufactured capital*: "Manufactured physical objects (as distinct from natural physical objects) that are available to an organization for use in the production of goods or the provision of services, including: buildings; equipment; infrastructure (such as roads, ports, bridges, and waste and water treatment plants). Manufactured capital is often created by other organizations, but includes assets manufactured by the reporting organization for sale or when they are retained for its own use."
- *Intellectual capital*: "Organizational, knowledge-based intangibles, including: intellectual property, such as patents, copyrights, software, rights and licenses; and, 'organizational capital' such as tacit knowledge, systems, procedures and protocols."
- *Human capital*: "People's competencies, capabilities and experience, and their motivations to innovate, including their: alignment with and support for an organization's governance framework, risk management approach, and ethical values; ability to understand, develop and implement an organization's strategy; and, loyalties and motivations for improving processes, goods and services, including their ability to lead, manage and collaborate."
- *Social and relationship capital*: "The institutions and the relationships within and between communities, groups of stakeholders and other networks, and the ability to share information to enhance individual and collective well-being. Social and relationship capital includes: shared norms, and common values and behaviours; key stakeholder relationships, and the trust and willingness to engage that an organization has developed and strives to build and protect with external stakeholders; and, intangibles associated with the brand and reputation that an organization has developed."

---

21 IIRC (2013). The International <IR> Framework. Page 12. This reference applies to all six definitions. Note: The IIRC merged with the Sustainability Accounting Standards Board (SASB) in 2020 to form the Value Reporting Foundation (VRF). The VRF subsequently merged within the IFRS Foundation to form the International Sustainability Standards Board (ISSB).

- *Natural capital*: "All renewable and non-renewable environmental resources and processes that provide goods or services that support the past, current or future prosperity of an organization. It includes: air, water, land, minerals and forests; and, biodiversity and eco-system health."

6.3.4.1.2 Capitals Coalition In their work, the Capitals Coalition identifies four capitals: human, social, natural, and produced capital. Definitions for these are:[22]

> *Human capital*: Represents the skills and expertise of every single individual in our network. Human capital is the sum of personal experiences that inspire each of us to work on making the economy more sustainable for nature and people, and the stories that we each can offer to inspire and motivate others to do the same. Another crucial aspect of our human capital is the diversity within our community which allows us to work in a way that is representative and accessible to everyone.
> *Social capital*: Represents the networks and relationships, values and understanding that facilitate cooperation within the Coalition community. It offers social norms such as reciprocity, trust and participation, and diversity of experience and opinion.
> *Natural capital*: Supports our operations through the provision of materials and space to work, as well as deep inspirational and emotional values that drive what we do. Underpinning all of this are fundamental ecosystem services such as sustenance, health and wellbeing.
> *Produced capital*: We rely on produced capital thanks to the freedom to travel, the use of meeting spaces, intellectual property, the provision of online technology and funding.

6.3.4.2 Capitals Coalition and Value

Continuing with the Capital Coalition's four shifts from Chapter 5 (§5.6.3), outcomes indicated for each of the shifts reflect their focus on value:

> *Change the math*: Reshaping how business, financial markets, and governments identify, measure, and value their relationship with nature, people, and the economy.
> *Change the conversation*: Embedding the value of a capitals approach in the global conversation.

---

22 Capitals Coalition (2020). Our value report 2020. p. 12. This reference applies to all four definitions.

Change the rules: Transforming incentives to deliver a new normal in decision-making.
Change the system: Reflecting the true value of all capitals in decision-making.[23]

6.4 Leadership and Participation in Frameworks

Early occupational health and safety management system (OHSMS) frameworks represented significant shifts in OHS management by highlighting leadership and participation. Although practitioners knew that they were cornerstones of the field, leadership and participation were not reflected in formal structures (e.g. regulations). This has changed over time, and, as organizational and business science has paid more attention to leadership, attention to frameworks and systems has increased as well. In OHSMSs there has also been an evolution in how participation is addressed. The term "consultation" has taken on expanded meaning, both with respect to requirements and/or recommendations, and with respect to whom should be included.

6.4.1 ISO 37000:2021

Stakeholder engagement and leadership are, respectively, 37001's sixth and seventh principles, and are categorized as "enabling" principles. Enabling principles are defined as "...principles [that] address the governance responsibilities pertinent to today's organizations – to meet evolving stakeholder expectations and the changing natural environment, social and economic context."[24] The standard defines a governing body as a:

> Person or group of people who have ultimate *accountability* for the whole *organization*. Note 1 to entry: Every *organizational entity* has one governing body, whether or not it is explicitly established. When the organization is not an organizational entity, the term *governing group* is applicable where 'governing body' is used throughout this document. Note 2 to entry: A governing body can be explicitly established in a number of formats, including, but not limited to, a board of directors, supervisory board, sole director, joint and several directors, or trustees. Note 3 to entry: ISO management system standards make reference to the term 'top management' to describe

---
23 Capitals Coalition (2020). Our value report. p. 13. https://capitalscoalition.org/wp-content/uploads/2021/08/Our-Value-Report_2020-21-.pdf (accessed 18 February 2022).
24 ISO 37000:2021, p. 12.

a role that, depending on the standard and organizational context, reports to, and is held accountable by, the governing body.[25]

The standard addresses leadership in terms of its role in setting the "tone for an ethical organizational culture." It states:

> The governing body should lead the organization ethically and effectively and ensure such leadership throughout the organization. In an organization, the governing body should set the tone for an ethical organizational culture. While all individuals contribute to this culture, what the governing body says, does and expects is critical in setting the tone for the whole organization. Leadership is therefore a critical issue for a governing body. Its own behaviours provide the model for the organization's behaviour. With the principles it establishes concerning the way stakeholders should be treated and the way goals should be pursued, the governing body creates standards and examples for others to follow. Visible, responsible and competent leadership ensures that the organization follows the expectations which have been set. In addition, effective leadership provides clarity in communication and an understanding of expectations across the whole organization.[26]

As it continues, it stresses the importance of leading "ethically and effectively." It states:

> The governing body should ensure ethical leadership across all areas. (1) Within the governing body: The members of the governing body should demonstrate that they are behaving in a manner consistent with the organizational values. (2) Within the organization: The governing body should ensure that the organization conducts itself in a manner consistent with its organizational values. (3) Within the organization's external context: The governing body should ensure that the organization treats stakeholders in a manner consistent with its organizational values.[27]

Examples are offered in terms of organizational culture and context. It states:

> Ethical leadership results in an organizational context and culture that: provides the individuals of an organization with a collective sense of belonging; assists in reconciling strategic dilemmas by creating

---

25 ISO 37000:2021, p. 6.
26 ISO 37000:2021, p. 25.
27 ISO 37000:2021, p. 26.

organizational alignment through the integration of opposites; contributes to the prevention of misconduct; provides competitive differentiation for stakeholders by providing clarity against which evaluators can assess the organization's behaviour, decisions and activities; and, provides increased certainty, which in turn, creates reputational value.[28]

The standard's ninth principle addresses risk management leadership and offers recommendations regarding decision-making behaviors. It states:

> The governing body should establish an organizational risk framework that ensures a formal, proactive and anticipative approach to the management of risk across the organization, including by the governing body. The governing body should ensure that this framework integrates risk management into all organizational activities. The governing body should ensure that the organizational risk framework, in respect to the management of risk: (b) guides decision-making behaviours and the impact of leadership actions, inactions or omissions on those behaviours.[29]

Stakeholder engagement recommendations are offered in ISO 37000:2021. The standard states that the governing body should ensure that the organization's stakeholders are appropriately engaged and their expectations considered."[30] It continues, stating:

> The governing body should ensure that the organization's stakeholders are identified, prioritized, appropriately engaged, consulted and their expectations understood. The governing body should do this to ensure that stakeholder relationships are effective and appropriate decisions about expectations are made to achieve the intended value generation objectives. When the governing body groups stakeholders, it should clarify its criteria for grouping, and for determining the relevance of, stakeholders. The governing body should also ensure that a stakeholder engagement process is devised on this basis.[31]

6.4.2 COSO's Enterprise Risk Management

Leadership and participation recommendations are touched on throughout the COSO ERM framework. In its fourth principle, "Demonstrates Commitment to

---
28 ISO 37000:2021, pp. 26–27.
29 ISO 37000:2021, p. 31.
30 ISO 37000:2021, p. 24.
31 ISO 37000:2021, p. 24.

Core Values [and] Embracing a Risk-Aware Culture," it states, "Maintaining strong leadership: The board and management places importance on creating the right risk awareness and tone throughout the entity. Culture and, therefore, risk awareness cannot be changed from second-line team or department functions alone; the organization's leadership must be the real driver of change."[32]

It continues within a context of enforcing accountability, it states, "Management provides guidance to personnel so they understand the risks. Management also demonstrates leadership by communicating the expectations of conduct for all aspects of enterprise risk management. Such leadership from the top helps to establish and enforce accountability and a common purpose."[33]

6.4.3 ISO 31000:2018

This standard's introduction begins with: "This document is for use by people who create and protect value in organizations by managing risks, making decisions, setting and achieving objectives and improving performance." Examples of "managing risk" are given, including: "Managing risk is part of governance and *leadership*, and is fundamental to how the organization is managed at all levels. It contributes to the improvement of management systems."[34]

As depicted in Figure 3.3, ISO 31000 is organized into three components, principles, framework, and process. The senior distinction in the framework component is Leadership and Commitment (5.2). The standard states:

> Top management and oversight bodies, where applicable, should ensure that risk management is integrated into all organizational activities and should demonstrate leadership and commitment by: customizing and implementing all components of the framework; issuing a statement or policy that establishes a risk management approach, plan or course of action; ensuring that the necessary resources are allocated to managing risk; and, assigning authority, responsibility and accountability at appropriate levels within the organization.[35]

And,

> Top management is accountable for managing risk while oversight bodies are accountable for overseeing risk management. Oversight bodies are

---

32 COSO, p. 33.
33 COSO, p. 34.
34 31000, Introduction, p. v.
35 31000:2018, p. 5.

often expected or required to: ensure that risks are adequately considered when setting the organization's objectives; understand the risks facing the organization in pursuit of its objectives; ensure that systems to manage such risks are implemented and operating effectively; ensure that such risks are appropriate in the context of the organization's objectives; and, ensure that information about such risks and their management is properly communicated.[36]

6.4.4 ISO MSS Examples – ISO 14001:2015 and ISO 45001:2018

Leadership is a major element in ISO's high-level management system structure. ISO 14001:2015's application of this contains three subsections: leadership and commitment (5.1); environmental policy (5.2); and organizational roles, responsibilities, and authorities (5.3). Relating to leadership and commitment, the standard states:

> Top management shall demonstrate leadership and commitment with respect to the environmental management system by: a) taking accountability for the effectiveness of the environmental management system; b) ensuring that the environmental policy and environmental objectives are established and are compatible with the strategic direction and the context of the organization; c) ensuring the integration of the environmental management system requirements into the organization's business processes; d) ensuring that the resources needed for the environmental management system are available; e) communicating the importance of effective environmental management and of conforming to the environmental management system requirements; f) ensuring that the environmental management system achieves its intended outcomes; g) directing and supporting persons to contribute to the effectiveness of the environmental management system; h) promoting continual improvement; i) supporting other relevant management roles to demonstrate their leadership as it applies to their areas of responsibility.[37]

ISO 45001:2018's application of the high-level structure includes "worker participation" in the section's title. It includes four subsections: leadership and commitment (5.1); OH&S policy (5.2); organizational roles, responsibilities, and

---

36 31000:2028, p. 5.
37 14001:2015, p.7.

authorities (5.3); and consultation and participation of workers (5.4). Refer to Chapter 3 (§3.4.2.4), where these subsections are addressed in greater depth.

6.5 Emerging Leadership Paradigms

What follows introduces two emerging leadership paradigms that offer ideas and tools for putting the Risk Matrix into action (Chapter 9) and in navigating the Matrix, leading to a new clearing and ways to escalate impact (Chapter 10). These are "System Leadership," and "Responsible Leadership." Each provides valuable perspectives that support the later chapters. The intent here is not to provide comprehensive overviews of these but rather to highlight their key points, and overall structure.

6.5.1 System Leadership – Senge, Hamilton, Kania

Leadership is a central piece in the system dynamics work that has evolved out of the MIT group mentioned in Chapter 5 (§5.6.2). I have been fortunate to participate over the years in several workshops conducted by the Society for Organizational Learning. In May 2018, I attended "Leading Across Boundaries for Systems Change." This workshop presented material from a paper by Peter Senge, Hal Hamilton, and John Kania titled "the Dawn of System Leadership,"[38] in which they introduced the characteristics and capabilities of "system leaders." They defined a system leader as "Someone who was able to bring forth collective leadership."[39] In the workshop, this notion was expanded to include fostering collective leadership across silos in organizations.

The presenters called out the impacts made by systems leaders: "Over time, their profound commitment to the health of the whole radiates to nurture similar commitment in others. Their ability to see reality through the eyes of people very different from themselves encourages others to be more open as well. They build relationships based on deep listening, and networks of trust and collaboration start to flourish. They are so convinced that something can be done that they do not wait for a fully developed plan, thereby freeing others to step ahead and learn by doing."[40] The authors continue by identifying three core capabilities of system leaders, gateways to becoming a systems leader, and guides for moving along the

---

38 Senge, P. et al. (2015). The Dawn of System Leadership. *Stanford Social Innovation Review*, Winter 2015.
39 The Dawn of System Leadership, p. 27.
40 The Dawn of System Leadership, p. 28.

path of system leadership. I encourage you to read this paper, as it provides valuable ideas and tools.[41]

Core system leadership capabilities are identified as:

1) *Ability to see the larger system*: Building a shared understanding of complex problems
2) *Ability to foster reflective conversation and more generative* conversations: Reflection means, thinking about our thinking. Challenging assumptions and mental models.
3) Ability to shift the collective focus from reactive problem solving to co-creating the future: Helping people articulate their deeper aspirations. Creating a space to face difficult truths about the current reality, and using the tension between vision and reality to inspire truly new approaches.[42]

The authors conclude their identification of system leadership capabilities with, "Much has been written about these leadership capabilities in the organizational learning literature and the tools that support their development. But much of this work is still relatively unknown or known only superficially to those engaged in collaborative systemic change efforts."[43] Aspects of system leadership qualities and capabilities in the context of generative fields are reviewed in Chapter 10.

6.5.2 Responsible Leadership – Accenture, World Economic Forum

> We are becoming increasingly aware that solutions to our global challenges must purposefully engage youth, at all levels – locally, regionally, nationally and globally. This generation has the passion, dynamism and entrepreneurial spirit to shape the future.[44]
>
> Prof. Klaus Schwab, Founder and Executive Chairman, World Economic Forum

A valuable body of leadership work has been initiated in collaboration with Accenture by the Forum of Young Global Leaders and the Global Shapers Community. A five-element model called Responsible Leadership evolved from

---

41 Download at https://ssir.org/articles/entry/the_dawn_of_system_leadership.
42 The Dawn of System Leadership, pp. 28–29.
43 The Dawn of System Leadership, p. 29.
44 https://www.globalshapers.org/partnerships/ (accessed 8 November 2023).

extensive research, including interviews with "people mostly born since 1980 – Generations Y and Z."[45] The model's elements are:

> *Stakeholder inclusion*: Safeguarding trust and positive impact for all by standing in the shoes of diverse stakeholders when making decisions – and fostering an inclusive environment where diverse individuals have a voice and feel they belong. Accountability, impact, and trust.
> *Emotion and intuition*: Unlocking commitment and creativity by being truly human, showing compassion, humility, and openness.
> *Mission and purpose*: Advancing common goals by inspiring a shared vision of sustainable prosperity for the organization and its stakeholders. Sensemaking, systems thinking, and integrity.
> *Technology and innovation*: Creating new organizational and societal value by innovating responsibly to solve problems using emerging technology. Tech vision, responsible innovation, and creativity.
> *Intellect and insight*: Finding ever-improving paths to success by embracing continuous learning and knowledge exchange.[46]

Ideas and issues identified in the Responsible Leadership are touched on in Chapters 9 and 10.

Suggested Reading

Bruch, H. and Vogel, B. (2011). *Fully Charged: How Great Leaders Boost their Organization's Energy and Ignite High Performance*. Boston: Harvard Business Review Press.

Cameron, K.S. et al. (2003). *Positive Organizational Scholarship, Foundations of a New Discipline*. San Francisco, California: Berrett-Koehler Publishing, Inc.

Capitals Coalition (2019). *Social and human capital protocol*. https://capitalscoalition.org/capitals-ap- proach/social-human-capital-protocol/.

Committee of Sponsoring Organizations of the Treadway Commission (2017). *Enterprise Risk Management – Integrating with Strategy and Performance*. Association of International Certified Professional Accountants.

---

45 Seeking New Leadership Responsible leadership for a sustainable and equitable world (2020). An initiative between the forum of Young Global Leaders and the Global Shapers Community, in collaboration with Accenture, to create a new framework for responsible leadership and to help organizations cultivate environments that can flourish. p. 6. https://www.accenture.com/content/dam/accenture/final/a-com-migration/pdf/pdf-115/accenture-davos-responsible-leadership-report.pdf#zoom=40 (accessed 1 May 2022).

46 *Seeking New Leadership* (2020), pp. 42–43.

Cross, R., Baker, W., and Parker, A. (2003). What creates energy in organizations? *MIT Sloan Management Review* 44 (4): 51–56.

IIRC (2013) The International <IR> Framework. Note: the IIRC merged with the Sustainability Accounting Standards Board (SASB) in 2020 to form the Value Reporting Foundation (VRF). The VRF subsequently merged within the IFRS Foundation to form the International Sustainability Standards Board (ISSB). https://integratedreporting.ifrs.org.

International Organization for Standardization (2015). *Environmental Management Systems – Requirements with guidance for Use, ISO 14001*. Geneva, Switzerland: International Organization for Standardization.

International Organization for Standardization (2018). *Risk Management – Guidelines, ISO 31000*. Geneva, Switzerland: International Organization for Standardization.

International Organization for Standardization (2018). *Occupational Health and Safety Management Systems – Requirements with Guidance for use, ISO 45001*. Geneva, Switzerland: International Organization for Standardization.

International Organization for Standardization (2021). *Governance of organizations – Guidance, ISO 37000*. Geneva, Switzerland: International Organization for Standardization.

Kaplan, R. and Mikes, A. (2012). Managing risks: a new framework, smart companies match their approach to the nature of the threats they face. *Harvard Business Review*, HBR Reprint R1206B: 1–13.

RWJF/GRI (2019). Culture of health for business, guiding principles to establish a culture of health for business. https://www.globalreporting.org/public-policy-partnerships/strategic-partners-programs/culture-of-health-for-business/ (accessed 17 July 2019).

Seeking New Leadership Responsible Leadership for a Sustainable and Equitable World (2020). An initiative between the forum of Young Global Leaders and the Global Shapers Community, in collaboration with Accenture, to create a new framework for responsible leadership and to help organizations cultivate environments that can flourish. https://www.accenture.com/content/dam/accenture/final/a-com-migration/pdf/pdf-115/accenture-davos-responsible-leadership-report.pdf#zoom=40 (accessed 18 October 2021).

Senge, P., Hamilton, H., and Kania, J. (2015). The dawn of system leadership. *Stanford Social Innovation Review* 13: 27.

7
Decision-Making – Expanding Perspective

CONTENTS
Awareness – Process, Paradox, and Tension, 203
Types of Decisions, 204
Organizational Learning, 204
Expanded Platform, 204
7.1 Background and Anchors, 205
7.1.1 Decision Science, 205
7.1.2 The Human, 206
7.2 Systems Perspective, 211
7.2.1 Systems 101, 212
7.2.2 Inputs and Processes, 214
7.2.3 Output, Outcome, and Impact, 215
7.2.4 Feedback, 216
7.2.5 Stocks and Flows, 216
7.2.6 Impact-Dependency Pathways, 218
7.3 Frameworks, 219
7.3.1 ISO 37000:2021, Governance of Organizations – Guidance, 219
7.3.2 ISO 31000:2018, Risk Management – Guidelines, 220
7.3.3 COSO Enterprise Risk Management – Integrating with Strategy and Performance, 222
7.3.4 ISO Management System Standards, 222
7.4 Key Considerations, 222
7.4.1 Carriers of Meaning, 223
7.4.2 Context, Framing, and Narrative – Or Is it the Number?, 223
7.4.3 Defining Risk, 224
7.4.4 Decision-Making Currency, 224
7.4.5 Rates, Cycles, and Time Horizon, 226
7.4.6 Delays and Buffers, 227
7.5 Risk Decision-Making Kernel, 227
Suggested Reading, 228

Organizational Risk Management: An Integrated Framework for Environmental, Health, Safety, and Sustainability Professionals, and their C-Suites, First Edition. Charles F. Redinger.
© 2025 John Wiley & Sons, Inc. Published 2025 by John Wiley & Sons, Inc.

With no awareness of the power of context, we continue to beat our heads against the same wall.[1]

Tracy Goss, et al.

Knowledge is anything that increases your ability to predict the outcome. Literally everything you do, you're trying to predict the right thing. Most people just do it subconsciously.[2]

Daryl Morey
Basketball Executive

Context shapes decision-making. Regulatory compliance has been the dominant context shaping EHSS/ORM thinking and action. The challenges touched on in Chapter 1 argue for the expansion of the decision-making perspective. Below, I repeat the four questions posed in Chapter 2 (§2.5).

- How aware are you and your team of your risk decision-making (RDM) process(es)?
- Is this something you all think about?
- To what extent, if at all, do your processes reflect integrated thinking?
- How would you describe or characterize the processes if asked?

These are valuable questions to consider. To Daryl Morey's observation, is awareness of this conscious or subconscious? And, with the risk management contexts and drivers presented in the risk field/matrix (regulatory/technical, organizational, social–human), is there awareness and understanding of how these interact and impact outcomes?

When I have asked clients, colleagues, and students these questions over the years, their responses more typically than not were circuitous or muddled. In many instances, they would say either "I've never thought of that," or "I don't know." This chapter dives into this inquiry and builds a foundation for moving on to Chapters 8, 9, and 10, where I present the Risk Matrix and ways to use it. What follows is not theoretical but rather pragmatic and highlights concepts and tools to put the Matrix into action. The focus here is on increasing awareness of the RDM process and then using this awareness to improve and routinize integrated thinking, decision-making, and action.

Whether you are a boots-on-the-ground (BOTG) worker in a plant, a mid-level manager, a director, or a C-suite executive, each role has its own unique accountabilities and different types of decision pressures. The Risk Matrix (Figure 8.3) presented in the next chapter provides an integrated framework that supports the

---

1 Goss T., Pascale R., and Athos A. (1993). The reinvention roller coaster: risking the present for a powerful future. *Harvard Business Review*, November–December, p. 101.
2 Lewis, M. (2017). *The Undoing Project*, 31. W.W. Norton.

range of each of these different organizational levels. The Matrix contains three dimensions, one of which is contexts and drivers. This dimension has three elements: (1) regulatory/technical, (2) organization, and (3) social–human. I draw your attention now to RDM in each of these from a stand-alone perspective and from an integrated perspective with emphasis on the social–human influence. I also draw your attention to decision-making nuances from the field/matrix Actors/Motive Force dimension, which has four elements: (1) individual; (2) team/department; (3) enterprise/company; and (4) community. The ways RDM plays out in the Matrix is explored in depth in Chapter 9.

There are a number of considerations in RDM, including maintaining regulatory compliance, company health, and profitability, and ESG/CSR.

Decision-making and conformity assessment are closely linked. It is hard to imagine any organizational decision that does not involve some sort of measurement activity (e.g. auditing), consideration of metrics or indicators, and possibly consideration of "who's the judge?" In Chapter 4, we looked at a range of conformity assessment-related topics, including the relationship among the human component of auditing (e.g. the auditor), the audit instrument, and audit outputs. With respect to decision-making, I draw your attention to the relationship between decision-makers (individual, team/department, etc.) and decision-making tools and processes. In decision-making processes, attention always needs to be given to types of data and information, measurement levels (e.g. nominal, ordinal, interval, and ratio), and the validity and reliability of each item or metric, or collection of items or metrics as in a scale or index.

It is valuable to consider how decision-making processes impact individual and organizational learning. I heard Dr. Sudip Bose[3] speak at a conference in 2021. He served in the U.S. Army as a frontline emergency physician. The doctor presented many useful tips, but one left an impression that is relevant here. He made a connection between decisions, habits, and character. He framed these in terms of making decisions under pressure and the value of paying attention to decisions we make. He asserted that, by paying attention, we begin to see that decisions impact habits, which in turn impact character. While he did not point this in a learning context, he was essentially pointing to a learning orientation.

Awareness – Process, Paradox, and Tension

Increasing awareness of RDM processes leads to more control over their outputs, outcomes, and impacts. As the social–human context/driver takes on greater prominence, along with the increased number of actors and motive forces,

---

3 https://www.docbose.com (accessed 3 May 2023).

(e.g., individuals and the broader community), RDM processes need to expand to accommodate this and to address shifts in the risk profile, as well as the added increases in paradox and tension.

Types of Decisions

In organizational life, we make decisions that range from the mundane and routine to ones that are mission-critical with significant consequences. Many of the former require little thought. Certainly, "automatic" decisions sometimes referred to as System 1 decisions, don't take much thought. Alternatively, System 2 decisions call for complex reasoning and deliberation.[4] I focus our attention on both of these (System 1 and 2) and the decision-making processes that impact them.

Organizational Learning

Decision-making and learning are, or at least should be, closely linked. I will be touching on systems-related topics in this chapter including (1) feedback loops, (2) impact-dependency causal pathways, and (3) stocks and flows. Awareness of these three combined with attention to how we learn is a requirement for developing an organization in which a culture of learning is established and sustainable.

Expanded Platform

The risk field/matrix construct provides a platform for expanding the RDM perspective. This is key to reducing uncertainty because the decision-making process is able to cast a wider net with greater capacity to identify risks that heretofore have gone unnoticed. Here and in the chapters that follow, I introduce impact-dependency pathways, and multidirectional flows, and the movements within them.

---

4 System 1 and 2 are addressed later in this chapter (§7.1.2.1). This notion of two systems has interested psychologists for decades. Keith Stanovich and Richard West are credited with formalizing it in 2000 in "Individual Differences in Reasoning: Implications for the Rationality Debate," *Behavior and Brain Science* 23: 645–665. The two-system model has been popularized in Daniel Kahneman's book, *Thinking, Fast and Slow* (2011).

7.1 Background and Anchors

7.1.1 Decision Science

Numerous terms are used to characterize the decision arena. "Decision science" and "decision theory" are terms used mainly in academic circles, while the term "decision making" is used mainly in business and organizational circles. Regardless of one's perspective, the decision arena is large and complex. It cuts across all domains of life, ranging from the personal, to the organizational, and to the societal, including public health, finance, and government. Decision science is anchored in multiple disciplines, such as mathematics (e.g. probability and statistics), psychology, economics, organizational behavior and artificial intelligence.

The first efforts to formalize the notion of decision-making in business are often credited to Chester Barnard and his seminal work in distinguishing between personal and organizational decision-making.[5] Up to the 1950s and the work of Nobel laureate Herbert Simon, a foundational concept in this arena was that humans approach decisions with rationality and logic. Simon challenged this assumption with the recognition that rationality in decision-making is limited and introduced the term "bounded rationality" to reflect these limits. In the 1970s, psychologists Daniel Kahneman and Amos Tversky began to identify specific, systemic cognitive biases that cause human decision-making to diverge from rationality in predictable ways. Their work helped launch and accelerate the decision science field.

Many of the advances in decision science (theory, analysis) have originated in psychological research with input from mathematics and applications to economics and business. Early work in this area began with researching how humans approach making judgments and how they make choices based on those judgments. Consideration of the role of probability dates back to the 1700s with the work of Daniel Bernoulli and has advanced with considerable input from mathematics and statistics. In the 1950s, the notion of "conservative Bayesian" was introduced by Ward Edwards. Embedded in Edwards's notion was the idea that humans are rational and seek to maximize utility. Early experiments done by Kahneman and Tversky debunked this idea and led them to explore the dynamics within humans that drive decision-making. From this came their characterization of "a two system brain," its use of heuristics (thinking shortcuts), and the seemingly irrational role of biases. Their work also demonstrated how errors are made without awareness of these dynamics. Table 7.1 compares a rational approach to judgment versus what Kahneman and Tversky concluded actually happens.[6]

---

[5] Barnard, C. (1938). *The Functions of the Executive*. Cambridge: Harvard University Press.
[6] Gino, F., Bazerman, M.H., and Shonk, K. (2016). *Decision-Making*. Harvard Business Publishing. p. 8.

Table 7.1 Normative versus observed decision processes.

Normative: Rational model assumptions	Observed: Organizational evidence
When making a decision, we consider all relevant alternatives and accurately assess and compare their probable outcomes.	Due to our limited information-processing capabilities, we typically only consider a small set of alternatives and gravitate towards rules-of-thumb, instincts, or gut feelings.
We use absolute standards and factual information to evaluate and choose among alternatives.	We often have an implicit favorite choice and bend the "facts" to meet this preference.
We evaluate all alternatives simultaneously using objective assessments and choose the one that has the highest payoff.	We frequently evaluate alternatives sequentially and choose the one that is "good enough."

The gaps observed in Table 7.1 between decision-making assumptions (rational model) and what is observed are important to consider in RDM. This phenomenon reinforces the assertion made in the introduction that "The historic ways we frame and practice ORM have diminished ability to meet post-2020 challenges." Narrowing such gaps and improving organizational RDM has been a driver in developing the Risk Matrix presented in this book. A valuable practice is to bring attention and awareness to gaps such as those highlighted here. Figure 2.1 offers a tool for this. It is in the outcome space depicted in this figure where the gaps can be observed.

7.1.2 The Human

It is important to consider briefly our (human) role in RDM as the designers, builders, operators, and participants of, and in, risk management systems, frameworks, and generative fields.[7] A handful of topics are presented here to help build the foundation that began in Chapter 5, which touched on internal-state topics – bias, heuristics, orientations, perspectives, and mental models. These topics fit into the arena of cognition, which The American Psychological Association defines as: "All forms of knowing and awareness, such as perceiving, conceiving, remembering, reasoning, judging, imagining, and problem-solving. Along with affect and conation, it is one of the three traditionally identified components of mind."[8]

---

7 Generative fields are addressed in Chapter 10. Our various roles as identified here were touched on Chapter 6 (6.2.2).
8 American Psychological Association, APA Dictionary of Psychology. https://dictionary.apa.org/cognition (accessed 5 September 2023).

7.1.2.1 Two-system Brain

> This is System 1 talking. Slow down and let your System 2 take control.[9]
> Daniel Kahneman, Princeton University

An increasingly popular way to characterize how humans think and make decisions is in terms of two distinctly different systems in the brain, simplified as System 1 and System 2.[10] This characterization is helpful when considering how we make decisions and ways to improve them. Kahneman defines these as:[11]

> System 1 operates automatically and quickly, with little or no effort and no sense of voluntary control as a result of the "well trained" subconscious mind. System 2 allocates attention to the effortful mental activities that demand it, including complex computations. The operations of System 2 are often associated with the subjective experience of agency, choice, and concentration.

Properties of the two systems are characterized in Table 7.2.[12]

Table 7.2 Properties of Systems 1 and 2.

System 1	System 2
Associative	Rule-based
Holistic	Analytic
Automatic	Controlled
Undermining cognitive capacity	Demanding cognitive capacity
Fast	Slow
Instantaneous	Takes time
Not aware	Consciously aware
Acquisition by biology, exposure, and personal experience	Acquisition by cultural and formal tuition

Both systems are operating simultaneously. "System 1 runs automatically and System 2 is normally in a comfortable low-effort mode, in which only a fraction of

---

9 Kahneman, D. (2011). *Thinking Fast and Slow*, 30. New York: Farrar, Straus, and Giroux.
10 This notion of two systems has interested psychologists for decades. Keith Stanovich and Richard West are credited with formalizing it in 2000 in "Individual Differences in Reasoning: Implications for the Rationality Debate," *Behavior and Brain Science* 23: 645–665. The two system model is also central in Daniel Kahneman's book, *Thinking, Fast and Slow* (2011).
11 Kahneman, D. (2011). *Thinking Fast and Slow*, 20–21. New York: Farrar, Straus, and Giroux.
12 Hassall, C.D. and Williams, C.C. (2017). The role of the amygdala in value-based learning. *Journal of Neuroscience*, 37 (28): 6601–6602.

its capacity is engaged. System 1 continuously generates suggestions for System 2: impressions, intuitions, intentions, and feelings. If endorsed by System 2, impressions turn into beliefs, and impulses turn into voluntary actions. When all goes smoothly, which is most of the time, System 2 adopts the suggestions of System 1 with little or no modification. You generally believe your impressions and act on your desires, and that is fine – usually."[13]

Awareness of these systems and their ramifications is important to understand when making risk-related judgments and choosing between an array of outcomes. Many of the quotidian decisions that humans make are driven by System 1. As complexity increases or significant new variables are introduced, System 2 is called on to kick in and engage. With increasingly complex risk profiles that extend beyond traditional EHSS risks, it is important to become more aware of which system (e.g. System 1 or System 2) is driving our judgments and resulting choices.

7.1.2.2 Perception

> The general law of perception... whilst part of what we perceive comes through our senses from the object before us, another part (and it may be the larger part) always comes out of our own head.[14]
>
> William James, American philosopher and psychologist

Perhaps the most vexing aspect of ORM is the tension between risk perception and science. People judge the consequences and outcomes of risk according to their emotions and perceptions, yet the calculation of risk is cognitive and scientific. Figure 7.1 presents the hot–cold decision triangle that provides a visual portrayal of tension between decision-making in the visceral state as opposed to the cognitive state.[15]

Emotions can be essential to decision-making, even when engaged in cold reasoning, because positive emotions are often the desired ends of our decisions. We choose well if we allocate attention more efficiently toward those things that will make us happier.[16] Stirring the emotions without context or facts yields a visceral response that can be more unproductive than the conclusion drawn when the time is taken to think through the problem. All humans are driven by emotions, but the place for an adrenaline rush may not be in the workplace, because we

---

13 Kahneman, D. (2011). *Thinking Fast and Slow*, 24. New York: Farrar, Straus, and Giroux.
14 James, W. (1890). The Principles of Psychology.
15 Yang, H., Carmon, Z., Kahn, B. et al. (2016). A framework for healthier choices: the hot–cold decision triangle. *Rotman Management Journal*, Winter 2016, pp. 49–54.
16 Dolan, P. (2014). *Happiness by Design: Finding Pleasure and Purpose in Everyday Life*. New York: Hudson Street Press (Penguin Group).

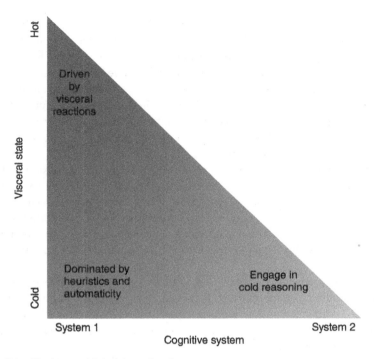

Figure 7.1 The hot–cold decision triangle.

want people healthy and safe at work. Recognizing this paradox and tension between on-the-job and off-the-job, as well as emotion and reasoning, can help the EHSS/ORM professionals communicate more effectively while engaging workers, management, and the public in innovative thinking about the efficacy and soundness of risk management considerations.

7.1.2.3 Brain Function

Perception of the physical world and the interpretation of that physical world, including risk, occur in multiple areas of the human brain. Perception is integrated both consciously and subconsciously and depends literally and figuratively on "where we are." The conscious part of the brain consists of the cortex and involves what we consider cognitive functions. The cortex is the part of the brain typically identified as being responsible for intellectual pursuits and rational decisions such as probability, calculus, and reasoning, as these pursuits require effort and deliberation, or System 2 slow thinking.

The subconscious part of the brain, equally important for cognition, consists of the brain stem and the limbic system which begin processing information even before the cognitive areas are aware that a risk exists. For example, because of subconscious spinal reflexes, humans may pull back their hand from a hot object even before the sensation of pain reaches the cortex. The cortex, however, also influences how the perceived information is integrated into a response to that information. The true extent of subconscious decisions made by the human brain has drawn parallels to artificial intelligence, where research has found that the average person is unaware of 99.68% of her or his decisions. Approximately 35,000 decisions are made per day by the human brain, although, on average, respondents believe they only make around 111.[17]

The brain stem and limbic system receive input from the senses such as touch, sight, and taste, as well as from internal body functions such as the digestive tract and the cardiovascular system.[18] Some of the information may never reach the cortex and never be integrated into conscious thought. Nonetheless, these incoming nerve impulses relay with effector neurons in the autonomic nervous system as well as the subcortical parts of the somatic (primarily voluntary) nervous system.

Subconscious effects on decision-making and risk-taking arise from several sources including physiological states related to, for example, hunger, thirst, and fatigue; remembered emotions such as anger, fear, reward, and punishment; and, emotions associated with the current situation such as pain, depravation, and disasters. In many cases, these subconscious effects operate quickly and effectively without any cortical or conscious recognition. Information pertaining to all three decision-making realms described earlier is conveyed and filtered by the cortex, the limbic system, and the brain stem as the subconscious perception interacts with conscious decision-making.

Trust between and among humans is strongly influenced by oxytocin, a polypeptide protein produced in the hypothalamus, another part of the limbic system. Oxytocin is released during such activities as positive social interactions like hugging. Experimental use of oxytocin significantly increases pro-social behavior, possibly by reducing activity in the limbic system (amygdala) associated with fear and anxiety.[19] Subjects given just a whiff of oxytocin are more likely to give their money to an investor than those not given a whiff of oxytocin.[20]

---

17 Lightspeed Research (2017) http://www.lightspeedresearch.com/; https://www.artificialintelligence-news.com/2017/12/05/research-ai-human-brains-subconscious/. Accessed November 2, 2018.
18 A. Venkatraman, B.L. Edlow, and M.H. Immordino-Yang, "The Brainstem in Emotion: A Review," *Front. Neuroanat.* 11:15. doi: 10.3389/fnana.2017.00015 (2017).
19 M. Kosfeld, M. Heinrichs, P.J. Zak, U. Fischbacher and E. Fehr, "Oxytocin increases trust in humans," *Nature*, 435: 673-676. doi:10.1038/nature03701 (2005).
20 T. Baumgartner, M. Heinrichs, A. Vontanthen, U. Fischbacher and E. Fehr, "Oxytocin Shapes the Circuitry of Trust and Trust Adaptation in Humans," *Neuron*, 58: 639–650 (2008).

Even though the subconscious, by definition, is not consciously perceived, it is an integral and powerful part of decision-making. It may be described as "gut" feelings or "instinct" and is independent of cognitive thought. The subconscious relies on somatic states, both remembered and currently being experienced or imagined. It is reactive, emotion-based, and, for better or worse, how humans make most decisions in their daily lives. When people are in stressful situations, like natural disasters, uncharacterized environmental contamination, and/or experiencing real or perceived injustice, the fast-thinking, subcortical system predominates. This is why establishing emotional connection and genuine trust is often more important than using charts, graphs, and cognitive reasoning when communicating about risk with people in stressful situations.

7.2 Systems Perspective

There are a number of ways decision-making processes can be addressed. A common response to the questions I posed at the beginning of the chapter – "How aware are you of your risk management contexts and decision-making processes?" – was posed in terms of management systems, primarily those of ISO such as ISO 14001(environment) and ISO 22301 (business continuity). People would mention ISO 31000 (risk management) and COSO's ERM framework. More operationally oriented EHSS professionals would mention process safety management and regulatory frameworks. Regarding context, meeting regulatory compliance has been the most common response, with some people mentioning organizational issues such as vision, mission, and organizational objectives.

I frame RDM here within a systems context. After introducing system basics, I turn our attention to stock-flow dynamics within RDM. These are reflected in the abovementioned approaches and frameworks. That is, in them you can identify inputs, RDM processes, and output considerations; and to some extent, there are considerations of outcomes and impacts. The notion of impacts is more prominent in ESG circles and certainly in ISO 37000 (Governance of Organizations).

Fundamental to decision-making are the relationships among judgment, choice, and decision, and the inputs to these, such as observation and sensemaking. While appearing to be elementary, these help frame thinking about RDM. In complex systems (e.g. building a refinery or nuclear power plant), there are of course many formal processes followed within numerous disciplines, not to mention things like regulations, building codes, and technology norms.

I have already drawn your attention to decision-making nuances from the field/matrix Actors/Motive Force dimension, which has four elements: (1) individual; (2) team/department; (3) enterprise/company; and, (4) community. Decision-making processes vary within each of these. We will dig into this in more depth in

Chapter 9 (Matrix in Action). Building on Figure 5.3, for each of these, I frame the decision-making sequence as (Figure 7.2):

> Awareness → Paying attention → See/observe → Sense making → Judgment → Choice → Decision.

Figure 7.2 Decision-making sequence.

Following this are considerations of decision implementation that include (1) motive force to move the decision along (e.g. action); and, (2) frameworks or structures to support it.

The attention that I have given to internal-state awareness and its associated factors or variables of perception, orientation, and mental models sets the stage for decision-making topics that include choice, judgment, context, framing, and narrative – as they are precursors to the RDM process. Decision-making and choice are often used synonymously; making judgments precedes and drives making a choice or decision. Every judgment is not necessarily followed by a decision, but every decision implies some judgment. As we proceed and dive into the decision-making process and our role in it, it is important to understand this difference between decision and judgment.

COSO's ERM framework points to the complexities of organizational decision-making. It states:

> Judgment has a significant role in defining the desired culture and management of risk across the culture spectrum. Judgment is often relied upon: when there is limited information or data available to support a decision; where there are unprecedented changes in the strategy, business objectives, performance, or risk profile of the organization; and, during times of disruption. Judgment is a function of personal experiences, risk appetite, capabilities and the level of information available, and organizational bias.[21]

7.2.1 Systems 101

In the simplest terms, a system is defined as "a set of things working together as parts of a mechanism or an interconnecting network, or a set of principles or procedures according to which something is done."[22] At a minimum, they have inputs, a process, outputs, outcomes and impacts as well as feedback channels/loops.[23]

---

21 COSO, p. 34.
22 New Oxford Dictionary, accessed in Apple Dictionary v.2.3.0.
23 Output and outcome are historic components of evaluation research. Impact is term more commonly used in ESG, referring to double materiality or the effect of reduced air emission on public health.

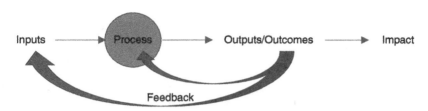

Figure 7.3 Depiction of system elements.

Systems can be characterized as open or closed. Open systems receive inputs and feedback from the external environments. A basic system schematic is depicted in Figure 7.3. Note that the system is embedded in the surrounding environment from which it receives feedback.

Characteristics of systems include: "They have a purpose; all parts must be present for a system to carry out is purpose optimally; the order in which the parts are arranged affects the performance of a system; and, systems attempt to maintain stability through feedback."[24] There is a wide range of system types, levels of complexity within them, and contexts and settings within which they occur. A common feature of systems is their tendency toward equilibrium, either intended or unintended. As system complexity increases, there is also a corresponding increase in the variables that impact the equilibrium and the necessary skills to measure outputs, outcomes, and impacts.

EHSS/ORM professionals are familiar with management systems. Common ones used in the EHSS space were reviewed in Chapter 3. ISO provides the following definition for a management system:

> A system to establish policy and objectives and to achieve those objectives. Management systems are used by organizations to develop their policies and to put these into effect via objectives and targets, using: an organizational structure where the roles, responsibilities, authorities, etc. of people are defined, systematic processes and associated resources to achieve the objectives and targets, measurement and evaluation methodology to assess performance against the objectives and targets, with feedback of results used to plan improvements to the system, and a review process to ensure problems are corrected and opportunities for improvement are recognized and implemented when justified.[25]

---

24 Kim, D. (1999). *Introduction to Systems Thinking*, 3. Waltham, MA: Pegasus Communication.
25 International Organization for Standardization (2001). *Guidelines for the Justification and Development of Management System Standards*, Guide 72, Geneva.

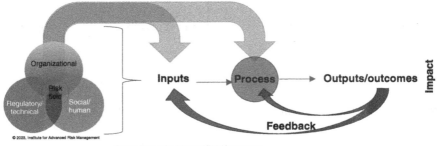

Figure 7.4 Depiction of RDM system elements.

7.2.2 Inputs and Processes

While not often distinguished as such, EHSS/ORM management systems provide an RDM framework. Within that framework are subsystems, often not identified or leveraged, such as communication, auditing, and risk assessment. When considering RDM from a systems perspective, the system elements are risk inputs, RDM processes, and risk-related outputs and outcomes. These are depicted in Figure 7.4.

This figure also depicts an integrated RDM process within the risk field (Risk Matrix) dimensions providing inputs and impacting the RDM process. It is valuable to consider here how the roles that the five evolution sequences presented in Chapter 2 (§2.1.1) play into RDM.

1) *EHSS management*: compliance, performance, impact
2) *Frameworks*: process, program, system, field
3) *Sustainability*: ESG, materiality, double materiality, value, capitals
4) *Object/foci*: shareholder, people/workers, stakeholders
5) *Organizational risk management (ORM)*: four generations – insurance, regulatory compliance, consensus standards, value and purpose

I revisit these five sequences in Chapter 9 when considering the Risk Matrix in action and integrated RDM.

7.2.2.1 Data and Measurement Consideration

Conformity assessment, measurement, and auditing are addressed in Chapter 4. It is important to be aware of these topics when making decisions. Regarding data input and their use in decision-making processes, pay attention to data characteristics, such as qualitative and quantitative, and measurement levels: nominal, ordinal, interval, and ratio – and how they can be analyzed.

Data used in risk decision-making can be classified as quantitative or qualitative. Quantitative input data have measurements that are at the ratio or interval level of measurement. Risk analyses that are familiar to EHSS/ORM professionals are driven primarily by quantitative data. Qualitative data, ordinal or nominal, can be analyzed quantitatively once they are assigned numerical values (1 = Female, 0 = Male). The quantitative analysis of qualitative variables would involve, for example, frequencies, chi-square analysis of the relationship between gender and political affiliation or the use of a dummy variable in a logistic regression model. Qualitative data have historically been used less in RDM, and notions of context, framing, and narratives has been remote, if not dismissed as unscientific and thus irrelevant by some.

7.2.3 Output, Outcome, and Impact

While these terms are often used interchangeably and considered synonymous, there are important nuances and differences between them. The definitions that follow are common in organizational science and program evaluation.

- *Output* is the result of an activity. Examples include the implementation and functioning of a risk management system as verified by an audit.
- *Outcome* is the short- to mid-term result. This would be the difference that the management system is making, e.g. fewer incidences, increased value, lower emissions. Sometimes effect and impact are used interchangeably here.
- *Impact* is the long-term result. It is the longer term benefit to which the outcome is logically expected to lead. In an emissions example, the installation and functioning of a control system is an output. Reduced emissions is the outcome. Reduced asthma in the community is the impact.

It is not uncommon for outcome and impact to be used interchangeably. In many cases, this is not a problem. However, clarity about what is happening at the impact level provides information that may show that different outputs and outcomes are needed to achieve desired impacts. Said differently, and sticking with community asthma, if reduced emissions are not producing the desired level of asthma reduction, then other things need to be considered. For example, are repairs on the original controls needed; or, are different controls needed? Is the right emission being addressed? Should a different (or additional) initiative be considered, such as vermin control?

These three terms are central to shifting and upgrading ORM. Are you getting what you need or want from your risk management efforts? In many instances, you would say yes. Things are fine from a compliance perspective, shareholders/stakeholders are generally satisfied, etc. But the imperatives identified in Chapter 1 suggest this might be otherwise. ORM systems generate outputs. It is vital to monitor the outputs, the initial results of a system or intervention.

In many instances, basic risk-related systems carefully monitor and report the outputs. It is also vital to measure and report the outcomes and their impacts. This sequence can also be framed in terms of cause-and-effect and with metrics, including leading and trailing indicators. A key piece of Daniel Kahneman's quotation in Chapter 5 (§5.5) related to decision-making in organizations is that "Learning medicine consists in part of learning the language of medicine. A deeper understanding of judgments and choices also requires a richer vocabulary than is available in everyday language."[26] The attention I give here to these three terms reflects Kahneman's point.

In Chapter 10, I return to this trio and discuss escalating ORM impact in the context of organizational purpose, a topic I have been weaving throughout the book. It is important to exercise rigor with these three terms when considering generative organizational risk fields and organizational purpose.

7.2.4 Feedback

A fundamental component of a system is feedback, sometimes referred to as feedback loops or feedback channels. This concept evolved in the 1950s with respect to biological systems. In order to survive and grow, biological (living) systems need feedback from the external environment on how things are going.

A defining feature of systems is their tendency toward equilibrium. This is accomplished through feedback loops or channels. A common example is a home heating system where a thermostat sends signals to the heat source. When the desired temperature is reached, a signal to turn off is sent to the heat source; when the temperature drops, a signal is sent to the heat source to turn on. In an organization there are many feedback loops/channels that carry information, such as audits, incident investigations, management reviews, suggestion boxes, and community forums. The program evaluation discipline provides feedback channels to agencies on the efficacy of public programs so that organizations can learn important lessons and make necessary mid-course corrections to improve performance.

7.2.5 Stocks and Flows

Stocks and flows are important considerations in making decisions in complex organizational systems. A brief introduction to these is offered here to help set the stage for when we look at capitals and impact-dependency pathways. When teaching this subject, it is common to use a very simple analogy of water in a bathtub as a stock and a faucet and drain as flows, entities that increase or decrease the level of stock (water) in the bathtub.

---

26 D. Kahneman, *Thinking Fast and Slow*, Farrar, Straus and Giroux, New York, 2011. page 3.

7.2 Systems Perspective

When considering ORM within a stock and flow context, the various risks contained in an organization's risk profile can be viewed as a stock, or I would say "risk stock." The flow for this stock is multifaceted and is the domain of well-established risk management paradigms (control, substitution, transfer, insurance, etc.). EHSS professionals are highly trained in these paradigms, say with chemical, health, and compliance risks.[27] Generally, there is less expertise when it comes to sustainability risks, mainly due to longer time horizons, and even less with risks related to the social–human dimension, mainly due to the recency of its emergence in organizational culture.

When considering the interconnected nature of capitals and the pathways through which value flows between them, this concept of stocks and flows is an important consideration in RDM. In one of its initial publications (2013), the International Integrated Reporting Council (IIRC) defined capital as "stocks of value on which all organizations depend for their success as inputs to their business model, and which are increased, decreased or transformed through the organization's business activities and outputs. The capitals are categorized in this framework as financial, manufactured, intellectual, human, social and relationship, and natural."[28] A factor to be considered when evaluating stocks and flows is their non-static nature. The IIRC identifies this by stating "The overall stock of capitals is not fixed over time. There is a constant flow between and within the capitals as they are increased, decreased or transformed."[29]

The Capitals Coalition addressed social and human capitals in 2019 in their Social and Human Capitals Protocol:

> You may consider social and human capital in terms of stocks and flows in a manner similar to financial capital stocks and flows; however, while businesses account for financial capital performance in balance sheets and profit and loss (P&L) statements, to date there is no equivalent mechanism for evaluating non- financial capital performance. Such accounting would go beyond the measurement of the ways business impacts social and human capital to also consider the ways in which business depends on social and human capital. This would help businesses understand how social and human capital relate to their risks and opportunities and how effective management of these capitals underpin sustainable performance.[30]

---

27 These are not mutually exclusive; all three of these quick examples can be, and in many instances are, connected.
28 International Integrated Reporting Council (2013). The International <IR> Framework. p. 33.
29 International Integrated Reporting Council (2013). The International <IR> Framework. p. 11.
30 Capitals Coalition (2019). Social and Human Capitals Protocol. p. 9.

MIT Professor John Sterman offers valuable advice on the importance of this topic. He writes, "Developing facility in identifying, mapping, and interpreting the stock and flow of networks of systems is a critical skill for any modern systems modeler."[31] While it is beyond the scope of this book to dive deeply into stock/flow at this level, I consider impact-dependency pathways in this chapter and in Chapter 10 (Escalate Impact) when considering generative fields.

7.2.6 Impact-Dependency Pathways

Aspects of systems thinking, system dynamics, and system components have been discussed. Feedback loops, networks, channels, etc. are the fundamental components of systems. Flow in these is one-way, from impacts, outcomes and outputs to process and inputs, and support the growth or improvement of the system. The term pathway is used in a similar fashion in capital-oriented decision-making.

The impact-dependency pathway is a central component in the Capitals Coalition protocols (Natural Capital, and Social and Human Capital Protocols). "Pathways draw links between [natural, social or human] capital issues identified and the business activities that affect or rely on them."[32] These can include issues that are "positive or negative; impacts or dependencies; known or potential issues; and risks or opportunities."[33] The *Social and Human Capital Protocol* (SHCP) states, "These pathways (also called logical frameworks, results chains or theories of change) outline the potential and empirically testable relationships between your business' activities and social and human capital creation, destruction or reliance."[34] A distinction is made between an impact driver and an impact. A single driver can be associated with multiple impacts.

The impact-dependency pathway model offers a valuable tool in organizational RDM. By looking at linking risks and opportunities in decision-making, the ability to expand the systems perspective addressed above (§7.2) is increased. The SHCP states that these pathways can:

- Help you understand the tracing of business activities all the way through *outputs* to *outcomes* and *impacts*.
- Highlight unintended consequences or indirect effects of a business activity that might occur despite not being the primary intention of the activity.

---

31 Sterman, J. (2000). *Business Dynamics, Systems Thinking, and Modeling for a Complex World*. p. 191.
32 Capitals Coalition (2019). Social and Human Capitals Protocol. p. 33.
33 Capitals Coalition (2019). Social and Human Capitals Protocol. p. 33.
34 Capitals Coalition (2019). Social and Human Capitals Protocol. p. 33.

- Articulate the causal links between a business's activity or product and downstream impacts. This can be particularly useful when you want to demonstrate the societal value of the use of your products.[35]

This concept is revisited in Chapter 9 when we begin to look at the Risk Matrix in action.

7.3 Frameworks

The frameworks included here are RDM frameworks that organizations tailor and populate for their unique internal and external contexts and needs. While there is a definitional nuance between a decision-making framework and a risk management framework, many practitioners treat them as synonymous. I introduce them both to help broaden your perspectives on how these can support your efforts in developing RDM processes.

COSO's ERM and ISO 31000 can be viewed as organizational RDM frameworks. ISO 14001 and 45001 can respectively be considered environmental and occupational health and safety RDM frameworks. In a similar manner, ISO 37000 offers robust support for framing ORM, especially when considering risk profiles and risk stocks and flows.

7.3.1 ISO 37000:2021, Governance of Organizations – Guidance

In its introduction, ISO 37000;2021 quickly makes a connection between decision-making and good governance. It states: "Good governance means that decision-making within the organization is based on the organization's ethos, culture, norms, practices, behaviours, structures and processes."[36] The standard includes decision-making transparency, improvement, structures, processes, stakeholder inclusion, resources, and agility.

ISO 37000's Section 6.8 (Data and decisions) offers detailed decision-making guidance. It is framed here in terms of data and decisions, recognizing "data as a valuable resource for decision-making by the governing body, the organization and others."[37] The standard states:

---

35 Capitals Coalition (2019). Social and Human Capitals Protocol. p. 35.
36 37000, p. vi.
37 37000, p. 28.

The value of data for decision-making can be considered from different perspectives, such as the following.

a) Decision-making internal to the organization:
   1) Decision-making within the governing body. The viability of an organization depends on the data on which the governing body relies to make decisions.
   2) Decision-making across the organization. The operation of an organization depends on structures and practices to ensure effective decision-making. Such decision-making is based on trusted information and decisions being made with the level of authority and responsibility appropriate for:
      i) the decisions;
      ii) the nature of the data on which the decisions rely.
b) Decision-making by others external to the organization: Because data are used to make decisions, they are valuable not only for the organization itself but also as a resource that can be bought, sold or otherwise distributed. For example, data are a resource that is used in products and their design, market and customer insights, as well as supply chain and product usage information.[38]

As you consider upgrading your audit programs or measurement endeavors, refer to this section in ISO 37000:2021. This guidance reinforces the importance of having awareness about decision-making inputs, and "the nature of the data on which the decisions rely."

7.3.2 ISO 31000:2018, Risk Management – Guidelines

Is this a risk management or RDM framework? As suggested above, either description can be used. I prefer the term "risk management framework" which includes three components: principles, framework, and process. For the process component, the standard states: "The risk management process should be an integral part of management and decision-making and integrated into the structure, operations and processes of the organization. It can be applied at strategic, operational, programme or project levels."[39] The process component is depicted in Figure 7.5.

---

38 37000, p. 28.
39 31000, p. 9.

7.3 Frameworks | 221

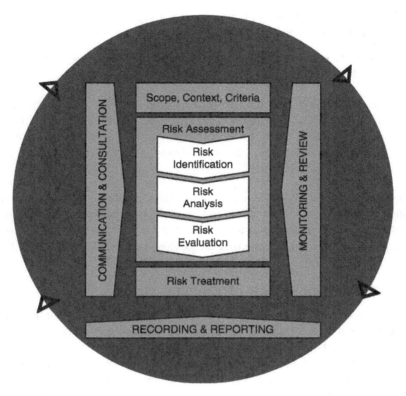

Figure 7.5 ISO 31000's process component.

The standard states:

- This document can be used throughout the life of the organization and can be applied to any activity, including decision-making at all levels.[40]
- The purpose of the risk management framework is to assist the organization in integrating risk management into significant activities and functions. The effectiveness of risk management will depend on its integration into the governance of the organization, including decision-making. This requires support from stakeholders, particularly top management.[41]
- Successful implementation of the framework requires the engagement and awareness of stakeholders. This enables organizations to explicitly address uncertainty in decision-making, while also ensuring that any new or subsequent uncertainty can be taken into account as it arises.[42]

---
40 31000, p. 1.
41 31000, p. 4.
42 31000, p. 8.

7.3.3 COSO Enterprise Risk Management – Integrating with Strategy and Performance

This framework's first section is "The Changing Risk Landscape," and its first paragraph states:

> Our understanding of the nature of risk, the art and science of choice, lies at the core of our modern economy. Every choice we make in the pursuit of objectives has its risks. From day-to-day operational decisions to the fundamental trade-offs in the boardroom, dealing with risk in these choices is a part of ***decision-making***.[43]

Decision-making is highlighted throughout the COSO framework. Some of the highlights include awareness of the role that decision-making has in: strategy development; alignment of resources with mission and vision; development of operating structures; improvement of capabilities; and, the elimination of silos.

7.3.4 ISO Management System Standards

As with the above three frameworks, I think of ISO MSSs as RDM frameworks. These are geared toward specific areas such as business continuity (ISO 22301), energy (ISO 50001), and environment (ISO 14001).

As an example, ISO 14001 states, "Top management can effectively address its risks and opportunities by integrating environmental management into the organization's business processes, strategic direction and decision making, aligning them with other business priorities, and incorporating environmental governance into its overall management system."[44] In ISO 45001, decision-making is addressed in terms of worker participation in decision-making (3.4) and consultation in making decisions (3.5).[45]

7.4 Key Considerations

There are many complexities in organizational decision-making, including ones related to ORM. A handful of key considerations in EHSS/ORM decision-making is offered here.

---

43 Committee of Sponsoring Organizations of the Treadway Commission (2017): *Enterprise Risk Management – Integrating with Strategy and Performance, Executive Summary.* Association of International Certified Professional Accountants. P. 1.
44 14001 (2015), p. vi.
45 ISO 45001, p. 2.

7.4.1 Carriers of Meaning

Carriers of meaning was introduced in Chapter 5 (§5.5.2). It is a term used in the social sciences to describe "fundamental mechanisms that allow us to account for how ideas move through space and time, who or what is transporting them, and how they may be transformed by their journey."[46] These are typically framed in terms of language and communication. With decision-making, it is important to be aware of these words (stock) and how they travel (flow).

Language and words are often thought of as the primary carriers, both oral and written. Others include contexts of language and words (e.g. regulations, legal documents, policies), processes (e.g. decision-making itself, hiring practices, environmental management practices, auditing), organizational structure (e.g. organizational charts, business unit relations), and risk management (e.g. relationship to risk, risk appetite, risk tolerance).

The risk kernel is introduced below (§7.5) as a useful construct that includes risk-related carriers of meaning.

7.4.2 Context, Framing, and Narrative – Or Is it the Number?

> No one ever made a decision because of a number. They need a story.[47]
>
> Daniel Kahneman

Context, framing, and narrative are essential dynamics in RDM. Addressing these may seem counterintuitive to EHSS/ORM professionals, as our historical orientation has been primarily toward the analytical and the quantitative – toward numbers and ratio/interval levels of management. This makes sense in many ways. But, as Kahneman observes, maybe it is not just about "the number." As mentioned throughout this book, RDM is rife with complexity, tension, and paradox that arise from its values-dependent and subjective nature. As Paul Slovic observed in his quotation in Chapter 2 (§2.2), risk is not absolute, but rather context-dependent. And as touched on in that chapter, lurking here is the question, "to whom or what is something a risk?" Is it the organization, the worker, the community, the environment, or the salamanders? Depending upon the context and the "who or what," a risk may be acceptable or not.

It is important to be aware of context, framing, and narrative in your RDM processes. As you become more attentive to these, you will notice that narratives and carriers of meaning are everywhere. You will see contexts when divergent values, points of view, opinions, etc. merge in your risk field (matrix). Context and

---

46 Scott, W.R. (2014). Institutions and Organizations. Sage Publications, Inc., p. 95.
47 Lewis, M. (2017). *The Undoing Project*, 250. W.W. Norton.

> **Box 7.1   Definitions – Context, Frame, Narrative**[48]
>
> *Context (n)*, the circumstances that form the setting for an event, statement, or idea, and in terms of which it can be fully understood and assessed.
> *Frame (v)*, to create or formulate a concept, plan, or system.
> *Narrative (n)*, a spoken or written account of connected events; a story; a representation of a particular situation or process in such a way as to reflect or conform to an overarching set of aims or values.
>
> (*Source:* Adapted from New Oxford American Dictionary.)

Framing are related to the material introduced in Chapter 5 (§5.3 and §5.4.2) in terms of "internal state" and things like orientation, perspectives, and mental models. Here I am shifting a bit and applying these to the RDM process, your RDM process, your teams, your organization, and your stakeholders. Awareness of these elements in the internal-state is a prerequisite to understanding the context, framing, and narratives that influence the RDM process.

The elements of the Contexts/Drivers dimension (regulatory/technical, organizational, social–human) influence the decision-making context and how a decision is framed and made. The decision-making process is influenced by narratives and stories put forth by stakeholders. As you become aware of these dynamics, you increase your ability to impact the process and quality of outputs, outcomes, and impacts.

7.4.3 Defining Risk

Attention to this topic is given in Chapter 2 (§2.2). A key piece within a decision-making context is Slovic's insight that risk is an "invented concept," and there is "no such thing as real risk or objective risk." There is a range of views on risk, and understanding this range in your settings and contexts is recommended, particularly when considering decisions in the social-human element of the contexts/drivers dimension in the Risk Matrix. In considering integrated ORM, the integration of different risk perspectives is essential. The Risk Matrix's Actors/Motive Force dimension's four elements (individual, team/department, enterprise/company, community) offer a construct for doing this (§9.4).

7.4.4 Decision-Making Currency

The term currency implies a medium or system of exchange. Used here, it refers to that which is central to RDM, specifically inference guidelines, residual and acceptable risk, and materiality, value, and purpose.

---

48 New Oxford American Dictionary, Apple OS 13.5.2.

7.4.4.1 Inference Guidelines

Inference guideline was introduced in Chapter 4 (4.1.3). It is a term used in the National Research Council's (NRC) *Red Book*[49] that relates to quantitative performance goals or standards such as those related to air and water quality, greenhouse gases, soil clean up levels, and occupational exposure limits. Such guidelines are analogous to speed limits. They are values that set numeric thresholds against which decisions are made. Risk management's historical orientation has been toward the quantitative analysis of ratio/interval data. It is important to begin to see how nominal and ordinal data can also be used as inference guidelines. This is critical when considering the social–human dimension of ORM and its four actors/motive force elements. An example is DE&I and policies related to gender and related issues in this area (e.g. bathrooms and pronouns).

7.4.4.2 Residual and Acceptable Risk

Residual risk is the amount of remaining risk or danger associated with an action or event after natural or inherent risks have been reduced by risk controls/treatments. Residual risk considerations are always present, whether expressed or not, when an inference guideline is used.

An issue with residual risk considerations are tensions within and between the risk field's subfields. Residual risk and uncertainty considerations are at the core of RDM. A conundrum in RDM occurs when there is pressure to achieve zero-risk goals. We recognize that zero risk is impossible to achieve. At best, it can be managed to achieve the residual level that is acceptable and thus is defined as safe. Even when residual risk is low (for example, 1 in 10^{-3}) or extremely low (1 in 10^{-53}), the risk for the population cannot be zero, and humans will judge a risk according to the expected or actual consequences of its occurrence, not its probability of occurring.

From a policy-making perspective, all policy choices ultimately involve balancing additional risk reduction and incremental costs with society's willingness to pay for achieving the expected benefits. Based on implicit tradeoffs between risk and money, economists have developed estimates of the "value of statistical life" (VSL), which provide a reference point for assessing the costs and benefits of risk reduction efforts and government risk policies. These can be as disparate as the draining of swamps near ancient Rome to suppress malaria, or limiting air pollution in developed countries over the past decades to reduce respiratory disease.[50]

---

49 National Research Council (1983).
50 Environmental Protection Agency, "Value of Statistical Life Analysis and Environmental Policy: A White Paper with Appendices for Presentation to Science Advisory Board – Environmental Economics Advisory Committee (2004)," www.epa.gov/environmental-economics/value-statistical-life-analysis-and-environmental-policy-white-paper (accessed 30 January 2019).

Clear guidance in determining residual and acceptable risk can at times be fleeting, and decisions are ultimately informed within the intersection of the risk field's three subfields presented in Figure 2.7. A dynamic at this intersection is a tension between evidence-based decision-making approaches and situations where evidence is lacking and intuition or "gut" instincts are drivers.

The Risk Matrix (Chapter 8) has been designed to address the decision-making challenges posed by the residual and acceptable risk reflected here.

7.4.4.3 Materiality, Value, and Purpose

Materiality, value, and purpose are relatively new concepts in organizational RDM, each having levels of maturity that are context-dependent and that can be considered RDM currency. Inference guidelines and issues related to residual and acceptable risk are well established in the regulatory/technical dimension, but less so in the organizational dimension, and even less in the social–human dimension. It is important to be aware of these three topics in ORM decision-making processes.

Materiality became a central concept in ESG as recently as the early 2000s. Outside of an ESG context, it is a well developed concept in finance and investment. Materiality, along with double materiality, is addressed in Chapter 3 (§3.6.1.4 and 3.6.1.5).

Value generation and preservation is central to the evolution of organizational governance and risk management, as reflected in ISO 37000:2021 and Capital Coalition Protocols. It is addressed throughout this book, specifically in Chapter 1 (§1.1.1) and Chapter 6 (§6.3).

Organizational purpose has also been addressed throughout the book, specifically in Chapter 1 (§1.1.1) and Chapter 8 (§8.5.2.1).

Decision-making currency is revisited in Chapters §9 and §10 in the context of decision-making within the Risk Matrix and generative fields.

7.4.5 Rates, Cycles, and Time Horizon

Possibly the most important dynamics to consider in organizational RDM are how fast things are happening, how close a decision point is, whether there is a repetitive nature to the process, and the decision's time horizon. The S1/S2 construct from Chapter 5 is central to this. Responding to an accident is very different from responding to, for instance, climate change. It is valuable to think about these variables as the organization's risk profile increases in complexity, e.g. responding to social media and citizen science.

7.4.6 Delays and Buffers

In systems thinking, delays and buffers are related to rates. The more complex a system is, the more delays and rates come into play. Tight coupling was mentioned in Chapter 1. As risk profiles have increased in complexity, including an increasing number of stakeholders and the advent of social media, there are fewer and fewer delays and buffers against the constant information flow and the expectation of a response. It is important to consider the speed at which information related to RDM flows and where there are or might need to be delays or buffers in the system.

7.5 Risk Decision-Making Kernel

Key aspects of RDM related to information that flows through feedback loops, channels, and impact-dependency pathways are: (1) source; (2) impact; (3) tolerance; (4) acceptability; and, (5) control/transfer. These topics were introduced in Chapter 2 (§2.3) and are grouped here as a decision-making heuristic that I refer to as the RDM kernel. The heuristic concept was introduced in Chapter 5 (§5.3.1) "as rules-of-thumb for the mind." These can also be framed in terms of mental or decision-making shortcuts that can impact the speed at which decisions are made. Aspects of the kernel's five elements have been touched on throughout the book.

Seasoned EHSS/ORM professionals have internalized these five key aspects, or some form of them, when making decisions about risk in their regulatory/technical element of the Contexts/Drivers dimension, and less so with organizational, and social–human dimensions. Combining RDM elements, as offered in the decision-making kernel, provides a tool for improving RDM.

Embedded in the kernel's source element is the risk itself, as touched on above in §7.4.3. That is, what is the risk, and how is it defined? This is addressed in Chapter 2 (§2.2). Risks in the regulatory/technical dimension originate primarily from regulatory noncompliance. In the organizational risk dimension, they arise from a host of sources, including nonconformance with policies and procedures and issues related to breakdowns in cultural norms. In the social–human risk dimension, they also arise from a host of sources, including those in the organizational dimension, as well as external stakeholder sources, including customers, supply chains, NGOs, and communities. These five elements of the risk decision making kernel can be used in any cell in the Risk Matrix. For example, Figure 2.5 (in §2.3.3) depicts the relationship between risk profile, risk appetite, and risk capacity (tolerance). This depiction can be applied to individual cells and the RDM kernel mechanism.

Figure 7.6, depicts the risk kernel's five elements in a box cell to reflect their use in the Risk Matrix.

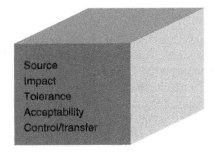

Figure 7.6 Risk decision-making kernel.

Suggested Reading

Anderson, V. and Johnson, L. (1997). *Systems Thinking Basics: From Concepts to Causal Loops*. Waltham, MA: Pegasus Communication.

Kehneman, D. (2011). *Thinking, Fast and Slow*. Farrer, Straus and Giroux. New York.

Kahneman, D., Lovallo, D., and Sibony, O. (2019). A structured approach to strategic decisions: reducing errors in judgment requires a disciplined process. *MIT Sloan Management Review* 60 (3): 67–73. With Dan Lovallo & Olivier Sibony.

Kahneman, D., Rosenfield, A.M., Gandhi, L., and Blaser, T. (2016). Inconsistent decision making is a huge hidden cost for many companies. Here's how to overcome what we call noise. *Harvard Business Review* 94: 38–46. With Andrew Rosenfield (Greatest Good Group); Linnea Gandi and Tom Blaser of the GGG.

Kahneman, D., Sibony, O., and Sunstein, C.R. (2021). *Noise – A Flaw in Human Judgment*. With Olivier Sibony and Cass Sunstein.

Kim, D. (1999). *Introduction to Systems Thinking*. Waltham, MA: Pegasus Communication.

Sterman, J. (2000). *Business Dynamics, Systems Thinking and Modeling for a Complex World*. Irwin McGraw-Hill.

Part III

Integrating Eras

8

Risk Matrix

An Integrated Framework

CONTENTS

    Generating Organizational Energy – The Engine, 232
    New Language and Dimensionality, 233
Tailoring, 233
8.1 Risk Field to Risk Matrix, 234
8.2 Matrix Structure, 236
    8.2.1 Nomenclature, 236
    8.2.2 Cells, Rows, and Columns, 238
8.3 Contexts/Drivers (y-Axis), 239
    8.3.1 Regulatory/Technical, 240
    8.3.2 Organizational, 240
    8.3.3 Social–Human, 240
8.4 Actors/Motive Force (z-Axis), 240
    8.4.1 Individual, 242
    8.4.2 Team/Department, 242
    8.4.3 Enterprise/Company, 242
    8.4.4 Community, 242
8.5 Risk Management Elements (x-Axis), 243
    8.5.1 Foundational Five, 245
    8.5.2 Trim Tabs, 254
    8.5.3 Operational Elements, 268
Suggested Reading, 277

Organizational Risk Management: An Integrated Framework for Environmental, Health, Safety, and Sustainability Professionals, and their C-Suites, First Edition. Charles F. Redinger.
© 2025 John Wiley & Sons, Inc. Published 2025 by John Wiley & Sons, Inc.

> A rules-based risk management system may work well to align values and control employee behavior, but it is unsuitable for managing risks inherent in a company's strategic choices or the risks posed by major disruptions or changes in the external environment. These types of risks require systems aimed at generating discussion and debate.[1]
>
> Kaplan & Mikes (2012)
Harvard University

Matrix (n): *An environment or material in which something develops, a surrounding medium or structure. An organizational structure in which two or more lines of command, responsibility, or communication may run through the same individual.*[2]

Organizational Risk Management (ORM) is commonly conceived, framed, and executed as a one-dimensional phenomenon characterized by a narrow focus on rules, regulations, and norms. ORM frameworks and practices have been playing catch-up to meet the challenges identified in Chapter 1, namely the challenges of increased complexity in general, a wider stakeholder net that leads to tighter coupling, increased tension, and seeming paradoxes. The three-dimensional Risk Matrix framework presented here represents an operationalization of the risk field construct (Chapter 2, §2.6). This and the remaining chapters, take it from concept to its boots-on-the-ground use as a framework to support integrated thinking, decision-making, and action. I have introduced aspects of these three dimensions in previous chapters. In Chapter 2, I laid the foundation for ORM contexts and drivers. This continued in Chapter 3 with a review of common frameworks used in ORM, such as ISO management systems, ISO's risk management standard (31000), and COSO's Enterprise Risk Management (ERM) framework. Core issues in risk science that impact ORM and risk decision-making (RDM) were also addressed. The content of Chapter 3 laid the foundation for the Matrix's Risk Management Elements dimension. The matrix's final dimension, Actors/Motive Force was introduced in Chapter 6 within the context of leadership and motive force.

Generating Organizational Energy – The Engine

Motive force and organizational energy was introduced in Chapter 6. The inclusion of a social-human dimension in the Matrix, and a focus on health

---

1 Kaplan, R. and Mikes, A. (2012). Managing risks: a new framework, smart companies match their approach to the nature of the threats they face. *Harvard Business Review*, June, 2012.
2 New Oxford American Dictionary, Apple OS 13.5.2.

(organizational and human), provide a foundation for shifting (evolving) ORM and generating organizational energy.

In his robust body of work on organizations, Jim Collins applies the conservation of energy concept in flywheels to organizational transformations.[3] He observes that transformations do not happen overnight, rather, they are the result of a series of actions that build and maintain momentum. The Risk Matrix provides a structure that supports this concept and generates organizational energy. In addition to Collins's work, there is a plethora of ideas, concepts, and approaches in the organizational science literature on ways to increase performance, increase efficiency, create value, etc. Relatively new to this liturature is the concept of "organizational energy." This concept is offered here as a "lever" to create new clearings and capacities that increase the ability to generate and preserve value, resilience, and fulfillment for an organization and its stakeholders, as discussed in Chapter 1 (1.2).

New Language and Dimensionality

At the beginning of §6.2 in Chapter 6, Otto Scharmer introduces us to the idea of social grammar. He points to the need for: "A social grammar: a language; a social technology: methods and tools; and a new narrative of societal evolution and change."[4] The Risk Matrix provides a new risk management construct and new language that will help meet the challenges identified in Chapter 1.

EHSS and ORM are historically executed in a single dimension containing a number of risk management elements. These are similar to programmatic elements found in regulatorily defined programs, as well as the elements found in ISO management systems, and COSO's ERM framework. With two additional dimensions explicitly identified (contexts/drivers; and actors/motive force), the matrix provides a framework and tool to: (1) promote and foster integrated thinking, decision-making, and action; and, (2) impact organizational culture, specifically in generating a learning culture, and a culture of. It does this in ways that the historical single dimension does not.

Tailoring

The notion of tailoring is touched on throughout the book, specifically in Chapter 3 (§3.1.4). As mentioned above, there are numerous ways the Matrix can be used. I encourage you to consider tailoring it to your specific needs. There might also be different elements that you would like to include in the three dimensions.

---

3 Collins, J. (2001). *Good to great*. Random House Business Books.
4 Scharmer, O. (2018). How to cultivate the social field. Garrison Institute conference, 1–3 October 2018. Conference introduction document, p. 11. Excerpt from *The Essentials of Theory U*, Berrett-Koehler.

8.1 Risk Field to Risk Matrix

The risk field concept was introduced in Chapter 2 (§2.6) and defined as: "the relational and contextualized space creates the interactions and collective behavior, which in turn produces the organization's risk management governance, strategy, execution, and outcomes."

There are any number of ways a risk field can be depicted. The three subfields chosen here evolved from my years of working in organizations, doing public policy work, and teaching. Conceptual models are valuable in many ways. They can provide platforms for generating ideas and helping with decision-making. Once operationalized, one can operationalize variables and develop a data collection and reporting plan tools.[5] An illustration of this process follows in a Risk Matrix. A starting point for this is depicted in Figure 8.1 with three dimensions of ORM identified as:

- Contexts/Drivers (y-axis),
- Actors/Motive Force (z-axis), and
- Risk Management Elements (x-axis).

In this operationalization, the social-human, organizational, and regulatory/technical subfields can now be viewed as elements of the Contexts/Drivers dimension (y-axis) in Figure 8.1. To fully operationalize the Risk Field in this three-dimensional matrix, we add the other two dimensions, 1) Risk Management Elements (x-axis), and 2) Actors/Motive Force (z-axis), elements that are found in ISO management systems and COSO's ERM frameworks. The second step, depicted in Figure 8.2,

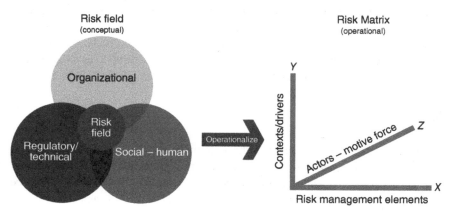

Figure 8.1 Field to matrix.

---

5 Specifying an operational definition is the process of turning abstract concepts or ideas into observable and measurable phenomena. This process is often used in the social sciences to quantify vague or intangible concepts and study them more effectively.

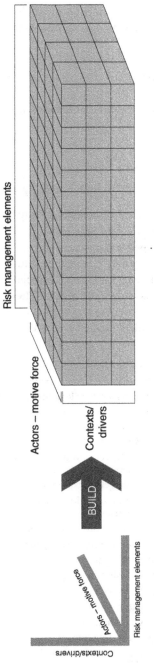

Figure 8.2 Building a Risk Matrix.

involves building out the three dimension by adding their respective elements producing the full three dimensional matrix of 168 cells.

8.2 Matrix Structure

The Risk Matrix's 3 dimensions and 21 elements are depicted in Table 8.1 and Figure 8.3.

8.2.1 Nomenclature

The following terms are essential for understanding and using the Risk Matrix:

- **Dimensions**. The Matrix has 3 dimensions. They also have been refered to as risk field subfields, as well as ORM elements.
- **Elements**. These are grouped within the 3 Matrix dimensions and are listed in Table 8.1. There are 21 elements: 3 associated with contexts/drivers, 14 associated with risk management elements, and 4 associated with actors/motive force.
- **Cells**. These are basic units within the Matrix. There are 168 cells. These are at the intersection of the 21 elements ($3 \times 14 \times 4 = 168$).
- **Slice**. Refers to groups of cells found in either Matrix rows or columns.

Table 8.1 Risk Matrix dimensions and their elements.

Contexts/drivers	Risk management elements	Actors/motive force
• Regulatory/technical • Organizational • Social-human	1) Purpose and scope 2) Risk assessment 3) Emergency preparedness and response 4) Management of change 5) Communication: systems and practices 6) Competency and capabilities 7) Social-human engagement 8) Leadership 9) Decision-making 10) Frameworks 11) Auditing/metrics 12) Operation 13) Escalating impact 14) Future-ready strategy	• Individual • Team/department • Enterprise/company • Community

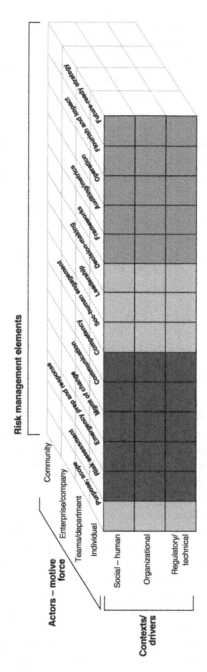

Figure 8.3 Risk Matrix.

There are several nuances to keep in mind are. First, the term social-human is used to refer an element in both the Contexts/Drivers and Risk Management Element dimension. Second, the dimension Risk management element contains 14 elements, and, therefore, can seem redundant or awkward when refering to elments within this dimension.

8.2.2 Cells, Rows, and Columns

Going forward, I draw your attention to the Risk Matrix rows, columns, and cells. These are at the core of using the Risk Matrix to support integrated thinking, decision-making, and action. Individual cells can be the foci when a specific issue needs attention within a specific risk management element, associated with a specific risk contexts/drivers element. For example, the focus could be on risk assessment within the social-human element, which could be an issue with any one of the four elements in the Actors/Motive Force dimension.

The concept of a risk decision-making kernel was introduced in the previous chapter. Cells and slices serve as units of analysis for RDM. It is within these cells and slices that I'll be focusing your attention in Chapters 9 and 10 on using the kernel's core concepts.

- Risk source/owner
- Impact,
- Risk tolerance,
- Acceptable risk, and
- Risk control/transfer.

The Risk Matrix provides a framework for more granular examination of risk using the heuristics for each of these five areas for any given cell of the Risk Matrix. This ability is central to increasing overall resilience and fulfillment for the organization and the people associated with it. It is at this micro level that organizational energy is generated because it involves individuals and groups who will be responsible for mitigating the risk.

Focusing on cells and slices associated with the social-human provides a way to increase skills and confidence in navigating in this turbulent environment. For example, focusing on risk assessment issues at the individual level, and addressing the five risk aspects listed above (risk owner, tolerance, etc.). The Matrix allows for templating what you know about, say chemical risk assessment, or the assessment of strategic risks, and using that knowledge within any given cell. The Matrix allows for this level of analysis and action. Individual cells are depicted in Figure 8.4 at the intersection of risk dimension elements.

Templating. A key aspect of the rationale for focusing on the Matrix's cells and slices is to leverage established skills, competencies, processes, programs, and systems in a particular part of the Matrix to ones that are less developed. For

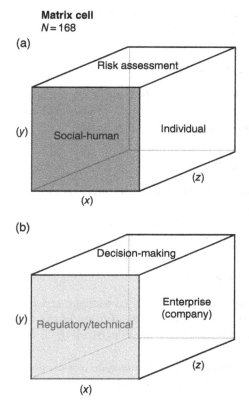

Figure 8.4 Cell examples – (a) Foundational cell, and (b) Trim tab cell.

example, risk assessment, in most enterprises, is well established and strong in the Risk Matrix's regulatory/technical element, but perhaps not as much in the organizational dimension, and perhaps not at all in the social–human element. Strengths in one area can be "templated" for use in another. This concept is developed further later in this chapter and in Chapters 9 and 10.

In the following sections, I describe each of elements in each of the three dimensions.

8.3 Contexts/Drivers (*y*-Axis)

The Contexts/Drivers dimension contains three elements: 1) regulatory/technical, 2) organizational, and 3) social–human. The social–human element expands the ORM purview such that social–human era issues that impact an organization's risk profile can be addressed in a uniform and structured manner. It also provides an anchor for integration with traditional EHSS/ORM issues reflected in this dimension's other two elements – regulatory/technical and organizational.

8.3.1 Regulatory/Technical

This element includes public policy (including risk science) constructs, laws, regulations, and consensus standards. This element is the most familiar to EHSS/ORM professionals and serves as the primary driver.

8.3.2 Organizational

This element includes historic governance and management ideas and theories, mission, vision, purpose, operations, frameworks, and systems. This element is more familiar to EHSS/ORM management, C-suite, and board-level decision-makers.

8.3.3 Social-Human

This element includes people, society, consumers, and the commons (environment). It is evolving and many aspects of it are not as familiar to EHSS/ORM professionals. It is a focus of this book, particularly in its integration with the other two elements.

Figure 8.5 shows an example of contexts/drivers element (social-human). The use of this is elaborated on in Chapter 9 (§9.4.2).

8.4 Actors/Motive Force (z-Axis)

This risk dimension's elements were introduced in Chapter 6 (§6.2). The term "actor" is used to depict individuals, or groups of individuals such as reflected in teams and departments, the organization itself, and the community. The community element is not developed extensively here but is included as it plays a critical role in ORM. The ORM object/foci evolution sequence, identified in Chapter 2 (§2.1.1.4), includes stakeholders at the end of its continuum. In Chapters 9 and 10, where Risk Matrix applications are presented, I include stakeholders in the community element. Explicitly identifying this dimension, and it's four elements is a significant advancement for EHSS/ORM. Doing this (1) reflects the increase in stakeholder impacts on organizational governance (this is embedded in the community element), (2) highlights dynamics beyond the enterprise/company, (3) highlights the role of individuals and teams/departments. The term "gravitational pulls" is introduced in Chapter 9 (§9.2.5) and is an important

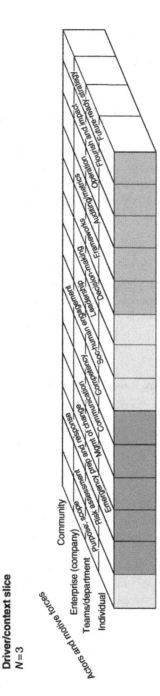

Figure 8.5 Contexts/drivers row example.

dynamic to consider in this dimension, especially "pulls" between individuals and the enterprise/company.

8.4.1 Individual

This includes people in an organization and those associated with it. The focus here is primarily on people who design, build, and operate programs and systems. Also included are senior management and boots-on-the-ground personnel. A valuable aspect of this element is to draw attention to the various "hats we wear" (discussed in §6.2), and how this dynamic can or does impact the outcome space depicted in Figure 2.1, particularly in terms of decision-making.

8.4.2 Team/Department

The primary foci here are the EHSS and risk management functions. Also included are organizational functions (departments) and teams within the enterprise/company, such as human resources, engineering, legal, and production. Business units in large organizations can be considered in the element.

8.4.3 Enterprise/Company

The term, enterprise/company, is used as the primary unit for an entity. Clarity on boundaries is important. In complex entities, attention should be given to this as risk profiles, and particularly accountabilities for them, may vary depending on how boundaries are defined.

8.4.4 Community

Community refers primarily to and community that is in close proximity to an enterprise's operations. More broadly, any external entity can be included, including stakeholders (generally), consumers, customers, supply chains, and suppliers.

Figure 8.6 shows an example of an actors/motive force element. A single slice as depicted here can be used for individuals in training and performance reviews, by teams/departments in developing programs and systems, by enterprises in developing KPIs (key performance indicators), dashboards, and audit tools. For the community element, the slice can be used by the other actors (including relevant stakeholders) in their decision-making actions to address community issues or concerns.

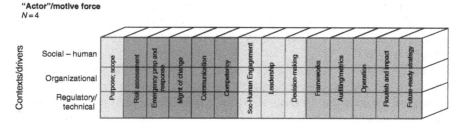

Figure 8.6 Actors/motive force row example.

The actors/motive force row can also aid actors in developing and strengthening their decision-making skills by providing a structure within which to address a particular problem or need for action. This example is expanded on in Chapter 9 (§9.4.3), particularly in terms of representing an enterprise/company-wide scorecard.

8.5 Risk Management Elements (*x*-Axis)

Fourteen risk management elements make up the Risk Matrix backbone. These are familiar elements in common ORM frameworks such as COSO's ERM, ISO 31000, and ISO management system standards. These are broken into three groupings:

- Foundational Five,
- Trim Tabs, and
- Operational elements.

The sorting of elements within these groupings is not rigid. This ordering is offered to provide a means to prioritize efforts to emphasize elements that (1) require 24/7/365 attention, (2) provide leverage, and (3) are operationally oriented. As mentioned earlier, I encourage you to arrange the elements within these groupings to align with your organizational frameworks, thinking, governance, etc. The point is to distinguish a topography around which to organize your thinking. The Foundational Five points to processes and activities that are frontline in ORM. This is highlighted first to emphasize activities that must be in the foreground at all times, i.e. 24/7/365.

The Trim Tab activities provide leverage in achieving ORM goals and are central to generating motive forces and organizational energy. These are associated with "→" (interventions) and "B" (goals, objectives, fulfilling purpose, etc.) in the A → B notation introduced in Chapter 1.

8 Risk Matrix

Graphically, a risk element column (e.g. risk assessment) is depicted generically in Figure 8.7. This depiction offers a powerful tool that will helps as we move forward in considering individual elements. Examples of this are presented in Chapter 9 (§9.4.1).

Fourteen (14) risk management elements are listed in Table 8.1. I have developed this set and groupings based on years of working in with oganizations standards-development. In these activities, I have seen a wide range of defining elements and their groupings. Some of these are depicted in the frameworks covered in Chapter 3. These 14 elements also reflect the work I have done in developing the Awareness-Based Risk Management™ framework described in Chapter 5. Element 1 – purpose and scope – is presented first in Table 8.2 (and Risk Matrix figures) due to its critical role in establishing context and in defining boundaries. It is followed by the Foundation Five elements, then by the remaining four Trim Tab elements.

A brief overview of the elements (E1 – E14) follows here, along with language from, (1) ISO 31000:2018, (2) COSO's ERM framework, (3) ISO MSSs, and (4) ISO 37000:2021. Regarding ISO MSSs, their environmental management MS (ISO 14001:2015), and occupational health and safety MS (ISO 45001:2018) are used as examples. The knowledge gained from these two examples can be applied to the ISO MSS generally.

It is important to understand that there are significant differences in the four frameworks, and how each of them address the 14 elements. There are different foci, and there are differences in what was taking place during the period in which

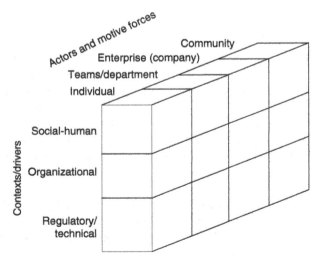

Figure 8.7 Risk management element column example.

8.5 Risk Management Elements (x-Axis) | 245

Table 8.2 EHSS risk management elements.

1	Purpose and scope
2	Risk assessment
3	Emergency preparedness and response
4	Management of change
5	Communication: systems and practices
6	Competency and capabilities
7	Social-human engagement
8	Leadership
9	Decision-making
10	Frameworks
11	Auditing/metrics
12	Operation
13	Escalating impact
14	Future-ready strategy

they were developed. For instance, ISO 37000:2021, the most recent version, contains the language of evolved concepts in the field of risk management.

8.5.1 Foundational Five

This grouping contains five elements: (E2) risk assessment; (E3) emergency preparedness and response (EPR); (E4) management of change (MOC); (E6) communication, systems and practices; and, (E6) competency and capabilities. In organizational engagements and teaching, I often ask clients and students to describe the bundle of "elements" that comprise what they do. Some point to elements of their management systems. Others point to regulatory compliance programs and technical aspects associated with them. Depending on the setting, I would ask if they do, or could group the elements they identified, such as I have done here. Then, I would ask about this notion of "foundational" elements. The five elements here are ones that rose to the top. You may find that there are others. You may have a Foundational Six or a Foundational Four. The point is to have a grouping like this to anchor the rationale for your groupings.

8.5.1.1 Risk Assessment (E2)

Risk assessment is at the core of risk management. There are different ways of thinking about this element and a range of terms used to describe it. For instance, risk profile is often embedded in risk assessment activities, and in some instances, it is considered as a separate, but parallel element with equal importance. Some common definitions are:

> SRA (2018) – Risk assessment
> "Systematic process to comprehend the nature of risk, express and evaluate risk, with the available knowledge."[6]

> ISO 31000 (2018) – Risk assessment
> "Risk assessment is the overall process of risk identification, risk analysis, and risk evaluation. [It] should be conducted systematically, iteratively, and collaboratively, drawing on the knowledge and views of stakeholders. It should use the best available information, supplemented by further enquiry as necessary."[7] In ISO 31000, three risk assessment components are identified: risk identification (6.4.2); risk analysis (6.4.3); and, risk evaluation (6.4.4).

> COSO ERM (2017) – Assessing risk
> "Risks identified and included in an entity's risk inventory are assessed in order to understand the severity of each to the achievement of an entity's strategy and business objectives. Risk assessments inform the selection of risk responses. Given the severity of risks identified, management decides on the resources and capabilities to deploy in order for the risk to remain within the entity's risk appetite."[8]

8.5.1.2 Emergency Preparedness and Response [EPR] (E3)

8.5.1.2.1 ISO 14001 (2015) and 45001 (2018), 8.2 Requirement Both standards state "The organization shall establish, implement and maintain the process(es) needed to prepare for and respond to potential emergency situations."[9]

ISO 45001:2018 includes workers and related communications and actions. The standard states:

> The organization shall establish, implement and maintain a process(es) needed to prepare for and respond to potential emergency situations, as identified in 6.1.2.1, [hazards] including: e) communicating and providing relevant information to all workers on their duties and responsibilities; f) communicating relevant information to contractors, visitors, emergency

---

6 SRA Glossary of Terms, p. 8.
7 ISO 31000:2018, p. 11.
8 COSO ERM (2017). Principle 11: Assesses Severity or Risk, p. 72.
9 ISO 14001, p. 13.

response services, government authorities and, as appropriate, the local community; g) taking into account the needs and capabilities of all relevant interested parties and ensuring their involvement, as appropriate, in the development of the planned response.[10]

In 45001:2018, reference is made to hazard identification, including: "how work is organized, social factors (including workload, work hours, victimization, harassment and bullying), leadership and the culture in the organization."[11] As you consider aspects of the Matrix in Action (§9), consider this ISO 45001 element in developing the matrix's social-human element within the context of templating.

8.5.1.3 Management of Change (E4)

This is a key dynamic in ORM frameworks, and organizational governance in general. The term "change management" is commonly used in organizational/business science. It is included as a risk management element given its importance to sound ORM. Of interest in the evolving social-human era is how existing policies and procedures can be used to address how "change" impacts the contexts/drivers, and social–human element (templating).

8.5.1.3.1 ISO 45001:2018 Related to MOC, this standard states:

> The organization shall establish a process(es) for the implementation and control of planned temporary and permanent changes that impact OH&S performance, including: a) new products, services and processes, or changes to existing products, services and processes, including: workplace locations and surroundings; – work organization; working conditions; equipment; work force; b) changes to legal requirements and other requirements; c) changes in knowledge or information about hazards and OH&S risks; d) developments in knowledge and technology. The organization shall review the consequences of unintended changes, taking action to mitigate any adverse effects, as necessary.[12]

While MOC is not explicitly addressed in the auditable portion of 14001:2015, its annex A states:

> Management of change is an important part of maintaining the environmental management system that ensures the organization can achieve the

---

10 ISO 45001, §8.2.e-g, p. 19.
11 ISO 45001, p.12.
12 45001, p. 18.

intended outcomes of its environmental management system on an ongoing basis. Management of change is addressed in various requirements of this International Standard, including maintaining the environmental management system (see 4.4), environmental aspects (see 6.1.2), internal communication (see 7.4.2), operational control (see 8.1), internal audit programme (see 9.2.2), and management review (see 9.3). As part of managing change, the organization should address planned and unplanned changes to ensure that the unintended consequences of these changes do not have a negative effect on the intended outcomes of the environmental management system. Examples of change include planned changes to products, processes, operations, equipment or facilities; changes in staff or external providers, including contractors; new information related to environmental aspects, environmental impacts and related technologies – changes in compliance obligations.[13]

8.5.1.4 Communication – Systems and Practices (E5)

It is important to consider the difference between communication systems and communication practices. Both are important. Communication systems are identified and defined in frameworks (left side of Figure 2.1). Communication practices are identified in the figure's rightside. Promoting communication practices, some of which are identified in Chapter 5 (§5.5 and §5.6), are, of course, critical to promote.

8.5.1.4.1 ISO 37000:2021 Some communication-related examples from this standard are found in its "principles" as follows:

> *6.1.3.2 – Define the organizational purpose*
>
> "The governing body should ensure that the essence of the organizational purpose is documented in a summary statement to promote effective communication and to assess and determine organization-wide actions and success."[14]
>
> *6.6.3.e – Key aspects of practice*
>
> "The governing body should ensure that an open and transparent communication culture within the organization is created and maintained to help bridge the gap between diverse stakeholder groups and varying perspectives based on, for example, gender, age, belief systems or cognitive abilities."[15]

---

13 14001, p. 18.
14 37000, p. 14.
15 37000, p. 25.

6.9.3.2.a – Set the tone for management of risk

"The governing body should ensure that the organizational risk framework, in respect to the management of risk: establishes the desired risk culture across the organization that encourages the reporting and communication of new and emerging risks and ensures that every person in the organization understands their risk management responsibility."[16]

6.9.3.4.h – Oversee risk management

"The governing body should oversee the organization's management of risk (see 6.4), ensuring that effective risk reporting and communication of risk are practised and promoted throughout the organization."[17]

8.5.1.4.2   COSO ERM   Communication is embedded throughout the framework. It is the focus of its fifth "interrelated component," *Information, Communication, and Reporting*. This component is characterized as:

> Communication is the continual, iterative process of obtaining information and sharing it throughout the entity. Management uses relevant information from both internal and external sources to support enterprise risk management. The organization leverages information systems to capture, process, and manage data and information. By using information that applies to all components, the organization reports on risk, culture, and performance.[18]

In the development of operating structures (Principle 2), related to communication, COSO states: "Factors to consider when establishing and evaluating operating structures may include the: type of reporting lines (e.g. direct reporting/solid line versus secondary reporting) and *communication channels*."[19]

Related to continual improvement efforts, communication is suggested as an area where improvement efficiencies maybe available. It states: "Communications: Reviewing performance can identify outdated or poorly functioning communication processes. For example, in reviewing performance an organization discovers that emails are not successfully communicating its initiatives. In response, the organization decides to highlight initiatives through a blog and instant message feed to appeal to its changing workforce."[20]

---

16   37000, p. 31.
17   37000, p. 32.
18   COSO, p. 54.
19   COSO, p. 64.
20   COSO, p. 45.

8 Risk Matrix

Principle 19: Communicates Risk Information states:

- "The organization uses communication channels to support enterprise risk management."[21]
- "Communication channels enable management to convey: the importance, relevance, and value of enterprise risk management; the characteristics, desired behaviors, and core values that define the culture of the entity; the strategy and business objectives of the entity; the risk appetite and tolerance; the overarching expectations of management and personnel in relation to enterprise risk and performance management; and, the expectations of the organization on any important matters relating to enterprise risk management, including instances of weakness, deterioration, or nonadherence."[22]
- *Methods of communicating.* "Communication methods can take the form of electronic messages (e.g. emails, social media, text messages, instant messaging); external/third-party materials (e.g. industry, trade, and professional journals, media reports, peer company websites, key internal and external indexes); informal/verbal communications (e.g. one-on-one discussions, meetings); public events (e.g. roadshows, town hall meetings, industry/technical conferences); training and seminars (e.g. live or online training, webcast and other video forms, workshops); written internal documents (e.g. briefing documents, dashboards, performance evaluations, presentations, questionnaires and surveys, policies, and procedures, FAQs)."[23]

8.5.1.4.3 ISO 31000:2018 This standard addressed communication in concert with consultation. There is value in considering this standard's notion of consultation in developing an integrated EHSS/ORM framework. Some examples are Clause 5.4.5 Establishing communication and consultation. For example:

The organization should establish an approved approach to communication and consultation in order to support the framework and facilitate the effective application of risk management. Communication involves sharing information with targeted audiences. Consultation also involves participants providing feedback with the expectation that it will contribute to and shape decisions or other activities. Communication and consultation methods and content should reflect the expectations of stakeholders, where relevant.[24]

---

21 COSO, p. 102.
22 COSO, p. 102.
23 COSO, p. 104.
24 ISO 31000, p. 7.

This clause asserts that communication and consultation should be done in a timely manner, with relevant information collected. It states: "Communication and consultation should be timely and ensure that relevant information is collected, collated, synthesised and shared, as appropriate, and that feedback is provided and improvements are made."[25]

The standard frames communication and consultation in a decision-making and integration context by stating in clause 6.2, that:

> The purpose of communication and consultation is to assist relevant stakeholders in understanding risk, the basis on which decisions are made and the reasons why particular actions are required. Communication seeks to promote awareness and understanding of risk, whereas consultation involves obtaining feedback and information to support decision-making. Close coordination between the two should facilitate factual, timely, relevant, accurate and understandable exchange of information, taking into account the confidentiality and integrity of information as well as the privacy rights of individuals. Communication and consultation with appropriate external and internal stakeholders should take place within and throughout all steps of the risk management process. Communication and consultation aims to: bring different areas of expertise together for each step of the risk management process; ensure that different views are appropriately considered when defining risk criteria and when evaluating risks; provide sufficient information to facilitate risk oversight and decision-making; build a sense of inclusiveness and ownership among those affected by risk.[26]

8.5.1.4.4 ISO MSS – ISO 45001:2018, as an Example[27] This standard contains requirements for establishing, implementing, and maintaining communication processes. It states:

> The organization shall establish, implement and maintain the process(es) needed for the internal and external communications relevant to the OH&S management system, including determining: a) on what it will communicate; b) when to communicate; c) with whom to communicate – 1) internally among the various levels and functions of the organization; 2) among contractors and visitors to the workplace; and 3) among other interested parties; and d) how to communicate.

---

25 31000, p. 7.
26 31000, p. 9.
27 45001 (2018), pp. 15–16.

The standards also include requirements to address diversity in (e.g. gender, language, culture, literacy, and disability) considering communication needs and views of external parties. Criteria for this related to internal and external communications is stated.

8.5.1.5 Competency and Capabilities (E6)

This refers to level of skill and the ability to do something well. Frameworks refer to competence, competency, and capabilities. These are closely related terms. Competence and competency are often considered synonymous. In some circles, competence is considered baseline, while competency reflects greater skill or mastery. A way to simplify this is to think of Jim Collin's notion of having "the right people on the bus, in the right seat."[28]

Box 8.1   Definitions, Competence, and Capability

Competence (n): "The ability to do something successfully or efficiently."
Capability (n): "The power or ability to do something."

8.5.1.5.1 ISO 37000:2021 This standard addresses competence as part of the expectations that an organization is well governed with accountability residing with the governing body. The standard states,

> Governing body members should continuously improve their competency regarding the organization's activities, legal requirements and, more broadly, the organization's context. This improving capability, together with regular reviews of governance practices, should ensure a continually improving governance environment.[29]

Related to assurance (6.4.3.3.c), the standard states, "The governing body should ensure that assurance providers have the necessary competency and capacity and that their efforts are appropriately focused."[30]

8.5.1.5.2 COSO ERM The fifth principle – "attracts, develops, and retains capable individuals" – contains human resource-related criteria. The first criterion indicated is, "establishing and evaluating competence." It states,

> Management, with board oversight, defines the human capital needed to carry out strategy and business objectives. Understanding the needed

---

28 Collins, Jim (2001) *Good to Great*.
29 ISO 37000, p. 9.
30 ISO 37000, p. 21.

competencies helps in establishing how various business processes should be carried out and what skills should be applied. Management considers the following factors when developing competence requirements: knowledge, skills, and experience with enterprise risk management; nature and degree of judgment and limitations of authority to be applied to a specific position; the costs and benefits of different skill levels and experience.[31]

An improvement worth considering here is the inclusion of integrated thinking, decision-making, and context, especially related to the Risk Matrix's contexts/drivers elements.

The COSO Framework's "front end," addresses integrating enterprise risk management and emphasizes the importance of integration. It states,

> An entity's success is the result of countless decisions made every day by the organization that affect the performance and ultimately, the achievement of the strategy or business objectives. Most of those decisions require selecting one approach from multiple alternatives. Many of the decisions will not be simply either 'right' or 'wrong', but will include trade-offs: time versus quality; efficiency versus cost; risk versus reward. When making such decisions, management and the board must continually navigate a dynamic business context, which requires integrating enterprise risk management thinking into all aspects of the entity, at all times. The Framework, therefore, views enterprise risk management in just that way. It is not simply a function or department within an entity, something that can be 'tacked on'. Rather, culture, practices, and capabilities are, together, integrated and applied throughout the entity.[32]

8.5.1.5.3 ISO 14001:2018 As part of ISO's High Level Structure, this standard contains an element that addresses competence (7.2). It states,

> The organization shall: determine the necessary competence of person(s) doing work under its control that affects its environmental performance and its ability to fulfil its compliance obligations; ensure that these persons are competent on the basis of appropriate education, training or experience; determine training needs associated with its environmental aspects and its environmental management system; where applicable, take actions to acquire the necessary competence, and evaluate the effectiveness of the actions taken.[33]

In ISO's conformity assessment-related standards, competence and competency are addressed in the ISO 17000 series of standards.

---

31 COSO, p. 41.
32 COSO, p. 17.
33 14001 (2015), p. 11.

8.5.2 Trim Tabs

This grouping contains four elements: (E1) purpose; (E7) social-human engagement; (E8) leadership; and, (E9) decision-making. These elements are ones that, over the years, clients and students have pointed to as providing the most leverage in growing and improving their EHSS/ORM activities. In Chapter 10, I introduce the term "escalate impact" to develop this idea). The notions of leverage and trim tabs were introduced in Chapter 1. Here and in the next chapter (§9), these are put into action with the Risk Matrix. References to how the Matrix's Trim Tabs are addressed in historic EHSS/ORM frameworks follow.

8.5.2.1 Purpose and Scope (E1)

Aspects of "purpose" were introduced in Chapter 2 (Risk Logics). The notion of organizational purpose is an important topic with many facets. The focus here is on its alignment with ORM and EHSS management, how ORM and EHSS management can impact it, and how it can impact ORM and EHSS management.

Box 8.2 Definitions, Purpose and Scope

Purpose (n): "The reason for which something is done or created or for which something exists."[34]

Scope (n): "The extent of the area or subject matter that something deals with or to which it is relevant."[35]

A dynamic of organizational purpose is materiality, both in the traditional sense of that which has a material impact on an organization and in terms of "double materiality" in looking at the impacts the organization has on others.

Implicit in the ORM frameworks presented in Chapter 3, and highlighted in this chapter, is their scope. With ISO 37000, the scope is organizational governance; with COSO's ERM, the scope is enterprise risk management; with ISO 31000, the scope is managing the risk faced by organizations; with ISO management system standards, scope is defined within each standard; with ISO 14001, it is environmental management; and, with ISO 45001, it is occupational health and safety management.

Scope (n): "The extent of the area or subject matter that something deals with or to which it is relevant."[36]

---

34 Apple OS dictionary, OS 13.5.2.
35 Apple OS dictionary, OS 13.5.2.
36 Apple OS dictionary, OS 13.5.2.

In addition to the content or topics included in its scope, are its boundaries. The scope includes the entire enterprise, or specific parts of it, such as individual business units or physical locations.

When applying the ideas presented in this book, consideration of scope (content and boundaries) is important. Facets of this include the dynamics of conformity assessment, such as measurement (auditing), and the stakeholders who are impacted by its scope, e.g., the "who's the judge?" question posed in Chapters 4 and 7.

The alignment of organizational structures and processes with purpose and scope is not static; rather, the assessment of alignment is an ongoing practice.

8.5.2.1.1 37000:2021   ISO 37000:2021's first principle of governance – purpose (6.1) – states,

> The [organization's] governing body should ensure that the organization's reason for existence is clearly defined as an organizational purpose. This organizational purpose should define the organization's intentions towards the natural environment, society and the organization's stakeholders. The governing body should also ensure that an associated set of organizational values is clearly defined.[37]

The ORM/EHSS functions play a central role in helping define the organization's purpose, but strictly speaking, in most instances, it is beyond their purview to take the lead in generating it. With the increasing attention to sustainability and ESG issues, the influence of EHSS in supporting organizational purpose is also increasing.

8.5.2.1.2 COSO ERM   Purpose is addressed throughout COSO's ERM framework in terms of value, and structures that support creating, preserving, and realizing value. It states: "Enterprise risk management is not a function or department. It is the culture, capabilities, and practices that organizations integrate with strategy-setting and apply when they carry out that strategy, with a purpose of managing risk in creating, preserving, and realizing value."[38]

8.5.2.1.3 31000:2018   ISO 31000 states that, "The purpose of risk management is the creation and protection of value. It improves performance, encourages innovation and supports the achievement of objectives." Eight elements (principles) are offered to "provide guidance on the characteristics of effective and efficient risk management, communicating its value and explaining its intention and purpose."[39]

---

37 ISO 37000:2021: *Governance of organizations – Guidance*, p. 13. Geneva, Switzerland.
38 COSO, p. 12.
39 31000, p. 2.

This standard reinforces the importance of scope and defines it. Section 6.3.1 states: "The purpose of establishing the scope, the context and criteria is to customize the risk management process, enabling effective risk assessment and appropriate risk treatment. Scope, context and criteria involve defining the scope of the process, and understanding the external and internal context."[40]

Regarding guidance in developing scope, Section 6.3.2 states:

> The organization should define the scope of its risk management activities. As the risk management process may be applied at different levels (e.g. strategic, operational, programme, project, or other activities), it is important to be clear about the scope under consideration, the relevant objectives to be considered and their alignment with organizational objectives. When planning the approach, considerations include: objectives and decisions that need to be made; outcomes expected from the steps to be taken in the process; time, location, specific inclusions and exclusions; appropriate risk assessment tools and techniques; resources required, responsibilities and records to be kept; relationships with other projects, processes and activities.[41]

8.5.2.1.4 ISO MSS Example – 45001:2018 ISO's management system standard related to occupational health and safety states:

> The purpose of an OH&S management system is to provide a framework for managing OH&S risks and opportunities. The aim and intended outcomes of the OH&S management system are to prevent work-related injury and ill health to workers and to provide safe and healthy workplaces; consequently, it is critically important for the organization to eliminate hazards and minimize OH&S risks by taking effective preventive and protective measures.[42]

Purpose is also addressed in Section 4.1 in terms of developing an understanding of the organization and its context. The standard states: "The organization shall determine external and internal issues that are relevant to its ***purpose*** and that affect its ability to achieve the intended outcome(s) of its OH&S management system."[43]

Regarding the development of scope, in Section 4.3, ISO 45001 states:

> The organization shall determine the boundaries and applicability of the OH&S management system to establish its scope. When determining this

---

40 31000, p. 10.
41 31000, p. 10.
42 45001, p. vi.
43 45001, p. 8.

scope, the organization shall: consider the external and internal issues referred to in 4.1; take into account the requirements referred to in 4.2; and take into account the planned or performed work-related activities. The OH&S management system shall include the activities, products and services within the organization's control or influence that can impact the organization's OH&S performance. The scope shall be available as documented information.[44]

8.5.2.2 Social-Human Engagement (E7)

The roots of social-human engagement go back to environmental, health, and safety management systems (EHSMS) and their focus on environmental aspects, and occupational health and safety. With the development of ORM frameworks, e.g. COSO's ERM and ISO 31000, a wider perspective was introduced. ISO 37000 widens the perspective to include stakeholders engaged with an organization, including the impact of risk governance.

In the management system world, going back to ISO 9001 (quality), the focus was on customer specification; with ISO 14001 (environment), the focus was on environmental aspects; with ISO 45001 (OHS), the focus was on the workers; and, with COSO's ERM, it is was on generating and protecting value.

The inclusion of a social-human engagement element is unique in the Risk Matrix. It reflects attention on the evolving social-human era, and a need to address it in EHSS/ORM. This is included as a risk management dimension element (E7), and as a contexts/drivers dimension element. I included it in these two places to reflect its role in impacting and driving EHSS/ORM. If the emphasis on this element is not strong now, I anticipate that it will likely be in the future. And, it is included as a risk management element to reflect the importance of integrating it with other risk management elements.

Box 8.3   Definitions, Human, Person, Social, and Society

Human (adj): "Relating to or characteristic of people or human beings."
Human (n): "A human being, especially a person as distinguished from and animal or (in science fiction) an alien."
Person (n): A human being regarded as an individual.
Social (adj): "Relating to society or its organization."
Social (n): "An informal social gathering, especially one organized by the members of a particular club or group."
Society (n): "The aggregate of people living together in a more or less ordered community."

---

44 45001, p. 8.

8.5.2.2.1   37000:2021   Social is addressed in this standard primarily in terms of environmental, social, and economic contexts. Human is addressed as a resource – as a capital – in terms of human rights, and in terms of human-based systems. These are framed in terms of value generation and stakeholders.

Stakeholders are addressed throughout ISO 37000, including providing stakeholders with an understanding of the organization's identity (6.1.2.b); identifying the stakeholders for which the organization exists to serve and the negative impacts that are to be avoided (6.1.3.2); and, ensuring that relevant stakeholders are engaged with the organization when defining organizational values (6.1.3.3).

The standard addresses value generation (6.2). It states,

> The focus for all organizations should be to fulfil their organizational purpose by generating value over time. To achieve this, organizations need to generate value which represents something of worth to its stakeholders. The ultimate value an organization is trying to generate (articulated in the organizational purpose) can only be achieved through collaboration with stakeholders. Appropriate value needs to be generated for stakeholders so that they are willing, and able, to support the organization in fulfilling its organizational purpose over time. The value stakeholders expect can take different forms and can impact the natural environment and society, as well as the stakeholders themselves. The governing body's function in this value generation includes stewardship – to ensure the organization not only creates but also protects value over time.[45]

Stakeholders are addressed as: (1) stakeholder, (2) member stakeholder, and (3) reference stakeholder. The standard defines these as:

- *Stakeholder* (3.3.1) person or *organization* (3.1.3) that can affect, be affected by, or perceive itself to be affected by a decision or activity. Note 1 to entry: Depending on the nature of the organization, stakeholders can include *member stakeholders* (3.3.2) and other stakeholders, including customers, regulators, suppliers and employees. Note 2 to entry: In ISO management system standards, a stakeholder can be referred to as an "interested party".
- *Member stakeholder* (3.3.2), [is a] *stakeholder* (3.3.1) who has a legal obligation or defined right to make decisions in relation to the *governing body* (3.3.4) and to whom the governing body is to account. Note 1 to entry: These rights or obligations are often recorded in the *organization's* (3.1.3)

---

45 ISO 37000, p. 15.

8.5 Risk Management Elements (x-Axis) | 259

constituting documents (3.1.5), laws and/or regulations. Note 2 to entry: These decisions can include, for example, the determination of the composition of the governing body or the parameters within which the governing body is to make decisions. Note 3 to entry: Governing bodies account to these stakeholders for the organization's outcomes as well as the governing body's performance. Note 4 to entry: Member stakeholders are often referred to, and can include, shareholders and members of an organization.

- *Reference stakeholder* (3.3.3), [is a] *stakeholder* (3.3.1) to whom the *governing body* (3.3.4) has decided to account to when making decisions pertaining to the *organizational purpose* (3.2.10). Examples, [a] scientific advisory board to a research organization, parents of the pupils in a school, community advisory boards for companies. Note 1 to entry: In some cases, a *member stakeholder* (3.3.2) can also be a reference stakeholder.[46]

ISO 37000 asserts that, when its guidance is applied, it enables an organization to realize a number of outcomes, including "generating value for stakeholders", and demonstrating "fairness in the treatment of, and engagement with, stakeholders."[47]

Regarding stakeholder engagement, the standard states,

> The governing body should ensure that the organization's stakeholders are identified, prioritized, appropriately engaged, consulted and their expectations understood. The governing body should do this to ensure that stakeholder relationships are effective and appropriate decisions about expectations are made to achieve the intended value generation objectives.[48]

As mentioned, ISO 37000:2021 addresses ORM in its ninth principle – risk governance. As a key aspect of risk governance practice, the standard states that, "The governing body should assume accountability for the organization's continual sensing of, and responding to, risk and communicating the chosen approach with relevant stakeholders as necessary."[49]

Possibly the most significant recommendation in this standard is related to risk governance that is characterized as setting the tone for the management of risk (6.9.3.2). Select clauses state:

> The governing body should ensure that the organizational risk framework, in respect to the management of risk: (d) considers the impact of changes

---

46 37000, p. 5.
47 ISO 37000, 5 Overview, a.3 and c.3, pp. 10–11.
48 ISO 37000, p. 24.
49 ISO 37000, 6.9.3, Risk governance, key aspects of practice, p. 31.

to and dependencies on the external and internal context of the organization, including: 1) stakeholders.; and h) mandates that relevant stakeholders are engaged responsibly and accurately, and considers the organization's positive and negative risk impacts on them.[50]

Regarding related practices, it states,

The governing body should consider and manage risk associated with its own activities in accordance with the organizational risk framework. For example, the governing body should: d) disclose substantive risk, limits and associated expectations to relevant stakeholders as a demonstration of its accountability; e) make clear to relevant stakeholders the nature and extent of accepted risk along with assurance that the organization will operate within the defined risk limits and take corrective action where necessary.[51]

8.5.2.2.2 ISO Example – 45001:2018 The focus of this standard is on workers. In its introduction, where general principles and recommendations are stated, as opposed to requirements (e.g. "shall" statements), it states: "The implementation and maintenance of an OH&S management system, its effectiveness and its ability to achieve its intended outcomes are dependent on a number of key factors, which can include: d) consultation and participation of workers, and where they exist, workers' representatives."[52] Of the eleven key factors included, "d" is highlighted because of its focus on engaging (participation) or workers (e.g. human).

Definitions are provided for, worker, participation, and consultation:[53]

- *Worker (3.3)* person performing work or work-related activities that are under the control of the *organization*. Note 1 to entry: Persons perform work or work-related activities under various arrangements, paid or unpaid, such as regularly or temporarily, intermittently or seasonally, casually or on a part-time basis. Note 2 to entry: Workers include *top management* (3.12), managerial, and nonmanagerial persons. Note 3 to entry: The work or work-related activities performed under the control of the organization may be performed by workers employed by the organization, workers of external providers, contractors, individuals, agency workers, and by other persons to the extent the organization shares control over their work or work-related activities, according to the context of the organization.

---

50 ISO 37000, 6.9.3.2, Set the tone for the management of risk, p. 31.
51 ISO 37000, 6.9.3.3, Practise effective risk management, p. 32.
52 ISO 45001, 0.3 Success factors, p. vi.
53 Iso 45001, 3 Terms and definitions, p. 2.

8.5 Risk Management Elements (x-Axis) | 261

- *Participation (3.4)* involvement in decision-making. Note 1 to entry: Participation includes engaging health and safety committees and workers' representatives, where they exist.
- *Consultation (3.5)* seeking views before making a decision. Note 1 to entry: Consultation includes engaging health and safety committees and workers' representatives, where they exist.

In terms of developing an "understanding [of] the needs and expectations of workers" the standard states:

> The organization shall determine: a) the other interested parties, in addition to workers, that are relevant to the OH&S management system; b) the relevant needs and expectations (i.e. requirements) of workers and other interested parties; and c) which of these needs and expectations are, or could become, legal requirements and other requirements.[54]

The inclusion of leadership and worker participation in OHSMS standards from their onset in the 1990s, was a significant advancement. ISO 45001, the standard's Section 5 shows this, as follows:

> Top management shall demonstrate leadership and commitment with respect to the OH&S management system by: k) protecting workers from reprisals when reporting incidents, hazards, risks and opportunities; and l) ensuring the organization establishes and implements a process(es) for consultation and participation of workers.[55]

In establishing policy, it states: "Top management shall establish, implement and maintain an OH&S policy that: f) includes a commitment to consultation and participation of workers, and, where they exist, workers' representatives."[56]

Related to this in terms of roles, responsibilities, and authorities, the standard states:

> The organization shall establish, implement and maintain a process(es) for consultation and participation of workers at all applicable levels and functions, and, where they exist, workers' representatives, in the development, planning, implementation, performance evaluation and actions for improvement of the OH&S management system.[57]

---

54 ISO 45001, 4.2, p. 8.
55 ISO 45001, 5.1.k-l, p. 9.
56 ISO 45001, 5.2.f, p. 9.
57 ISO 45001, 5.4, p. 10.

The standard contains criteria for training; access to information; and, emphasis on consultation and participation of nonmanagerial workers. Significant attention is given to engagement of workers in hazard identification and assessment of risks and opportunities. It also states:

> The organization shall establish, implement and maintain a process(es) for hazard identification that is ongoing and proactive. The process(es) shall take into account, but not be limited to: e) people, including consideration of: 1) those with access to the workplace and their activities, including workers, contractors, visitors and other persons; 2) those in the vicinity of the workplace who can be affected by the activities of the organization; 3) workers at a location not under the direct control of the organization.[58]

The standard also contains criteria for the assessment of opportunities. It states,

> The organization shall establish, implement and maintain a process(es) to assess: a) OH&S opportunities to enhance OH&S performance, while taking into account planned changes to the organization, its policies, its processes or its activities and: 1) opportunities to adapt work, work organization, and work environment to workers; and, 2) opportunities to eliminate hazards and reduce OH&S risks.[59]

Regarding OHS objectives and the planning done to achieve them, the standard states:

> The organization shall establish OH&S objectives at relevant functions and levels in order to maintain and continually improve the OH&S management system and OH&S performance. The OH&S objectives shall: c) take into account: 3) the results of consultation with workers and, where they exist, workers' representatives.[60]

Engagement is also reflected as well in the standards management review requirements. It states

> Top management shall review the organization's OH&S management system, at planned intervals, to ensure its continuing suitability, adequacy and effectiveness. The management review shall include consideration

---

58 ISO 45001, 6.1.2.1.e, p. 12.
59 ISO 45001, 6.1.2.3.a, p. 13.
60 ISO 45001, 6.2.1.c.3, p. 14.

of: d) information on the OH&S performance, including trends in: 5) consultation and participation of workers."[61] And, "Top management shall communicate the relevant outputs of management reviews to workers, and, where they exist, workers' representatives.[62]

Requirement related to participation in "improvement" activities, including incident and nonconformity investigations and associated corrective actions, state:

> The organization shall establish, implement and maintain a process(es), including reporting, investigating and taking action, to determine and manage incidents and nonconformities. When an incident or a nonconformity occurs, the organization shall:... b) evaluate, with the participation of workers and the involvement of other relevant interested parties, the need for corrective action to eliminate the root cause(s) of the incident or nonconformity, in order that it does not recur or occur elsewhere, by: 1) investigating the incident or reviewing the nonconformity; 2) determining the cause(s) of the incident or nonconformity; and, 3) determining if similar incidents have occurred, if nonconformities exist, or if they could potentially occur.[63]

Finally, the requirements related to involvement in continual improvement efforts, state,

> The organization shall continually improve the suitability, adequacy and effectiveness of the OH&S management system, by: c) promoting the participation of workers in implementing actions for the continual improvement of the OH&S management system; and, d) communicating the relevant results of continual improvement to workers, and, where they exist, workers' representatives.[64]

8.5.2.2.3 COSO ERM The COSO ERM Framework reinforces value in, and engagement by people and stakeholders. Principle 5 – attracts, develops, and retains capable individuals – underscores the value of human capital by stating, "The organization is committed to building human capital in alignment with the strategy and business objectives."[65]

---

61 ISO 45001, 9.3.d.5, p. 21.
62 ISO 45001, 9.3, p. 22.
63 ISO 45001, 10.2.b, p. 22.
64 ISO 45001, 10.3.c-d, p. 2.
65 COSO, p. 41.

Social-human engagement is reflected in the framework's encouragement to escalate issues of concern. It states, "Instilling more transparency and risk awareness into an entity's culture requires actions such as: Encouraging people to escalate issues and concerns without fear of retribution."[66]

Stakeholders are defined in COSO's ERM, as "Parties that have a genuine or vested interest in the entity."[67] Their primacy is reflected by, "A discussion of enterprise risk management begins with this underlying premise: every entity – whether for-profit, not-for-profit, or governmental – exists to provide value for its **stakeholders**."[68] That is reflected in linking value realization to stakeholder benefit. The Framework states, "The value of an entity is largely determined by the decisions that management makes – from overall strategy decisions through to day-to-day decisions. Those decisions can determine whether value is created, preserved, eroded, or realized. Value is realized when stakeholders derive benefits created by the entity. Benefits may be monetary or non-monetary."[69]

COSO's Principle 19 – Communicates Risk Information – begins with reinforcing communication with stakeholders. Internal and external communication channels are identified. The Framework states:

> Various channels are available to the organization for communicating risk data and information to internal and external stakeholders. These channels enable organizations to provide relevant information for use in decision-making. Internally, management communicates the entity's strategy and business objectives clearly throughout the organization so that all personnel at all levels understand their individual roles. Management also communicates information about the entity's strategy and business objectives to shareholders and other external parties. Enterprise risk management is a key topic in these communications so that external stakeholders not only understand the performance against strategy but the actions consciously taken to achieve it.[70]

8.5.2.2.4 31000:2018 This standard addresses the social-human element in its introduction. It states, "This document is for use by people who create and protect value in organizations by managing risks, making decisions, setting and achieving objectives and improving performance." Examples of "managing risk"

---

66 COSO, p. 18.
67 COSO, p. 110.
68 COSO, p. 3.
69 COSO, p. 3–4.
70 COSO, p. 102.

related to the social-human includes: "Managing risk considers the external and internal context of the organization, including human behaviour and cultural factors."[71]

Eight principles are offered to support value creation and protection, the seventh principle states: "Human and cultural factors – Human behaviour and culture significantly influence all aspects of risk management at each level and stage."[72]

The standard provides a risk management process that, "...involves the systematic application of policies, procedures and practices to the activities of communicating and consulting, establishing the context and assessing, treating, monitoring, reviewing, recording and reporting risk."[73] The role of culture and humans in the process is identified. The standard states. "The dynamic and variable nature of human behaviour and culture should be considered throughout the risk management process."[74] These dynamic and variable factors are present in the outcome space depicted in Figure 2.1. Tools for impacting these is are offered in Chapter 5 (§5.5, and §5.6).

ISO 31000:2018 defines stakeholder as, "person or organization that can affect, be affected by, or perceive themselves to be affected by a decision or activity."[75]

Risk treatment is part of the EHSS/ORM process. In selecting risk treatment options (6.5) stakeholders need to be addressed. The standard states,

> Justification for risk treatment is broader than solely economic considerations and should take into account all of the organization's obligations, voluntary commitments and **stakeholder** views. The selection of risk treatment options should be made in accordance with the organization's objectives, risk criteria and available resources. When selecting risk treatment options, the organization should consider the values, perceptions and potential involvement of **stakeholders** and the most appropriate ways to communicate and consult with them. Though equally effective, some risk treatments can be more acceptable to some stakeholders than to others.[76]

The standard reinforces the importance of engaging stakeholders in the implementation of the EHSS/ORM framework implementation. It states, "Successful

---

71 31000, Introduction, p. v.
72 31000, Principles, p. 4.
73 31000, p. 8.
74 31000, Process, p. 9.
75 31000, Terms and definitions, p. 1.
76 31000, Risk Treatment, p. 13.

implementation of the framework requires the **engagement** and awareness of stakeholders. This enables organizations to explicitly address uncertainty in decision-making, while also ensuring that any new or subsequent uncertainty can be taken into account as it arises."[77]

8.5.2.3 Leadership (E8)
Chapter 6 is devoted to leadership. The way this topic is addressed in various frameworks is reviewed. Leadership is also addressed in Chapters 9 and 10 in terms of risk field/matrix leverage (§9.4) and creating generative fields (§10.3).

8.5.2.4 Decision-Making (E9)
Decision-making is one of the central anchors presented in this book. It has been covered in its own stand-alone chapter (7), as well as in Chapter 5. Its inclusion as a stand-alone element in the risk management elements dimension, is unique. A core premise, put forth in Chapters 1 and 2 (depicted in Figures 1.1 and 2.1), is that integrated thinking, decision-making, and action occurs at the intersection of systems/frameworks (tangible) and "the Human" – social and human capital (nontangible realm), in a space where EHSS/ORM outcomes are generated.

A key aspect in the frameworks addressed in Chapter 3 are the decision-making processes embedded in them. In a sense, each framework can be viewed as a decision-making framework that an organization tailors and populates according to its unique internal and external context. For instance, COSO's ERM and ISO 31000 can be viewed as organizational risk decision-making frameworks. ISO 14001 and 45001 can also be considered environmental, and occupational health and safety risk decision-making frameworks. For this element (decision-making), I have hesitated whether to parse within the frameworks as I have with the other 13 elements. Specific decision-making elements within the frameworks follows, but keep in mind the larger (macro) decision-making function that each one offers.

8.5.2.4.1 37000:2021 In its introduction, this standard quickly makes a connection between decision-making and good governance. It states: "Good governance means that decision-making within the organization is based on the organization's ethos, culture, norms, practices, behaviours, structures and processes."[78] Attention to decision-making transparency, improvement, holistic nature, structures, processes, stakeholder inclusion, resources, and agility, for example, are addressed throughout the standard.

Section 6.8 of 37000 offers detailed decision-making guidance. It is framed here in terms data and decisions, recognizing "data as a valuable resource for

---

77 31000, p. 8.
78 37000, p. vi.

decision-making by the governing body, the organization and others."[79] Greater detail on value from 37000 with respect to decision-making was presented in Chapter 7.

8.5.2.4.2 *COSO ERM* This framework's first section is "The Changing Risk Landscape," and its first paragraph states:

> Our understanding of the nature of risk, the art and science of choice, lies at the core of our modern economy. Every choice we make in the pursuit of objectives has its risks. From day-to-day operational decisions to the fundamental trade-offs in the boardroom, dealing with risk in these choices is a part of decision-making.[80]

Decision-making is highlighted throughout the framework. Some of the highlights include: the role that decision-making has with; strategy development; alignment of resources with mission and vision; development of operating structures; improvement of capabilities; and, elimination of silos.

8.5.2.4.3 *31000:2018* Decision-making is a prominent topic in this standard. In its scope, it states: "This document can be used throughout the life of the organization and can be applied to any activity, including decision-making at all levels."[81]

Decision-making and integration are reinforced in introducing the standard's framework. It states,

> The purpose of the risk management framework is to assist the organization in integrating risk management into significant activities and functions. The effectiveness of risk management will depend on its integration into the governance of the organization, including decision-making. This requires support from stakeholders, particularly top management.[82]

Related to the standard's process, it states: "The risk management process should be an integral part of management and decision-making and integrated into the structure, operations and processes of the organization. It can be applied at strategic, operational, programme or project levels."[83]

---

79 37000, p. 28.
80 COSO, p. 9.
81 31000, p. 1.
82 31000, p. 4.
83 31000, p. 9.

8.5.2.4.4 ISO MSS Example - 14001 and 45001 Decision-making is addressed but not as directly as in ISO 37000, COSO's ERM, or ISO 31000. ISO 14001 states generally,

> Top management can effectively address its risks and opportunities by integrating environmental management into the organization's business processes, strategic direction and decision making, aligning them with other business priorities, and incorporating environmental governance into its overall management system.[84]

In 45001, decision-making is touched on in terms of worker participation in decision-making processes.

8.5.3 Operational Elements

This grouping contains five elements: (E10) frameworks; (E11) auditing/metrics; (E12) operation; (E13) escalating impact; and, (E14) future-ready strategy. Over the years, I have wrestled with how to label this group and their part of the EHSS/ORM topography. That I refer to them as operational elements points to their role in the routine functioning of EHSS/ORM. There is an individual element called *operation* (E12). It addresses specific activities with roots in ISO MSSs.

8.5.3.1 Frameworks (E10)

This term refers to processes, programs, systems, and fields. These are presented as a continuum in Chapter 2 (§2.1.1.2), within which there have been evolutions, particularly from programs to systems, and from systems to fields. Historic EHSS/ORM frameworks require or recommend that organizations define – or show that processes exist that are related to – that which the framework addresses. For instance, if the framework addresses a management system (e.g. ISO 14001), then there needs to be evidence of processes that, in aggregate, demonstrate the existence of an environmental management system that is designed to conform with ISO 14001.

8.5.3.1.1 ISO MSS While this may seem circular, these management system standards require that the organization have a management system that conforms to the MSS that is being followed. In ISO MSSs (e.g. 45001 and 14001), this is addressed in Section 4.4 xx management system. With ISO 14001, xx refers to

---

84 14001 (2015), p. vi.

environmental, and, with ISO 45001, it refers to occupational health and safety. These standards state:

> *45001:2018.* "The organization shall establish, implement, maintain and continually improve an OH&S management system, including the processes needed and their interactions, in accordance with the requirements of this document."[85]
>
> *14001:2015.* "To achieve the intended outcomes, including enhancing its environmental performance, the organization shall establish, implement, maintain and continually improve an environmental management system, including the processes needed and their interactions, in accordance with the requirements of this International Standard. The organization shall consider the knowledge gained in Understanding the Organization and its Context (4.1) and Understanding the Needs and Expectations of Interested Parties (4.2) when establishing and maintaining the environmental management system."[86]

8.5.3.1.2 ISO 37000:2021 The standard addresses organizational governance and in one of its 11 principles addresses risk governance (6.9), and provides ORM framework guidance.

For governance frameworks in general, the standard states in its introduction:

> Good governance means that decision-making within the organization is based on the organization's ethos, culture, norms, practices, behaviours, structures and processes. Good governance creates and maintains an organization with a clear purpose that delivers long-term value consistent with the expectations of its relevant stakeholders. The implementation of good governance is based on leadership, values, and *a framework of mechanisms, processes and structures that are appropriate for the organization's internal and external context.*[87]

Organizational governance framework is defined as (3.1.2): "strategies, *governance policies* (3.2.9), decision-making structures and *accountabilities* (3.2.2) through which the *organization's* (3.1.3) governance arrangements operate."[88]

---

85 ISO 45001, p. 8.
86 ISO 14001, p. 6.
87 ISO 37000, p. vi.
88 ISO 37000, p. 1.

The notion of "integrated governance" is woven throughout ISO 37000. In relation to frameworks, it states:

> The governing body is accountable for establishing and maintaining an integrated organizational governance framework across the organization that coordinates these governance activities such that the organization realizes effective performance, responsible stewardship and ethical behaviour. This organizational governance framework should ensure that decision-makers have appropriate authority, competence and resources for the responsibilities given to them. Effective delegation and transparent decision-making empower personnel to act appropriately, resulting in a more resilient and agile organization. Controls and subsequent improvement actions should be planned and implemented to ensure that the governance system remains adequate for the organization's purpose.[89]

Principle 6.9 addresses risk governance in general, and specifically sets the tone for the management of risk in 6.9.3.2. In this section it states,

> the governing body should establish an organizational risk framework that ensures a formal, proactive and anticipative approach to the management of risk across the organization, including by the governing body. The governing body should ensure that this framework integrates risk management into all organizational activities.[90]

The standard offers criteria related to establishing risk culture, and decision-making behaviors.

ISO 37001 does not explicitly address MOC as is done in ISO MSSs. However, related to ORM, it offers a valuable MOC recommendation. It states that the risk framework should consider:

> ... the impact of, changes to and dependencies on the external and internal context of the organization, including: stakeholders; short-, medium- and long-term trends including social responsibility and sustainability trends; organizational purpose; organizational values; the value generation model; [and], intended strategic outcomes.[91]

---

89 ISO 37000, p. 7.
90 ISO 37000, p. 31.
91 ISO 37000, p. 31 (6.9.3.2.d).

8.5.3.1.3 COSO ERM The COSO ERM Framework document states in its cover page:

> This project was commissioned by the Committee of Sponsoring Organizations of the Treadway Commission (COSO), which is dedicated to providing thought leadership through the development of *comprehensive frameworks* and guidance on internal control, enterprise risk management, and fraud deterrence designed to improve organizational performance and oversight and to reduce the extent of fraud in organizations.[92]

8.5.3.1.4 ISO 31000:2018 This standard presents a three-pillar model (principles, framework, process). Regarding its framework pillar, it states:

> The purpose of the risk management framework is to assist the organization in integrating risk management into significant activities and functions. The effectiveness of risk management will depend on its integration into the governance of the organization, including decision-making. This requires support from stakeholders, particularly top management.[93]

The framework contains six elements: (1) leadership and commitment; (2) Integration; (3) Design; (4) Implementation; (5) Evaluation; and, (6) Improvement.

Reinforcing the notion of tailoring that I have raised earlier, this standard states, "The components of the framework and the way in which they work together should be customized to the needs of the organization."[94]

8.5.3.2 Auditing and Metrics (E11)

The role and importance of measurement cannot be overstated. Aspects of measurement are addressed generically in Chapter 4 (Conformity assessment), and specifically with respect to the Risk Matrix and integrated EHSS/ORM in Chapters 9 (Matrix in Action) and 10 (Escalating Impact). I chose to label this element as I have because of prominence of auditing and metrics in EHSS/ORM, and the obvious meaning they have. Likewise it could have also been labeled "performance measurement" or "performance assessment" or – as is done in ISO management systems – "performance evaluation" with sub-labels, or sub-elements such as: monitoring, measurement, analysis, and evaluation; internal audit; and, management review. Historic EHSS/ORM frameworks require or recommend different aspects of measurement. Examples follow.

---

92 COSO, coverpage.
93 31000, p. 4.
94 31000, p. 4.

It is important to understand the range and types of audits (e.g. conformity assessment activity) that a given standard or framework addresses (e.g. recommendation versus requirement). Accordingly, as the Risk Matrix presented here is put into action you should be clear about the range and types of audits needed.

8.5.3.2.1 ISO 31000:2018 This standard addresses evaluation (section 5), stating:

> In order to evaluate the effectiveness of the risk management framework, the organization should: periodically measure risk management framework performance against its purpose, implementation plans, indicators and expected behaviour; and, determine whether it remains suitable to support achieving the objectives of the organization.[95]

8.5.3.2.2 ISO MSS – 14001 and 45001 Examples ISO 14001 Section 9 addresses "Performance evaluation." This section has three subsections: 9.1, Monitoring, measurement, analysis, and evaluation; 9.2, Internal audit; and, 9.3, Management review. The section provides a range of requirements related to environmental performance. Its sister standard, ISO 45001 states, related to OH&S:

> The organization shall determine: a) what needs to be monitored and measured, including: 1) the extent to which legal requirements and other requirements are fulfilled; 2) its activities and operations related to identified hazards, risks and opportunities; 3) progress towards achievement of the organization's OH&S objectives; and, 4) effectiveness of operational and other controls.[96]

Internal audits are required at "planned intervals." Criteria are included for the establishment of the defined internal audit programs.

8.5.3.2.3 37000:2021 This standard frames the issues in terms of oversight and assurance. Principle 6.4 contains assurance criteria to "oversee performance (6.4.3.2)," and to "obtain assurance (6.4.3.3)." Overseeing performance is framed in terms of the assessment and taking corrective action on a range of issues, among which are values and governance polices, and engagement with stakeholders. Assurance criteria include the use of entities other than the governing body to verify assurance of internal audit protocols, reports on risk and compliance management.

---

95 31000, p. 8.
96 45001, pp. 19–20.

8.5.3.2.4 COSO ERM This framework offers guidance in defining and establishing measurement structures. It offers a number of these structures linked to performance management in its introduction, stating:

> Performance relates to actions, tasks, and functions to achieve, or exceed, an entity's strategy and business objectives. Performance management focuses on deploying resources efficiently. It is concerned with measuring those actions, tasks, and functions against predetermined targets (both short- and long-term) and determining whether those targets are being achieved.[97]
>
> Because a variety of risks – both known and unknown – may affect an entity's performance, a variety of measures may be used (for example): Financial measures, such as return on investments, revenue, or profitability; operating measures, such as hours of operation, production volumes, or capacity percentages; obligation measures, such as adherence to service-level agreements or regulatory compliance requirements; project measures, such as having a new product launch within a set period of time; growth measures, such as expanding market share in an emerging market; and stakeholder measures, such as the delivery of education and basic employment skills to those needing upgrades when they are out of work.[98]

The framework's ninth principle (*Formulates Business Objectives*) addresses establishing measurable performance standards, where it states:

> The organization sets targets to monitor the performance of the entity and support the achievement of the business objectives. For instance: an asset management company seeks to achieve a return on investment (ROI) of 5% annually on its portfolio; a restaurant targets on-line home delivery orders to be delivered within forty minutes; and, a call center endeavors to minimize missed calls to 2% of overall calls received.[99]

Examples of performance measurements and targets, for business objectives are provided in the framework's Example 7.9, reproduced here (with permission) as Table 8.3.

---

97 COSO, p. 5.
98 COSO, p. 5.
99 COSO, p. 61.

Table 8.3 COSO performance measures example.

	Business objective	Performance measure and target
Business objectives (entity)	• Continue to develop innovative products that interest and excite consumers • Expand retail presence in the health food sector	• 8 products in R&D at all times • 5% growth year over year
Business objectives for North America (division)	• Increase shelf space in leading stores that share our core values • Continue to source products in local markets	• 7% increase in shelf space • 92% local source rate
Business objectives for Confectionary (operating unit)	• Develop high-quality and safe snack products that exceed consumer expectations	• 4.8 out of 5 in customer satisfaction survey
Business objectives for Human Resources (function)	• Maintain favorable annual turnover of employees • Recruit and train product sales managers in the coming year	• Turnover less than 10% • Recruit 50 sales managers • 95% training rate for sales staff

Performance measurements have a range of applications in EHSS/ORM. The COSO ERM Framework highlights their use in assessing performance against established tolerances. It states,

> Performance measures related to a business objective help confirm that actual performance is within an established tolerance (see Example 7.10). Performance measures can be either quantitative or qualitative. Tolerance also considers both exceeding and trailing variation, sometimes referred to as positive or negative variation. Note that exceeding and trailing variation is not always set at equal distances from the target.[100,101]

---

100 COSO, pp. 104–105.
101 COSO, Example 7.10 (page 62). "Trailing Target Variation. "A large beverage bottler sets a target of having no more than five lost-time incidents in a year and sets the tolerance as zero to seven incidents. The exceeding variation between five and seven represents greater incidents and potential for lost time and an increase in health and safety claims, which is a negative result for the entity. In contrast, the trailing variation up to five represents a benefit: fewer incidents of lost time and fewer health and safety claims. The organization also needs to consider the cost of striving for zero lost-time incidents."

Another application relates to impact measures that can be framed in terms of hazard or risk. The framework's eleventh principle (assesses severity of risk) addresses selecting severity measures:

> Management selects measures to assess the severity of risk. Generally, these measures align to the size, nature, and complexity of the entity and its risk appetite. Different thresholds may also be used at varying levels of an entity for which a risk is being assessed. The thresholds used to assess the severity of a risk are tailored to the level of assessment – by entity or operational unit. Acceptable amounts of risk to financial performance, for example, may be greater at an entity level than an operating unit level. Management determines the relative severity of various risks in order to select an appropriate risk response, allocate resources, and support management decision-making and performance.[102]

8.5.3.3 Operation (E12)

The term "operation" is a common element in ISO MSSs. In ISO's high level structure, it is MSS element 8. Other historic EHSS ORM frameworks embed operational recommendations or requirements.

8.5.3.3.1 ISO MSS – 45001:2018 Example This element is a major component of ISO MSSs. Section 8 in these standards addresses operational planning and control. ISO 45001:2018 states:

> The organization shall plan, implement, control and maintain the processes needed to meet requirements of the OH&S management system, and to implement the actions determined in Clause 6, by: establishing criteria for the processes; implementing control of the processes in accordance with the criteria; maintaining and retaining documented information to the extent necessary to have confidence that the processes have been carried out as planned; and, adapting work to workers.[103]

There is variation in how specific ISO MSSs tailor section 8. In 45001:2018, there are subelements that address eliminating hazards and reducing OH&S (8.1.2) and management of change (8.1.3).

---

102 COSO, p. 74.
103 ISO 45001, p. 17.

8.5.3.4 Escalating Impact (E13)

In management system-speak, this element refers to continual improvement, management review, and corrective actions. The term "escalating impact" is used in this book to reflect the expanded dimensionality of the Risk Matrix.

ISO 37000 indirectly addresses this element in its principle 6.11, "Viability and performance over time." Criteria in this element address value generation, remaining viable over time and addressing stakeholder expectations. It suggests that, at a minimum, it cannot go backward, stating:

> The governing body should ensure that the organization protects and restores those systems on which it depends. In this regard, the governing body should consider and manage risk associated with those decisions it makes that can impact the natural environmental, social and economic systems (see 6.9). While doing so, the governing body should ensure that relevant stakeholders are consulted and engaged (see 6.6). This should provide clarity regarding the impact the governing body's decisions have, over time, on those aspects on which the organization is: directly dependent; not directly dependent but whose ability to be sustained will be affected by the governing body's decisions.[104]

ISO 31000:2018 states: "The purpose of risk management is the creation and protection of value. It improves performance, encourages innovation and supports the achievement of objectives."[105]

8.5.3.5 Future-Ready Strategy (E14)

A future-ready strategy is one that is focused on generating and preserving value. It is important that EHSS/ORM professionals align their activities with organizational strategies, and where needed and appropriate, help develop and guide that strategy.

ISO 37000 principle 6.3 addresses strategy. It states: "The governing body should direct and engage with the organizational strategy, in accordance with the value generation model, to fulfil the organizational purpose."

The COSO ERM framework includes "strategy" in its title – "Enterprise Risk Management: Integrating with **Strategy** and Performance." (emphasis added). As such, strategy is addressed throughout the framework.

---

104 37000, p. 35.
105 ISO 31000:2018, p. 2.

Suggested Reading

Committee of Sponsoring Organizations of the Treadway Commission (2017). *Enterprise Risk Management – Integrating with Strategy and Performance.* Association of International Certified Professional Accountants.

Goss, T., Pascale, R.T., and Athos, A.G. (1993). The reinvention roller coaster: risking the present for a powerful future. *Harvard Business Review* 71: 97.

International Organization for Standardization (2021). *Governance of Organizations – Guidance, ISO 37000.* Geneva, Switzerland: International Organization for Standardization.

International Organization for Standardization (2018). *Occupational Health and Safety Management Systems – Requirements with Guidance for Use, ISO 45001.* Geneva, Switzerland: International Organization for Standardization.

International Organization for Standardization (2018). *Risk Management – Guidelines, ISO 31000.* Geneva, Switzerland: International Organization for Standardization.

International Organization for Standardization (2015). *Environmental Management Systems – Requirements with guidance for Use, ISO 14001.* Geneva, Switzerland: International Organization for Standardization.

Meadows, D. (2008). *Thinking in Systems.* White River Junction, VT: Chelsea Green Publishing Company.

Kaplan, R. and Mikes, A. (2012). Managing risks: a new framework, smart companies match their approach to the nature of the threats they face. *Harvard Business Review* June: 48–60.

Redinger, C.F. and Levine, S.P. (1998). Development and evaluation of the Michigan occupational health and safety management system: a universal OHSMS performance measurement tool. *American Industrial Hygiene Association Journal* 59: 572–581.

Redinger, C.F. and Levine, S.P. (2001). Evaluation of an occupational health and safety management system performance measurement tool – II: scoring methods and field study sites. *American Industrial Hygiene Association Journal* 63: 34–40.

Redinger, C.F. and Levine, S.P. (2002). Evaluation of an occupational health and safety management system performance measurement tool – III: measurement of initiation elements. *American Industrial Hygiene Association Journal* 63: 41–46.

9

Matrix in Action

CONTENTS
9.1 Risk Matrix Applications, 282
9.2 Matrix Dynamics, 283
9.2.1 Complexity and Tight Coupling, 283
9.2.2 Z-axis – Actors/Motive Force, 284
9.2.3 Y-axis – Contexts/Drivers, 285
9.2.4 X-axis – Risk Management Elements, 286
9.2.5 Gravitational Pulls, 287
9.2.6 Topographies/Ecosystems, 288
9.3 Integrate and Integration, 288
9.3.1 Integrating What?, 289
9.3.2 Templating, 290
9.3.3 Integrated Thinking, 291
9.4 Scorecards and Dashboards – Portals for Integration, 293
9.4.1 Risk Management Elements Dimension Example – Decision-Making Dashboard/Slice, 295
9.4.2 Contexts/Drivers Dimension Example – Social–Human Element Dashboard/Slice, 297
9.4.3 Actors/Motive Force Dimension Example – Enterprise/Company Element Scorecard/Slice, 299
Suggested Reading, 302

If you want to teach people a new way of thinking, do not bother trying to teach them. Instead, give them a tool, the use of which will lead to new ways of thinking.[1]

Buckminster Fuller

---

1 https://www.goodreads.com/author/quotes/44478.R_Buckminster_Fuller (accessed 16 November 2023).

Organizational Risk Management: An Integrated Framework for Environmental, Health, Safety, and Sustainability Professionals, and their C-Suites, First Edition. Charles F. Redinger.
© 2025 John Wiley & Sons, Inc. Published 2025 by John Wiley & Sons, Inc.

> *The phrase learning organization has become a handy label to talk about almost any company. The fact is, we do not know a lot about organizational learning. We do not know how to systematically intervene in the culture to create transformational learning across the organization. Transformational Learning requires something more than profound individual learning. Indeed, one of the greatest business challenges is to find some models for how a whole organization can learn.[2]*
>
> Edgar Schein
"The Anxiety of Learning"

In Chapter 1, I asserted that: The historic ways we frame and practice ORM have diminished ability to meet post-2020 challenges. To say that current ORM approaches are bankrupt may be too strong. But there is ample evidence that fresh thinking is needed, along with approaches and solutions that address changing risk profiles and the bundle of external challenges that organizations face.

Chapter 8 introduced the Risk Matrix, while Chapters 1 through 7 addressed topics, issues, tools, constructs, etc. that are associated with risk management and organizational dynamics. In this chapter, I draw on all of these chapters to examine fresh perspectives on ORM.

In Chapter 2 (§2.6), I introduced Richard Scott, a thought leader for decades in sociology and institutional theory regarding fields constructs. He points to their ability to examine "… the behavior of the objects under study … not by their internal attributes, but by their location in some physical or socially defined space. The objects, or actors, are subject to vectors of force (influences) depending on their location in the field and their relation with other actors as well as the larger structure within which these relations are embedded."[3] His insights will be helpful as we move forward in looking at the Risk Matrix in action.

The Risk Matrix represents an environment within which organizational risk management happens – it represents the relational and contextual space that creates the interactions and collective behavior, which in turn produce the organization's risk management governance, strategy, execution, and outcomes. The Risk Matrix provides a wide multifaceted lens of increased dimensionality through which to view risks. The dimensionality of the Risk Matrix is increased by making explicit the social–human as an element in the Contexts/Drivers dimension, and including the organizational field term "actors" and identifying them as the Actors and Motive Forces dimension containing four elements: – individual, team/department, enterprise, and community. Application and use of the Risk Matrix offer the possibility of addressing challenges posed by Ed Schein in "The Anxiety of Learning."

---

2 *HBR March 2002.*
3 Scott, R.W. (2014). *Institutions and Organizations – Ideas, Interests, and Identities*, 4e, 220.

While many valuable topics have been examined thus far, the points that Tracy Goss[4] and Donnella Meadows[5] add are the most significant in terms of impacting the outcome space depicted in Figure 2.1. Let us refresh on their key points here:

> Being alters action; context shapes thinking and perception. When you fundamentally alter the context, the foundation on which people construct their understanding of the world, actions are altered accordingly. Context sets the stage; being pertains to whether the actor lives the part or merely goes through the motions.[6]

> Paradigms are the source of systems. From them, from shared social agreements about the nature of reality, come system goals, and information flows, feedbacks, stocks, flows and everything else about systems.[7]

In the context of impacting ORM and the outcome space at the intersection of systems/frameworks and the human aspects (social–human capital), I have attempted to integrate their insights for use in evolving ORM generally, and specifically when we consider generative fields in Chapter 10 (Escalating Impact). This integration is:

> Awareness of risk contexts is a necessary and fundamental first step in transcending the limits of historic ORM approaches. Shifting the context to value creation and preservation, and focusing on social and human capitals and their strength/health alters being and action by changing internal states and perceptions. It provides a clearing whereby the outcome space represented in Figure 2.1 expands – and where outcomes such as increased resilience and fulfillment – for the organization and its stakeholders – can happen.

Risk context here refers to the Risk Matrix's y-axis (Contexts/Drivers). The Risk Matrix's additional dimensionality, namely with the z-axis (Actors/Motive Force), and the risk management elements (x-axis), approximates Meadows's highest leverage point that allows for the possibility of transcending historic ORM paradigms.

---

4 Chapter 2, introductory quote.
5 Chapter 1 (§1.2.3).
6 Goss, T. et al. (1993). The reinvention roller coaster: risking the present for a powerful future. *Harvard Business Review*. November-December 1993. p. 101.
7 Meadows, D. (1999). *Leverage Points: Places to Intervene in a System*, 18. Hartland, VT: The Sustainability Institute.

Early in discussing risk logics (Chapter 2), I asserted that "Organizational Risk Management (ORM) outcomes are a function of decision-making processes that happen within systems and frameworks that are driven by people and teams in an organization." The Venn diagram in that chapter (Figure 2.1) depicts the intersection of organizational systems and frameworks and social and human capital. The intersection depicts what I refer to as the outcome space where integrated thinking and decision-making occur. Circling back on this, what does Figure 2.1 look like now within the Matrix's environment, specifically in the outcome space? I will explore this with you in Chapter 10.

This chapter lays a foundation for discussing generative fields and their role in escalating impact in Chapter 10.

9.1 Risk Matrix Applications

The increased dimensionality of the Risk Field makes explicit the social–human as an ORM context/driver, including the organizational field distinction "actors," and identifying them as motive forces: individual, team/department, enterprise, and community. Its value lies providing a construct to build upon in order to navigate in the social–human era.

Buckminster Fuller points to the use of tools in teaching new ways of thinking. The Matrix is such a tool. Several applications are identified:

> *First*, at a base level, it offers a construct that can be used to help frame ORM design, decision-making generally, and, specifically, the consideration of social–human era dynamics.
>
> *Second*, it can be used to qualitatively assess existing ORM frameworks and structures. This can be done at a nominal level of "present" or "not present" for a given matrix cell, row or column. It can also be done at an ordinal/interval level with the development of scales and detailed metrics.
>
> *Third*, It can be used as a framework for revising existing ORM frameworks or developing an entirely new framework.

Embedded in these applications are (1) integrated risk decision-making, and (2) integrated action, which involves the integration of elements in each of the Matrix's three dimensions (Contexts/Drivers, Actors/Motive Force, Risk Management Elements). I envision this as meeting Edgar Schein's challenge to create a framework by which "a whole organization can learn."

For those who will consider using this Matrix with the second and third applications, I provide guidance intended to help you do so. Considering each of these within the A → B construct introduced in Chapter 1 (§1.2.2), the Risk Matrix is an intervention, represented by the arrow "→".

Use of the Risk Matrix depends on what you want to accomplish ("B"). At a minimum, useful ideas generated within the Risk Matrix are offered here. Examples follow of an enterprise scorecard and the use of two dashboards, one for the Social–Human element of the Context/Driver dimension, and one for the decision-making element of the Risk Management Element dimension.

9.2 Matrix Dynamics

> *Dynamics (pl. n)*: The branch of mechanics concerned with the motion of bodies under the action of forces; the branch of any science in which forces or changes are considered. The forces or properties which stimulate growth, development, or change within a system or process.[8]

We now turn our attention to some of the Risk Matrix's characteristics and dynamics, and their interplay in terms of the three ORM dimensions – Actors/Motive Force, Contexts/Drivers, and Risk Management Elements. The Matrix represents an environment where integrated thinking, decision-making, and actions arise and evolve; we can also call this environment a particular kind of field, a *risk field*. There are flows in the Risk Matrix such as the movement of energy, information, meanings; and, there are forces that generate and impact these flows.

9.2.1 Complexity and Tight Coupling

I have touched on the changing nature of ORM. Some of the dynamics with this can be attributed to the complexity of risk profiles, increased presence of social–human factors, and the nature of organizations themselves.

In their book, *Meltdown: Why Our Systems Fail and What We Can Do About It*, Chris Clearfield and Andres Tilcsik point to the role that complexity and tight coupling play in system breakdowns. They point to numerous examples of this dynamic ranging from Thanksgiving dinners to nuclear power plants. In a 2018 interview Tilcsik said, "In many cases, we make our systems more complex and tightly coupled for good reasons. Complex systems enable us to do all sorts of innovative things, and tight coupling often arises because we are trying to make a system leaner and more efficient. That is the paradox of progress: We often do these things for good economic reasons; but, at the same time, we are pushing ourselves into the danger zone without realizing it."[9] He also points to this as a

---
8 New Oxford American Dictionary, macOS 13.5.2.
9 Tilcsik, A. (2018). The Paradox of Progress: why our systems fail and what we can do about it. Interview conducted by Karen Christensen. *Rotman Magazine*, Spring 2018. p. 43.

phenomenon that is unavoidable in many instances, especially with an expanded network of stakeholders.

The ORM taxonomy depicted in the Risk Matrix offers a framework and tools to identify complexity and tight coupling. Tilcsik points to diagnosis as a "good first step" in navigating complex and tightly-coupled systems.[10]

The risk decision-making kernel (risk kernel) introduced in Chapter 7 (§7.5) has several uses as a tool for navigating the Risk Matrix, one of which is identifying complexity and tight coupling. This heuristic is based on five central elements of ORM: source, impact, tolerance, acceptability, and control/transfer. As suggested in §7.5, I invite you to tailor this heuristic to your needs. The point is to have this tool available for your use.

9.2.2 Z-axis – Actors/Motive Force

The Actors/Motive Force dimension and its four elements were introduced in Chapter 8 (§8.4). The four elements are: (1) individual, (2) team/department, (3) enterprise/company, and (4) community. These elements can be viewed in numerous ways in terms of the intersections with the Matrix's Risk other two dimensions (Context/Drivers and Risk Management Elements).

Figure 9.1 depicts a row of cells that represent this dimension's four elements. The Matrix contains 42 such rows. There are 3 Contexts/Drivers rows for each of the 14 Risk Management Elements (3 × 14 = 42). The row depicted in Figure 9.1 is generic for a Risk Management Element associated with the social–human

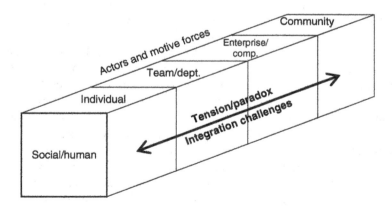

Figure 9.1 Actors/Motive Force dimension row.

---

10 Tilcsik, A. (2018). The Paradox of Progress: why our systems fail and what we can do about it. Interview conducted by Karen Christensen. *Rotman Magazine*, Spring 2018. p. 46.

element in the Contexts/Drivers dimension. Tension/paradox and integration challenges are indicated in the Figure to bring your attention to these dynamics when, for instance, using the RDM kernel (source, impact, tolerance, acceptability, and control/transfer) to help integrate the four cells depicted. For example, the row depicted in Figure 9.1 could represent the intersection of the social–human and risk assessment (E2) elements of their respective dimensions. In this case, using the RDM kernel, you can begin to identify dynamics (tension/paradox) within individual cells and between cells in the row.

Viewing individual rows in this way, allows one to begin considering decision-making issues from different viewpoints. Complexity and tight coupling have already been mentioned. In addition, different they facilitate the identification of things like tension, paradox, and other integration challenges. Especially with social–human element issues that play out in the 14 Risk Management Elements, gaps can be identified between the individual, team/department, enterprise/company, and community levels. Section 9.4.3 focuses on the enterprise/company element and its use in developing an integrated ORM auditing tool.

9.2.3 Y-axis – Contexts/Drivers

The Contexts/Drivers dimension and its three elements were introduced in Chapter 8 (§8.3). The three elements are: (1) regulatory/technical, (2) organizational, and (3) social–human. These elements can be viewed in numerous ways in terms of their intersections with the Risk Matrix's other two dimensions (Actors/Motive Force and Risk Management Elements).

Figure 9.2 depicts a column of cells that represent this dimension's three elements. The Matrix contains 62 such columns. There are 4 Actors/Motive Force columns for each of the 14 Risk Management Elements (4 × 14 = 62). As with Figure 9.1, the column depicted in Figure 9.2 is generic, and reflects tension/paradox and integration challenges.

Viewing individual columns in this way, allows one to begin considering decision-making issues from different viewpoints. This view also facilitates integration of the social–human element with the regulatory/technical and organizational elements in the Contexts/Drivers

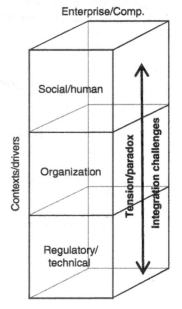

Figure 9.2 Contexts/Drivers dimension column.

dimension. There are varying levels of development and maturity (stage of evolution) of this dimension's three elements across the 14 elements in the Risk Management Elements dimension (e.g. risk assessment [E2], and operation [E12]). The term "templating" I introduced in Chapter 8 (§8.2.2) is applicable here in considering how developed/mature practices such as risk assessment at the regulatory/technical level can be used (templated) to develop practices at the social–human element level. This idea of templeting is developed further in §9.3.2. Section 9.4.2 focuses on using the social–human element as an example to develop a Contexts/Drivers dashboard.

9.2.4 X-axis – Risk Management Elements

The Risk Management Elements dimension and its 14 elements were introduced in Chapter 8 (§8.5). As with the other Risk Matrix dimensions, this dimension can be viewed in several ways, similar to that described above. The Risk Management Elements are the backbone of the Risk Matrix. These are well established in organizations at the regulatory/technical level, less so at the organizational level, and even less so at the social–human level. However, existing regulatory/technical level structures and practices can be used as templates to support development of the other two ORM dimensions.

Figure 9.3 depicts a column of cells that represents one of the elements in the Risk Management Elements dimension. I also refer this these columns as Risk

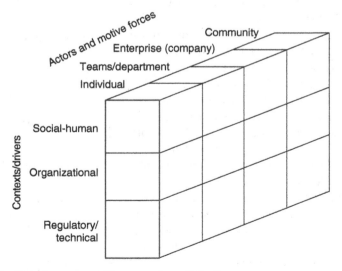

Figure 9.3 Risk Management Element column ("slice").

Matrix "slices" later in this chapter. Each column/slice contains 12 cells, 3 for the Contexts/Drivers dimension, and 4 for the Actors/Motive Force dimension ($3 \times 4 = 12$). Discussion of Figure 9.3 is expanded later in this chapter (§9.4.1).

9.2.5 Gravitational Pulls

We have all experienced gravitational pulls in our organizational lives. Call it power, energy, culture, or vibe. In considering how the Risk Matrix can be used in your organization to advance integrated thinking, decision-making, and action, it is valuable to identify these forces. While characterizing organizational context is a topic in the next chapter, a brief preview is presented here in terms of thinking about it in the context of the Matrix's three dimensions, and in terms of "topography/ecosystems."

Box 9.1 Definitions: Gravitational, Center of Gravity, Gravity

Gravitational (adj): "Relating to movement toward a center of gravity. Denoting a forceful attraction or movement toward something."[11]

Center of gravity (n): "A point from which the weight of a body or system may be considered to act. In uniform gravity it is the same as the center of mass."[12]

Gravity (n): "The force that attracts a body toward the center of the earth, or toward any other physical body having mass."

The idea of gravitational pull points to having awareness of energy and energy dynamics throughout the organization. Each organizational function, whether legal, human resources, engineering, operations, C-suite, etc. has a role. Within their roles, other dynamics can be characterized in terms of how things get done. There are "centers of gravity" that might not always be obvious, especially as dynamics are considered through the matrix dimensions. For example, with the Actors/Motive Force dimension (z-axis), there are individuals, teams, the organization itself, and the community. This is especially important when beginning to consider the social–human dimension.

---

11 New Oxford American Dictionary, Apple OS 13.5.2.
12 New Oxford American Dictionary, Apple OS 13.5.2.

9.2.6 Topographies/Ecosystems

This topic is useful when navigating within the Risk Matrix, the organization, and their interactions. In management-system-speak, this is akin to organizational context. There is texture in organizations. This distinction – topographies/ecosystems – brings attention and awareness to these interactions.

There are two primary considerations when bringing something new into existence in an organization. The first are those defining features of the organization such as culture, structure, mission, processes, etc. The second is the "intervention," defined as the actions that will bring about something new. In the simple A → B model, this is depicted by the arrow. For example, the implementation of something like an ISO management system, or COSO's ERM framework, is considered the intervention. Characterizing the organization's topography/ecosystem and the framework/system that is being used is important when considering an intervention.

There are any number of ways to characterize the Risk Matrix topography and ecosystems, whether by the three dimensions and their 21 elements (§9.2.2.4), or by the Risk Management Elements dimension's groupings – Foundational Five, Trim Tabs, or Operational Elements.

Box 9.2   Definitions: Topography and Ecosystem

Topography (n): "The arrangement of the natural and artificial features of an area. A detailed description or representation on a map of the natural and artificial features of an area."

Ecosystem (n): "A biological community of interacting organisms and their physical environment. (In general use) a complex network or interconnected systems."

9.3   Integrate and Integration

> Integrating enterprise risk management practices throughout an entity helps to accelerate growth and enhance performance.[13]
>
> *Committee of Sponsoring Organizations of the Treadway CommissionEnterprise Risk Management, Integrating Strategy and Performance*

---

[13] COSO (2017). Enterprise Risk Management, Integrating Strategy and Performance, Executive Summary. Committee of Sponsoring Organizations of the Treadway Commission, p. 1.

Integrate and integration are common terms in many arenas; with integrated thinking used in business, ESG, and risk management. The title of COSO's ERM framework includes "integrating" in its title. They are used throughout ISO standards in terms of integrating a standard's focus topic across an organization. We are familiar with integrating the E, H, S, and S of EHSS, and integrating it with business processes. A central aspect of COSO's second edition framework (2017) is evolving risk management from a siloed phenomenon to one that is integrated throughout an organization.

Whether used as a verb (integrate) or a noun (integration), these are important if not central distinctions in moving beyond historic ORM methods and practices, and to meeting social–human era challenges. The root of both integrate and integration is "to make whole."[14] I explore here what this means both conceptually and operationally in the ORM context in the social–human era. These words are commonly used conceptually, but less often operationally.

This topic is included here to highlight both the role it plays in organizational risk management and how risk management can be used to facilitate the development of integrated risk management thinking, decision-making, and action. The Risk Matrix offers a means to begin moving from conceptual to operational use. Breaking down silos, leading across boundaries, and thinking from the whole are popular notions in organizational practice. In the mix with these ideas is the integration of practices, approaches, methods, etc. In many cases, talking about integration is easier said than done. In many cases, whether it is integrating risk-based thinking with organization processes, or integrating social and human capital issues with ORM and risk decision-making, fundamental shifts in mindsets and paradigms are needed.

When present in a risk field (e.g. risk matrix), an integration/integrate context can meet the organizational learning challenge posed by Ed Schein. This context, when integrated within the Matrix's Context/Drivers and Actors/Motive Force dimensions, sets the stage for Donnella Meadows's highest leverage point, power to transcend paradigms.[15]

9.3.1 Integrating What?

There is seemingly no end to things being integrated in EHSS/ORM. Integration of the E, H, S, and S have been mentioned. Integration goals (in some cases requirements) within ISO standards and the COSO ERM framework have also been mentioned. Other examples include:

- Integrated audits (regulatory compliance, and management system conformance)
- Integrated capitals assessment (§1.2.4.3)

---

14 To refresh on definitions for integration, integrate, whole, and wholeness, revisit Box 1.2 in Chapter 1.
15 Chapter 1 (§1.2.3) and Chapter 5 (§5.6).

- Tangible v. nontangible (Figures 1.1 and 2.1)
- Integration of stakeholder perspectives
- Integration in matrix columns and rows.

9.3.1.1 Integrated Thinking → Integrated Decision-Making

A key point here is increasing awareness of ORM decision-making processes, and the role that integrated thinking plays in them. Figure 7.4 depicts an integrated risk decision-making system, by showing the Context/Drivers risk dimension elements as inputs to decision-making inputs themselves, and the decision-making process. This diagram could be expanded to show the Risk Matrix's other two dimensions and their elements as inputs as well.

9.3.1.2 Risk Field

To recapitulate, the risk field that is operationally expressed in the Risk Matrix integrates three dimensions of ORM: regulatory/technical, organizational, and social–human.

The risk field introduced in Chapter 2 (§2.6) and built out in Chapter 8 represents an integrated ORM framework.

9.3.2 Templating

Using templates is common when building or creating things. For example, templates are used in construction, in making clothing, and when creating documents in business.

As indicated above and in Chapter 8 (§8.2.2), it is helpful to characterize the use (templating) of matrix components (e.g. cells, rows, columns) in less developed or less mature components. This is especially the case with the social–human element, which in most organizations is neither mature nor well developed. There are a number of ways to embark on building out this element, one of which is to use what has been developed in other areas as a template. For instance, the tools and skills used to develop risk assessment (E2) or management of change (E4) at the regulatory/technical level can serve as a template for developing these elements at the social–human level.

Templating can serve as a first step in integrating/integration of Risk Matrix elements. In the example of risk assessment (E2) and management of change (E4), and after developing or upgrading social–human element policies and procedures, our attention can turn to integrating them with the other two Context/Drivers Elements for E2 and E3. This concept is developed later in this chapter (§9.4).

The Capitals Coalition's transformative change model of organizational decision-making (§5.6.3) – change the math, conversation, rules, and system – provides a different type of "templating" example that can shift ORM decision-making and outcomes. This includes developing new metrics (math), contexts for new conversations, and policies/procedures (rules). The multi-capital foundation (system) offers a template for building out the generative field construct discussed later in Chapter 10 (§10.2).

9.3.3 Integrated Thinking

Integrated thinking and integrated decision-making are taking place all the time but in many instances not distinguished. As touched on above, the term "integrated thinking" is used in: (1) business, (2) ESG (namely related to integrated reporting), (3) COSO's ERM, and (4) the Risk Matrix presented in Chapter 8.

9.3.3.1 Rotman School of Management – Integrative Thinking

Roger Martin and colleagues at the Rotman School of Management at the University of Toronto have pioneered integrative thinking ideas and constructs. In his 2007 book titled, *The Opposable Mind: How Successful Leaders Win Through Integrative Thinking*, Roger Martin writes:

> The leaders I have studied share at least one trait, aside from their talent for innovation and long-term business success. They have the predisposition and the capacity to hold two diametrically opposing ideas in their heads. And then, without panicking or simply settling for one alternative or the other, they're able to produce a synthesis that is superior to either opposing idea. Integrative thinking is my term for this process – or more precisely this discipline and synthesis – that is the hallmark of exceptional businesses and the people who run them.[16]

As a "working definition" for integrative thinking, he writes:

> The ability to face constructively the tension of opposing ideas, and instead of choosing one at the expense of the other, generate a creative resolution of the tension in the form of a new idea that contains elements of the opposing ideas but is superior to each.[17]

---

[16] Martin, R.L. (2007). *The Opposable Mind: How Successful Leaders Win Through Integrative Thinking*, 6. Boston, Ma: Harvard Business School Press.
[17] Martin, R.L. (2007). *The Opposable Mind: How Successful Leaders Win Through Integrative Thinking*, 15. Boston, Ma: Harvard Business School Press.

Martin offers several integrative thinking constructs and tools including a four-step decision-making model, a six dimension personal knowledge system construct, and three tools. The three tools used by integrative thinkers are: "Generative Reasoning rather than solely Declarative reasoning, Causal Modeling rather than Conventional Wisdom, and Assertive Inquiry rather than reliance on Advocacy."[18] The term generative reasoning is used to describe "creative" new solutions that arise from combining deductive, inductive, and abductive logic.

9.3.3.2 The International Integrated Reporting Council

The International Integrated Reporting Council (IIRC) was mentioned in Chapter 1 (§1.3.2) in the context of integrated reporting, which is often associated with integrated thinking. The IIRC has published numerous documents that address integrated thinking. In June 2021, they and the Sustainability Standards Accounting Board (SASB) formed the Value Reporting Foundation (VRF). The VRF subsequently merged with the International Financial Reporting Standards (IFRS) Foundation in August 2022. In 2021, the VRF published *Integrated Thinking: A Virtuous Loop, The business case for a continuous journey towards multi-capital integration*. It is described as:

> Rather than using the narrow focus on financial tools, today's best performing organizations are basing business decisions on interconnected information across multiple capitals, including natural, social and relationship, human, manufactured and intellectual, as well as financial. This is integrated thinking, or the active consideration by an organization or the relationships between various operating and functional units and the capitals that the organization uses or affects – inputs, outputs and outcomes. As such integrated thinking leads to integrated decision making and actions that consider the creation, preservation or erosion of value over the short, medium and long term.[19]

The VRF report continues, stating: "This integration delivers better decision-making, increased transparency and is forward looking. But it requires time to embed in an organization."

---

18 Martin, R. (2007). Becoming and Integrative Thinking. Rotman Management, Rotman School of Management, University of Toronto. Fall 2007, pp. 8–9.
19 Value Reporting Foundation (2021). Integrated thinking: a virtuous loop, the business case for a continuous journey towards multi-capital integration, p. 6. https://www.value reportingfoundation.org (accessed 23 January 2022). Note, the Value Reporting Foundation merged with the IFRS Foundation in 2022.

9.3.3.3 COSO Enterprise Risk Management (ERM) Framework

The COSO ERM framework was introduced in 2004, with a second edition published in 2017. The integration of ERM with organizational processes was a key emphasis of the revised framework. COSO states:

> Enterprise risk management is not a function or department. It is the culture, capabilities, and practices that organizations integrate with strategy-setting and apply when they carry out that strategy, with a purpose of managing risk in creating, preserving, and realizing value.[20]

The 2017 framework was introduced in Chapter 3 (§3.5). Figure 3.7 depicts the framework's five components as in a pair of intertwined ribbons:

> Illustrates these components and their relationship with the entity's mission, vision, and core values. The three ribbons in the diagram of Strategy and Objective-Setting, Performance, and Review and Revision represent the common processes that flow through the entity. The other two ribbons, Governance and Culture, and Information, Communication, and Reporting, represent supporting aspects of enterprise risk management. The figure further illustrates that when enterprise risk management is integrated across strategy development, business objective formulation, and implementation and performance, it can enhance value.[21]

9.4 Scorecards and Dashboards – Portals for Integration

> What gets measured gets done.
>
> *Peter Drucker*

This phrase is popular in business. It is often attributed to Peter Drucker, but many point out that what he actually said was "what gets measured gets managed." Regardless of which one is correct, they both point to the primacy of measurement. Performance measurement tools are indispensable in organizational risk management. They provide feedback on processes, programs, and systems. There are of course a lot of ways to look at this topic – there are tools, the use of

---

20 COSO (2017). Enterprise Risk Management, Integrating Strategy and Performance, Executive Summary. Committee of Sponsoring Organizations of the Treadway Commission, p. 1.
21 COSO (2017). Enterprise Risk Management, Integrating Strategy and Performance. Committee of Sponsoring Organizations of the Treadway Commission, p. 21.

them, and output reporting from their use – but more is at play here. This topic – and things like scorecards and dashboards – provide portals for seeing, transforming, and unleashing organizational capacities.

Risk Matrix dynamics and applications have been presented, followed by issues related to integration in ORM. Three examples – one for each matrix dimension – suggest ways dimension elements can be used within the applications listed above in §9.1. The elements used for this are:

- Decision-making (element of the Risk Management Elements dimension),
- Social–Human (element of the Contexts/Drivers dimension), and
- Enterprise (element of the Actor/Motive Force dimension).

The terms scorecard and dashboards are used to frame this discussion. These represent forms of reporting. We are familiar with scorecards from sports and playing games. We are familiar with dashboards from operating vehicles and the display of the important outputs related to the vehicle's performance. Both of these terms are commonly used in business. You will see that I use the term dashboard in the first two examples and scorecard for the third. This is somewhat arbitrary as either term – scorecard or dashboard – could be used for all three. The use of the term scorecard for the third example is due to the fact that this example is closely aligned with what an organization-wide (enterprise-wide) audit output (scorecard) could look like.

While this discussion is framed in terms of scorecards and dashboards, which imply reporting, the use of these matrix elements as tools is broader than performance measurement. They can be used as tools: (1) to identify gaps in or between matrix cells, rows, or columns; (2) in root-cause analysis when there are breakdowns; (3) to facilitate integration in or with matrix cells, rows, or columns; and, (4) integrated decision-making as depicted in Figure 7.3 and expanded above (§9.3.1.1). The word "slice" is used in the examples to reflect wider contexts than just measurement.

Seven postulates around which this work is organized are offered in Chapter 1. Two of these are: (1) a social–human era is evolving; and, (2) fresh ORM frameworks, orientations, and perspectives are needed to meet new challenges. The Risk Matrix represents a "fresh framework" that offers a way to integrate social–human era dynamics into an organization's risk management structures approach. The social–human element of the Contexts/Drivers dimension is used as an example because it addresses how organizations can begin to expand ORM context and meet social–human era challenges.

The three examples point to multifaceted uses for which Risk Matrix elements can be used. I suggest they offer solutions, or at least openings to solutions, to the challenges Ed Schein poses in the opening quotation from "The Anxiety of Learning." These examples are central to creating new clearings and capacities that increase an organization's ability to generate and preserve value and increase resilience.

9.4.1 Risk Management Elements Dimension Example – Decision-Making Dashboard/Slice

Decision-making is an element of the Risk Management Elements dimension. Figure 9.4 shows 12 matrix cells contained within this element, each of which falls within an intersection of the Actors/Motive Force and Contexts/Drivers risk dimensions. As there may be a need to examine the dynamics for any of the 14 Risk Management Elements, this 12-cell dashboard/slice provides a way to distinguish dynamics within a specific cell (as reflected in Figure 8.4a, b), or within a specific Actors/Motive Force row (as reflected in Figure 9.1), or within a Contexts/Drivers column (as reflected in Figure 9.2).

By seeing Risk Matrix dynamics at the granular level, you can begin to observe gravitational pulls within each of these (cell, row, column). The risk kernel (Figure 7.6) provides a tool (construct) for performing focused assessments; its five components (source, impact, tolerance, acceptability, control/transfer) can be used in aggregate to perform – as or when needed – a surgical ORM intervention. Individual kernel elements, can be used, for instance "[risk] source," to help guide the framing of risk communication related to the social–human element.

This dashboard/slice is at the heart of ORM integration. While there are numerous ways integration can be done within the Risk Matrix, it is here, with the 14 Risk Management Elements, that one should start.

Templating can serve as a first step in integrating/integration of Risk Management Elements. This can be done by using policies and procedures that are already established at the regulatory/technical level as templates for the organizational and social–human levels. In instances where there may not be existing policies or procedures, for instance, with decision-making, templates can be used from other sources.

With this dashboard/slice, you can also begin to see interconnections, such as communication channels, impact-dependency pathways, and decision-making feedback loops. This view can be useful, for instance, in beginning to drill down into areas found in an audit to be needing improvement. Flows were mentioned above in §9.2. Additional examples of flows are decision-making currency, including materiality, value and purpose. Each of these can affect those within the Actors/Motive Force dimension very differently. For example, the tradeoffs between those within the community, enterprise/company, team/department and individual elements can be very different.

This example, related to the third application identified in §9.1, provides a starting point for developing or evolving an integrated Risk Management System (RMS)

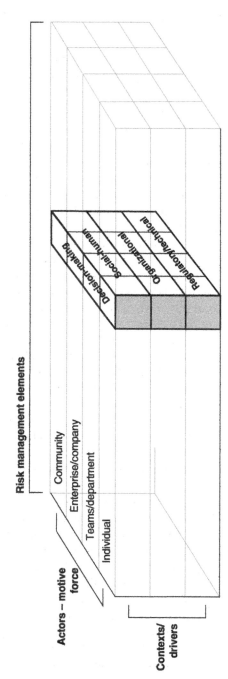

Figure 9.4 Risk Management Elements dimension "slice" example.

Table 9.1 Risk Management Elements dashboard example.

	Individual	Team/department	Enterprise/company	Community
Social–human				
Organizational				
Regulatory/technical				

framework in your organization. Element-by-element, you would work through all of the 14 Risk Management Elements.

A two-dimensional view of the 12 cells depicted in Figure 9.4 is presented in Table 9.1. The risk decision-making kernel's five elements, and the seven risk awareness elements (§5.4, and Box 5.2) can be used as tools to build-out cells, cell-by-cell.

9.4.2 Contexts/Drivers Dimension Example – Social–Human Element Dashboard/Slice

Social–human is an element of the Context/Drivers dimension. Figure 9.5 shows 56 matrix cells contained within this element, each of which falls within the intersection of the Actors/Motive Force and the 14 Risk Management Elements. As with the previous example (§9.4.1), there are any number of ways this dashboard/slice can be used to examine specific dynamics.

Figure 9.1 provides a generic example of the four cells at the intersection of risk management, and Actors/Motive Force elements. In the description offered for this figure, tension/paradox issues are identified, as well as integration challenges. If there is interest, for instance, in considering risk assessment (E2) integration within the four Actors/Motive Forces elements, in the social–human element in these four cells can be singled out as depicted in Figure 9.1. The risk decision-making kernel's five elements, or the seven risk awareness elements (§5.6, and table no. 5.3) can be used as tools to facilitate this integration.

Another way to use this dashboard/slice is to consider a specific Actors/Motive Force element. For instance, focusing on team/department, the 14 cells contained in this slice can be used to develop training programs to build social-human-related skills across the 14 Risk Management Element cells. In a similar fashion, focusing on the community element, the 14 cells can provide a foundation for understanding community issues, engaging with them, and responding as needed.

A two-dimensional view of the 56 cells depicted in Figure 9.5 is presented in Table 9.2. The risk decision-making kernel's five elements, and the seven risk awareness elements (§5.4, and Box 5.2) can be used as tools to build out cells, cell-by-cell. This can be used for any of the three elements in the Context/Drivers dimension. Of particular interest is using this to help build out the social–human

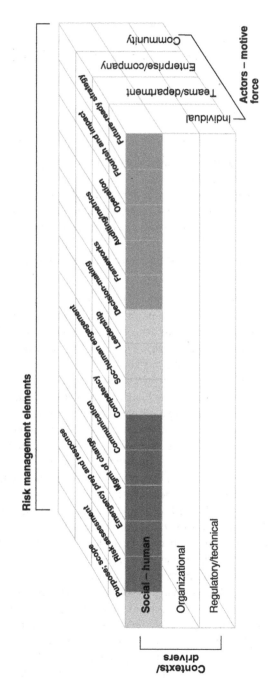

Figure 9.5 Contexts/Drivers dimension "slice" example.

Table 9.2 Contexts/Drivers dashboard example.

	Individual	Team/ department	Enterprise/ company	Community
Purpose and scope (E1)				
Risk assessment (E2)				
Emergency preparedness and response (E3)				
Management of change (E4)				
Communication: systems and practices (E5)				
Competency and capabilities (E6)				
Social–human engagement (E7)				
Leadership (E8)				
Decision-making (E9)				
Frameworks (E10)				
Auditing/metrics (E11)				
Operation (E12)				
Escalating impact (E13)				
Future-ready strategy (E14)				

element as it is more-likely-than-not the least mature, but in in many instances, the most pressing in the emerging social–human era. Ideas offered in Chapter 4 (Conformity Assessment), particularly ones related to levels of measurement (§4.2), are helpful in using this dashboard. For instance, a start would be simply to make the nominal level assessment (yes/no, present/not present) for each cell. With more mature or evolved cells, ordinal or interval scales could be developed.

This example is focused on the social–human element in the Contexts/Drivers risk dimension. The 56 cell dashboard can also be used in making stakeholder assessments. That is, to assess the status of the 14 Risk Management Elements – by individual, team/department, enterprise/company, community – for specific stakeholders.

9.4.3 Actors/Motive Force Dimension Example – Enterprise/ Company Element Scorecard/Slice

Enterprise/company is an element of the Actors/Motive Force dimension. Figure 9.6 shows 42 matrix cells contained within this element, each of which falls within the intersection of the Contexts/Drivers elements and the 14 Risk

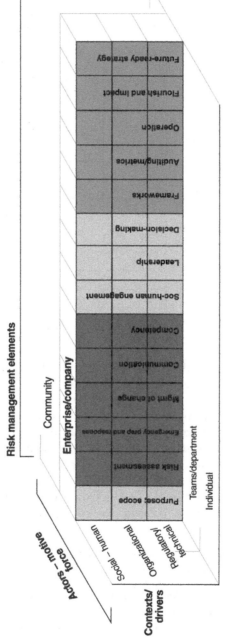

Figure 9.6 Actors/Motive Force dimension "slice" example.

9.4 Scorecards and Dashboards – Portals for Integration | 301

Management Elements. As with the previous examples (§9.4.1, 9.4.2), there are any number of ways this dashboard/slice can be used to examine specific dynamics.

The 42 cells depicted – especially those in the regulatory/technical and organizational element rows – reflect the foci in traditional audits, both of regulatory compliance and management system conformance. To build out and have an integrated ORM auditing tool, use templates (§9.3.2, and §8.2.2).

As in the example of the Contexts/Drivers risk dimension example (§9.4.2), regardless of whether you would consider it as a scorecard or dashboard, it can be used in training and development at the level of individual and teams/departments. It includes all three Contexts/Drivers elements which makes it more robust than the more narrow application in §9.4.2.

A two-dimensional view of the 42 cells depicted in Figure 9.6 is presented in Table 9.3. The risk decision-making kernel's five elements, and the seven risk awareness elements (§5.4, and Box 5.2) can be used as tools to build out cells, cell-by-cell. This process can be used for any of the four Actors/Motive Force elements. Of particular interest is using it to help build out the community element, and an integrated approach to interacting with external stakeholders

Table 9.3 Actors/Motive Force dimension dashboard example.

	Regulatory/Technical	Organizational	Social–human
Purpose and scope (E1)			
Risk assessment (E2)			
Emergency preparedness and response (E3)			
Management of change (E4)			
Communication: systems and practices (E5)			
Competency and capabilities (E6)			
Social–human engagement (E7)			
Leadership (E8)			
Decision-making (E9)			
Frameworks (E10)			
Auditing/metrics (E11)			
Operation (E12)			
Escalating impact (E13)			
Future-ready strategy (E14)			

as addressed, for instance, in ISO 37000:2021. As mentioned in §9.4.2, ideas offered in Chapter 4 (Conformity Assessment), particularly ones related to levels of measurement (§4.2), are helpful in using this dashboard. For instance, a start would be simply to make the nominal level assessment (yes/no, present/not present) for each cell. With more developed or evolved cells, ordinal or interval scales could be developed.

Suggested Reading

Clearfield, C. and Tilcsik, A. (2019). *Meltdown: Why Our Systems Fail and What We Can Do About It*. Atlantic Books.

Committee of Sponsoring Organizations of the Treadway Commission (2017). *Enterprise Risk Management – Integrating with Strategy and Performance*. Association of International Certified Professional Accountants.

Fritz, R. (1999). *The Path of Least Resistance for Managers – Designing Organizations to Succeed*. San Francisco, CA: Berrett-Koehler Publishers.

Goss, T. et al. (1993). The reinvention roller coaster: risking the present for a powerful future. *Harvard Business Review* 71: 97.

International Organization for Standardization (2021). *Governance of Organizations – Guidance, ISO 37000*. Geneva, Switzerland: International Organization for Standardization.

Kim, D.H. (1999). *Introduction to Systems Thinking*. Waltham, Massachusetts: Pegasus Communications.

Kotter, J. and Rathgeber, H. (2005). *Our Iceberg is Melting – Changing and Succeeding Under any Conditions*. New York: St. Martins Press.

Martin, R.L. (2007). *The Opposable Mind: How Successful Leaders Win Through Integrative Thinking*, 6. Ma. p: Harvard Business School Press, Boston.

Meadows, D. (2008). *Thinking in Systems*. White River Junction, VT: Chelsea Green Publishing Company.

Scott, W.R. (2014). *Institutions and organizations, ideas, interests and identities*, 4e. Thousand Oaks, California: Sage Publications, Inc.

Senge, P., Hamilton, H., and Kania, J. (2015). The dawn of system leadership. *Stanford Social Innovation Review* 13: 27.

Vogel, B. and Bruch, H. (2012). Organizational energy. In: *Oxford Handbook of Positive Organizational Scholarship* (ed. K. Cameron and G. Spreitzer), 691–702. New York, NY: Oxford University Press.

10

Escalate Impact

CONTENTS

A → B, 306
10.1 Transcending Paradigms, 306
    10.1.1 Evolutions, 307
    10.1.2 New Clearings, 307
    10.1.3 Shifts, 308
10.2 Generative Fields, 308
10.3 Leverage – Creating Generative Fields, 312
    10.3.1 Value Generation and Health, 313
    10.3.2 Interiority, 314
    10.3.3 Generative Field Engine – Social–Human Engagement (E7) and Leadership (E8), 316
    10.3.4 Pedagogy of Evaluation, 319
10.4 Carriers of a Field, 322
    10.4.1 Portal to Future-Ready, 323
    10.4.2 The Table and Its Seats, 324
    10.4.3 Trim Tab, 324
Suggested Reading, 325

Organizational Risk Management: An Integrated Framework for Environmental, Health, Safety, and Sustainability Professionals, and their C-Suites, First Edition. Charles F. Redinger.
© 2025 John Wiley & Sons, Inc. Published 2025 by John Wiley & Sons, Inc.

If we get health right, everything else will follow. Engagement is the key to the effective communication, understanding and management of hazards and risk.[1]

Dr. Alistair Fraser,
Vice President, Health at Royal Dutch Shell

Generative social fields – are where families, schools, businesses, governments, and the larger society in which we live can create relational interactions that support the nurturance of compassion, connection, curiosity, and well-being. Within such generative fields, a sense of belonging, meaning and effective action emerges.[2]

IPNB Pre-Conference Description

Dawn Awakening – there is a broad, though still largely unarticulated, hunger for processes of real change.[3]

Peter Senge, Hal Hamilton, and John Kania
"The Dawn of System Leadership"

Let us refresh on the opening paragraph in Chapter 1: "Fundamentally this book addresses how to create new clearings and capacities that increase ability to generate and preserve value, resilience, and fulfillment for an organization and its stakeholders. The focus is on using organizational risk management (ORM) as a platform, and environmental, health, safety, and sustainability (EHSS) management as a vehicle. Integrated thinking and decision-making are the fuel, generating organizational energy. This happens within a generative risk field. The possibility offered here is the transformation of ORM, and in turn, the transformation of organizational governance and purpose."

The Risk Matrix represents a new clearing, with a focus on the integrated thinking and integrated decision-making that are key to increasing ability to generate and preserve value, resilience, and fulfillment for an organization and its stakeholders. Logics for these are woven through the chapters, particularly in the previous chapter (§9) where I drilled down into various aspects of the Matrix, and Chapter 5 (§5.6) where I introduced aspects of shifting mindsets and paradigms as articulated in the work of Tracy Goss, Donnella Meadows, Peter Senge, and others cited throughout this book.

Considerable attention has been given to Figure 2.1 and what has been referred to as the ORM outcome space. We now look beyond ORM outcomes by turning our

---

[1] Warner Lecture, 2015 IOHA/BOHS Conference London, England; April 29, 2015. "Health – Who Cares?"
[2] April 26, 2019. "2019 IPNB Pre-Conference Workshop Awareness-Based Systems Leadership: Cultivating Generative Social Fields and Systems Change" Marina Del Rey, CA.
[3] Senge, P. et al. (2015). The Dawn of System Leadership. *Stanford Social Innovation Review*, Winter 2015, p. 33.

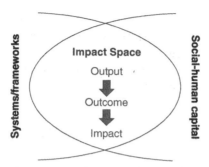

Figure 10.1 New clearing – impact space.

attention to their impacts. Figure 10.1 depicts the ORM impact space resulting from the intersection of a systems/frameworks orientation and the social–human capital Context/Driver.

The terms output, outcome, and impact are used in a range of settings, from simple, noncomplex decision-making and process/program interventions, up to complex enterprise-wide decision-making and system interventions. The 14 elements in the Risk Management Elements dimension depicted in the Risk Matrix have individual outputs as discussed in §9.4.1. The enterprise scorecard as discussed in §9.4.3 depicts the 14 elements as an aggregated output of the enterprise/company element in the Actors/Motive Force. Further, the inclusion of the other three Actors/Motive Force dimension elements yields the full Risk Matrix as depicted in Figure 8.3.

Regardless of level, the question here is about how impact can be escalated, in terms of their frequency, their magnitude, or both.

Box 10.1   Definition, Escalate, Impact[4]

Escalate (v): Increasing rapidly. Become or cause to become more intense or serious.
Impact (n): The action or one object coming forcibly into contact with another. The effect or influence of one person, thing, or action on another.

As discussed in Chapters 2 (§2.6) and 8 (§8.1), the Risk Matrix represents the first step in operationalization of an organizational risk field (Box 2.4). Ways to begin using the Risk Matrix as an integration and development tool were suggested in Chapter 9. While it has numerous uses, its primary use is to support (1) integrated risk decision-making, and (2) integrated action, which involves

---

[4] New Oxford American Dictionary, Apple OS 13.5.2.

integration of elements in each of the matrix's three dimensions (Contexts/Drivers, Actors/Motive Force, Risk Management Elements). The Matrix is based on an integrated ORM model, framed as a risk field. Its expanded perspective provides integrated inputs and processes for the overall risk decision-making (RDM) process as depicted in Figure 7.4.

A → B

In considering ORM impact, EHSS impact, or, more generally organizational impact, let us revisit the simple causal model A → B. Although a simple model, it offers a starting point to frame the various goals and objectives ("B") in EHSS/ORM. Setting goals and objectives, developing a plan to achieve them, faithfully implementing those plans, evaluating the results, and providing feedback can be complex. There is a range of time horizons (e.g. short-term, mid-term, and long-term) in play. There is a range of B's among the four elements (individual, teams/departments, enterprise/company, and community) within in the Actors/Motive Force dimension. It is important to distinguish among these and consider how they have evolved or changed along with the increased complexity and tight-coupling in risk profiles. I suspect that some of the topics raised in this book, namely related to value and purpose, will precipitate consideration of new B's. And, if this is the case, a question arises as to how might they be achieved (e.g. the →), and how quickly (e.g. time horizon, and rate)?

In Chapter 9, I framed the Risk Matrix as an intervention (→). What I have presented thus far, particularly the outcome/impact space depicted in Figures 2.1 and 10.1, has focused on how the Risk Matrix functions at the intersections depicted in these figures. This chapter continues to focus on these intersections and considers how they can evolve to escalate impact. The generative field construct is offered as a means to this end.

10.1 Transcending Paradigms

> **Box 10.2 Definitions, Transcend, and Paradigm**
>
> *Transcend* (v): "Be or go beyond the range or limits of (something abstract, typically a conceptual field or division. Surpass."
> *Paradigm* (n): "A typical example or pattern of something; a model. A worldview underlying the theories and methodology of a particular scientific subject."

10.1 Transcending Paradigms

In Chapter 1, I introduced you to the work of Donnella Meadows. In her inquiry into how to change systems and the leverage points to do so, she identified the notion of transcending the paradigm within which the system under consideration exists. Meadows frames this as "The power to transcend paradigms."[5]

Transcending points to the "why" question touched on earlier as it related to double-loop learning (§5.6.2), and the notion of context introduced in the work of Tracy Goss et al. Contexts do shift, and there can be, and often are, secondary shifts. Much like the outputs, outcomes and impacts from any intervention, this notion of transcending can be framed in numerous ways, such as change, evolution, or, at the deeper levels of double-loop learning and the "below the waterline" factors in the system dynamics iceberg (§2.5.2). In this book, I have been using the term "shift," which can be synonymous with transcending.

10.1.1 Evolutions

EHSS/ORM professionals are not strangers to transcending paradigms. Our familiarity with doing so is reflected in the five evolution sequences presented in Chapter 2 (§2.1.1). For example, when formal management systems began to proliferate in the 1990s (e.g. ISO 90001, 14001), the shift in orientation from programmatic and compliance to systems, represents an obvious paradigm shift in the frameworks but also in internal states (§5.3). We see this phenomenon currently with the advent of the capitals approach in organizational decision-making, and focus on stakeholders, value, and purpose.

With the Risk Matrix, we also see the phenomenon of transcending paradigms in the progression of the Contexts/Drivers dimension's three elements – regulatory/technical, organizational, and social–human.

Transcending paradigms can be viewed in terms of a continuum, as depicted in Figure 2.2, which show an EHSS/ORM evolution from compliance, to performance, to impact-driven orientations. Recall from Chapter 2 (§2.1.1.1) the decision-making characteristics (e.g. reactive, proactive, and generative) associated with each evolutionary stage. Observe and think about how your decision-making processes have evolved in these phases.

10.1.2 New Clearings[6]

A question worth considering is whether the result of the transcendence (the new paradigm) is fortuitous or the result of a plan. It is likely a bit of both. With large things (e.g., enterprise-wide interventions, such as implementing COSO's ERM

---

5 Meadows, D. (1999). *Leverage Points: Places to Intervene in a System*, 3. Hartland, VT: The Sustainability Institute.
6 Clearing, the state and context where frameworks, structures, policies, procedures, and culture arise.

framework) there is of course planning. In a sense, this points to problem-solving 101: identify the problem, consider options on how to solve it, pick an option, and implement it. And of course, from a systems perspective, measure the outputs, outcomes, and impacts, and, through feedback loops/channels, fiddle with the selected option as needed. This question points to "B" and to the complexities of achieving it.

Organizational shifts in focus to those of value generation (and of course protection/preservation), capitals thinking, and purpose are also creating new clearings and opportunities. Awareness of the social–human element's role in these shifts is increasing. The multidimensional risk field construct and the Risk Matrix, used along with the processes offered in this book, are powerful tools to bring about the possibility of real change by seeing new clearings, characterizing them, and successfully navigating within them.

10.1.3 Shifts

It is well-known how difficult it can be to impact complex systems (e.g. large, formal organizations). It is the nature of systems to seek equilibrium wherein their state, motion and internal energy tend to remain constant over time. Donnella Meadows's inquiry (§1.3) on leverage points for intervening in systems offers insights on how change can be realized. As I have highlighted numerous times, one of the highest leverage points is the shifting of mindsets or paradigms within a system. This is the topic in Chapter 5 (§5.6).

When beginning to look at escalating impact, and the generative field construct, there is value in looking at the kinds of "shifts" needed to create a generative field (§10.3). Shifts in the outcome/impact space depicted in Figures 2.1 and 10.1 involve both systems/frameworks and the seven risk awareness elements (§5.4), including internal states (§5.4.2).

10.2 Generative Fields

Risk field has been discussed at various points in this book. The framework evolution in Chapter 2 (§2.1.1.2) shows the field at the end of the evolutionary sequence: process → program → system → field. A brief field primer is provided in Chapter 2 (§2.6), and the operationalization of the risk field construct is the focus of Chapter 8 with the presentation of the Risk Matrix.

While the field concept is well developed in the social sciences, in particular in sociology and institutional theory, and in Kurt Lewin's social field theories, generative fields is a relatively new concept. Also in Chapter 2 (§2.5.3) was a reference to a 2018 event that focused on an evolving body-of-work called "Generative Social Fields."[7] A follow-up event in 2019 continued the evolution of this work. Complementary with to Daniel Kahneman's desire to develop a richer and more precise language for decision science, the conveners of these events have been trailblazing the development of a lexicon for investigating generative social fields (GSF). A definition for GSF that emerged from their work is included at the beginning of this chapter. In 2019, I began applying GSF concepts to my ORM work.

To help us as we move forward, let us refresh on some of the definitions offered in Chapter 2 (§2.6).

> **Box 10.3   Definition: Field, Organizational Field, Organizational Risk Field, Generative Organizational Risk Field**
>
> *Field (noun)*: "An area of open land, especially one planted with crops or pasture, typically bound by hedges and fences. A space or range within which objects are visible from a particular viewpoint or through a piece of apparatus. The region in which a particular condition prevails, especially one in which a force or influence is effective regardless of the presence or absence of a material medium."[8]
>
> *Organizational field*: "Those organizations that, in the aggregate, constitute an area of institutional life: key suppliers, resource and product consumers, regulatory agencies, and other organizations that produce similar services or products."[9] // "a community of organizations that partakes of a common meaning system and whose participants interact more frequently and fatefully with one another than with actors outside of the field."[10]
>
> *Organizational risk field*: The relational and contextualized space that creates the interactions and collective behavior, which in turn produces the organization's risk management governance, strategy, execution, and outcomes.[11]
>
> *Generative organizational risk field*: An organizational risk field is generative when there is collective energy and leadership creating and preserving value that includes social, human, and natural capital.

---

7 Generative Social Fields Conference, at the Garrison Institute, Garrison, New York, 1–3 October 2018.
8 Apple OS dictionary. New Oxford American Dictionary.
9 DiMaggio and Powell (1983), p. 148.
10 Scott, R.W. (1994), pp. 207–208.
11 Redinger, C. (2019). Organizational Risk as a Generative Social Field – New Promise for Increasing Risk Awareness, Resilience, and Improving Decision-Making. Poster presented at the Society for Risk Analysis Annual Meeting, 9–11 December 2019.

In their pioneering work on generative fields, the initial characteristics identified by Peter Senge, Otto Scharmer, and Mette Boell include:

- a field where phenomena consistently and intentionally arise that are beneficial within a larger context – i.e., contribute to well-being of a larger system;
- a field where learning and collective creativity – realizing new outcomes and building new capacities – takes place;
- a field where the boundary among individuals (self-other) collapses;
- a field where the boundary between what is (current) and what is possible (emerging future) opens up.[12]

Numerous speakers at the 2018 event offered and discussed definitions for GSFs. Two that I share here are by Mette Boell who has been championing this concept with Otto Scharmer, Peter Senge, and others. And second, by Diana Walsh, former president of Wellesley College (1993–2007).

> Boell: The way I have come to think about social fields is that they are an effect of the living. I think fields are constituted by the web of interconnectedness of all living beings over evolutionary time and we can cultivate our sensitivity to this field in different ways. This field is a "quasi" social field, or a potential social field, because it requires an orientation towards the other – fellow human or other species – to *actualize* the potential of the social field. When the field becomes social, there's always a relational space opening. Some social encounters are brief, without much awareness or intensity like the passing of strangers on the street while our minds are occupied elsewhere. When passing strangers on the street and noticing them, there's an enhanced density or intensity of the social field. And when we engage with each other in relational spaces we can do so with increasing intentionality and willingness to connect. When this happen, a social field can become generative. *By generative I mean self-reproducing in intensity of the connection.* In physics, we talk about how fields (other non-social fields) can propagate – and I think the same is true for generative social fields, where *there's an increase in intensity or density and a propagating phenomenon.* To generate means *"to reproduce or make more"*, it's a term widely used in biology – for example we can talk about *natural forces generating a particular effect.* The same seems true for social fields.[13]

---

12 Senge, Peter, Scharmer, Otto, Boell, Mette (2015). Towards a Lexicon for Investigating Generative Social Fields. Academy for Contemplative and Ethical Leadership. Reprinted in Garrison Institute event (2018) document, p. 15.
13 Mette Miriam Boell, Garrison Institute, program brochure, 1–3 October 2018, p. 5. Emphasis added.

Walsh: The concept of a "generative social field" *highlights the element of relationship*, the moments in which we, as members of *a social group*, begin to *sense and feel the power of an internal force*, an "awakened mind," pulling us into awareness of our own deeper nature, coaxing us out of our limitations, our confusion, our finitude, our fear, *bringing us into relationship with others and something larger than ourselves.*[14]

I have highlighted key ideas that Boell and Walsh raise that identify generative field characteristics:

- "... self-reproducing in intensity of the connection."
- "... there's an increase in intensity or density and a propagating phenomenon."
- "... to reproduce or make more."
- "... natural forces generating a particular effect."
- "... highlights the element of relationship."
- "... sense and feel the power of an internal force."
- "... bringing us into relationship with others and something larger than ourselves."

Think about the extent to which these characteristics are present generally in your organization, and specifically in your EHSS/ORM activities. They point to common aspirations I have seen in my work, such as increased engagement, motivated people and teams, alignment on common purpose, and focus on impacting people and natural resources. The Risk Matrix offers a multidimensional environment for characteristics such as these to flourish. As we move on to considering creating generative fields, think back to the nuggets of wisdom Boell and Walsh have provided.

The development of the risk field and Matrix and their applications (§8 and §9) have been influenced by these GSF events and research I have conducted on fields. Numerous reasons are offered on why academics and practitioners have explored and developed this area. A common reason is that field constructs and their operationalization provide access to impacting complex systems.

Impacting complex systems[15] is easier said than done. Regulatory compliance is a historic approach. Over time as business and organizational science have evolved, along with their methods and tools, they have produced such innovations

---

14 Diana Chapman Walsh, Garrison Institute, program brochure, 1–3 October 2018, p. 6. Emphasis added.
15 A Complex System is composed of a large number of parts that interact with each other. The interaction between the parts leads to the emergent collective behavior, which in turn influences the parts. Natural Capital Coalition (2015). Learning from a complex systems perspective. p. 6. https://rolandkupers.com/wp-content/uploads/2013/06/019750_NCC_White_Paper__Edinburgh.pdf (accessed 18 February 2022).

as systems thinking, Total Quality Management, and Six Sigma. A multidimensional risk field construct and associated operational matrix have been presented here as a fresh approach to the complexity of impacting ORM outcomes. In recent decades the risks organizations face have become more complex, more tightly coupled, and more frequent.

The aspiration to improve health and well-being at the human, community, environmental and global levels is universal, and yet the capacity to do so is untapped in most systems. The generative field construct is a way to unleash this vital capacity.

10.3   Leverage – Creating Generative Fields

As we have seen, the generative field construct is relatively new. Researchers who have been exploring social fields as a means to learn more about how to impact complex systems have begun to develop a lexicon and the concepts to characterize this construct. Mette Boell's and Diana Walsh's identification of generative field characteristics enables us to ask: "what is/are the environment and conditions(s) from which these characteristics arise?"

The outcome/impact space in Figures 2.1 and 10.1 offers a key, to this question. That is, it is the space within which systems/frameworks and people who design and operate within them interact, and from which outcomes and impacts originate. A primary orienting driver is a purpose that is focused on value generation. Organizational purpose was introduced in Chapter 2 (§2.1.2). Purpose is the primary governance principle in ISO 37000:2021. The standard states:

> The pursuit of purpose is at the centre of all organizations and of primary importance for the governance of organizations. Therefore, this principle is the primary consideration for governance and the central point of all the other principles in this document. All other principles are to be read in the context of the application of this principle.[16]

Building on Figures 2.1 and 10.1, and building on the primary governance principle in ISO 37001:2021, Figure 10.2 depicts a purpose-based, generative field construct for our consideration. Dynamics in this space include risk field/matrix → integrated decision-making → impact.

In an organization, purpose is typically created or defined at the C-suite or board levels; ISO 37000:2021 assigns responsibility for this to the organization's governing body. The Risk Matrix includes purpose as one (E1) of the 14 elements in the Risk Management Elements. Within the Matrix, consideration is given to purpose in terms of the four elements in the Actors/Motive Force dimension – individual, team/department, enterprise/company, and community; and the three Contexts/Drivers

---

16 ISO 37000:2021: *Governance of Organizations – Guidance*, p. 11. Geneva, Switzerland.

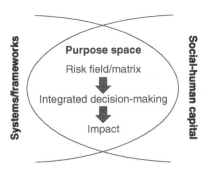

Figure 10.2 Purpose-based generative field construct.

elements – regulatory/technical, organizational, and social–human. Understanding purpose at all of the levels identified is obviously important. Purpose in each of the four Actors/Motive Force levels has unique Contexts/Drivers which need to be determined. ISO 37000:2021 states: "... organizational purpose should define the organization's intentions towards the natural environment, society and the organization's stakeholders. The governing body should also ensure that an associated set of organizational values is clearly defined."[17] As a place to start, I suggest the following definition of purpose: value generation focused on human, social, and natural capitals, and health and well-being within those capitals.

10.3.1 Value Generation and Health

Value generation and protection/preservation are primary aspects of ORM and organizational governance. In concert with health and engagement, they are also the primary components in creating generative fields. This is reflected in this chapter's opening statement by Alistair Fraser (Vice President, Health, Royal Dutch Shell) where he refers to the primacy of health as a driver for engagement. Many aspects of the generative field offered by Boell and Walsh are present at Shell.[18]

Value generation is central to what EHSS/ORM professionals do. This is typically framed, or thought of, as value protection. Our systems, training, orientations, etc. all revolve around value protection. While value protection is of course critical, much as the Foundational Five elements of the Risk Matrix are (§8.5.1), I suggest that, as with evolution and integration within the Risk Matrix (e.g. Foundational Five and Trim Tabs), EHSS/ORM professionals can "flip the script," so to speak, to view themselves as value generators as well as value protectors/preservers.

---

17 ISO 37000:2021: *Governance of Organizations – Guidance*, p. 13. Geneva, Switzerland.
18 I base this statement on Dr. Fraser's presentation at the 2015 IOHA/BOHS conference, conversations with colleagues who have worked at Shell, and Arie de Geus's book, *The Living Company* (2002).

10.3.2 Interiority

Interiority is a central concept in the generative field construct as well as in other phenomena essential to EHSS/ORM. These are engagement and participation, which are historically directed to "workers," and "employees." As related in Chapter 8 (§8.5.2.2), the inclusion of "participation" (and consultation) was a significant advancement in formal occupational, health, and safety management systems.

Interiority points to deeper and wider phenomena. The historic notions of engagement are focused on individuals, typically at the level of production or operations, colloquially at the "floor level," or "in the field." Here, interiority points to all who are part of a system, from the bottom to the top of organizational charts. This impacted the development of the Risk Matrix, particularly the Actors/Motive Force dimension. In historical single-dimensional ORM constructs, the focus participation, engagement, interiority, etc. is on the individual. The Risk Matrix expands this concept to include team/department, enterprise/company, and community. Purpose and value generation are the common denominators that link the four Actors/Motive Force dimension elements. Figure 10.2 depicts the role of integrated decision-making – really its precursor, integrated thinking – in distinguishing, navigating, and leveraging this link.

The relationship between fields and systems is a function of their degree of interiority.[19] The distinction between fields and systems is articulated by Senge et al.:

> In the physical sciences, fields are extensive forces that operate over space and time (like gravity or electromagnetism) and create contexts wherein particular living or non-living systems operate. In academic fields like system dynamics, social systems are conceived as a structure of relationships connecting individuals, groups, organizations, and larger systems in ways that give rise to collective behaviors and outcomes. Using the distinctions of 'first-person, second-person, and third-person,' we may say that, "At the moment we step inside a social system – that is, at the moment we begin to inquire into its interiority by 'turning the camera' around from the third-person to the first-person view – we switch the perspective from the social system to the social field.[20]

---

19 Garrison Institute document, p. 15. This distinction is referenced to, Scharmer, C.O. "The Blind Spot: Uncovering the Grammer of the Social Field," http://www.huff-ingtonpost.com/otto-scharmer/uncovering-the-grammar-of-the-socialfield_b_7524910.html, June 2015 and Theory U, 2006.
20 Senge, Peter, Scharmer, Otto, Boell, Mette (2015). Towards a Lexicon for Investigating Generative Social Fields. Academy for Contemplative and Ethical Leadership. Reprinted in Garrison Institute event (2018) document, p. 15.

10.3 Leverage – Creating Generative Fields

In his Theory U work, Scharmer points to the importance of paying attention to and increasing awareness of blind spots and "interior conditions" both individually and collectively.[21] In his Theory U writings and in workshop presentations, he often quotes Bill O'Brien, late CEO of Hanover Insurance: "The success of an intervention depends on the interior condition of the intervener."[22]

In the pre-event reader for the 2018 Garrison Institute event, each of the conveners offered insights into the generative field construct, some of which have already been shared here. Scharmer offered insights on "How to Cultivate the Social Field." In the reader and during the event, he referenced his foundational Theory U work, *The Essentials of Theory U* which reinforces several distinctions presented in previous chapters of this book, particularly internal state, awareness, perspective, language, attention, and energy. *(Emphasis included in original document.)* He writes:

> The essence of leadership is to become aware of our blind spot (these interior conditions) and then **to shift the inner place from which we operate** as required by the situations we face. This means that our job as leaders and change makers is **to cultivate the soil of the social field.** The *social field* consists of the relationships among individuals, groups, and systems that give rise to patterns of thinking, conversing, and organizing, and which then produce practical results.
>
> Social fields are like social systems – but they are seen from within, from their interior condition. To shift from a social *system* perspective to a social *field* perspective, we have to become aware of our blind spot, the source level from which our attention and our actions originate. That source level fundamentally affects the quality of leading, learning, and listening.
>
> The problem with leadership today is that most people think of it as being made up of individuals, with one person at the top. But if we see leadership as the capacity of a system to co-sense and co- shape the future, then we realize that all leadership is distributed – it needs to include everyone. To develop collective capacity, everyone must act as a steward for the larger eco-system. To do that in a more reliable, distributed, and intentional way, we need: A social grammar: a language; a social technology: methods and tools; and a new narrative of societal evolution and change.
>
> The reason why this matters is that **energy follows attention**. Wherever we put our attention as leader, educator, parent, etc. – that is where the energy of the team will go. The moment we see the quality of attention

---

21 Scharmer, C.O. (2016). *Theory U – Leading from the Future as it Emerges*. Oakland, California: Berrett-Koehler Publishers.
22 Theory U, p. 27.

shifting from ego to eco, from *me* to *we,* that is when the deeper conditions of the field open up, when the *generative social field* is being activated.

My work with these and other methods of change over the past two-plus decades boils down to this: The quality of results achieved by any system is a function of the quality of awareness that people in these systems operate from. In short: Form follows consciousness.[23]

Of particular interest at this juncture is the notion of generative field activation. The topics in Chapter 5 (Awareness in Risk Management) and Chapter 6 (Field Leadership) build toward this point, that is – activating a generative field.

10.3.3 Generative Field Engine – Social–Human Engagement (E7) and Leadership (E8)

Interiority is key in activating a generative field. As suggested, there is value in viewing this as more than an individual (social–human) phenomenon as reflected in the right side of the Venn diagrams in Figures 2.1, 10.1, and 10.2 and to consider more broadly the role systems and frameworks play in this activation. This consideration is possibly one of the most powerful benefits of developing the Risk Matrix.

The idea of leverage is addressed immediately in Chapter 1 with Buckminster Fuller's trim tab quote, and Professor Levine's simple depiction of an essentially cause-and-effect construct in his leverage diagram (Figure 1.2). The work of Donnella Meadows continues by looking at leverage within systems. Tracy Goss et al.'s ideas of context and being continue the thread into Chapter 2; and, in Chapter 3 Winston Churchill's opening quotation orients us to how structures shape us. The motive force and energy thread begins in Chapters 5 and 6 and is woven throughout with a consideration of decision-making (Chapter 7) and the multidimensional Risk Matrix (Chapter 8). These focus us on looking at how things are put into action (Chapter 9).

The interplay between systems/frameworks, and the people who operate them, and are "in them" – as suggested in Figures 2.1, 10.1, and 10.2 – has been a personal area of interest since graduate school. In the 1990s and the early days of management system work, it was important to highlight leadership and participation (e.g. worker, employee as mentioned above). At that time, the ideas of organizational energy had not been widely published, nor was Otto Scharmer's "energy follows attention" in the organizational development lexicon, let alone that of EHSS/ORM.

Within leverage and "field activation" thinking, Figure 10.3 and 10.4 offer two perspectives. The first looks at generative field activation from the perspective of two Risk Management Elements – Social–Human Engagement (E7) and

---

23 Garrison document pp. 10–12.

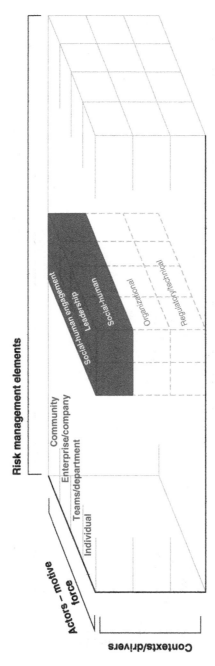

Figure 10.3 Generative field activation – social–human dimension and Risk Management Elements E7 and E8.

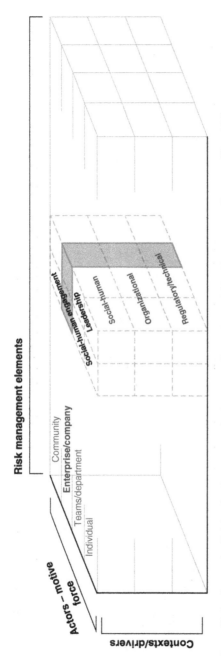

Figure 10.4 Generative field activation – enterprise/company dimension and Risk Management Elements E7 and E8.

10.3 Leverage – Creating Generative Fields | 319

Leadership (E8) – specifically highlighting the eight cells (four each) that are in the social–human element of the Contexts/Drivers dimension. This perspective reflects several evolutions. The first is consideration of the impacts the social–human era has had on ORM. The second is an evolution from the early management system period in the 1990s that highlighted concepts of leadership and interiority that included engagement and participation.

The second perspective (Figure 10.4) looks at the same two Risk Management Elements, but focuses more narrowly on the enterprise/company element. There are any number of foci here and depending on a particular situation or application, other parts of the Risk Matrix can be involved. This second perspective is offered as an example of focusing on the enterprise/company itself, and the integration of the cells highlighted.

Narrowing further the consideration of leverage and "field activation" in the Risk Matrix, two cells, the Risk Management Elements E7 and E8, highlighted in Figure 10.5, are at the intersection of the enterprise/company in the social–human elements.

While I was not conversant in the distinctions of risk fields, multiple risk dimensions, and generative fields at the time, I can now see that the characteristics identified by Senge et al. in 2015, and subsequently by Mette Boell and Diana Walsh, were present in my work with organizations and in the development of risk management standards.

They are also present in the three frameworks identified at the end of Chapter 3 (§3.7): 1) the National Institute for Occupational Safety and Health (NIOSH) Total Worker Health® initiative, 2) the Robert Wood Johnson Foundation (RWJF), and Global Reporting Initiative's (GRI) Culture of Health for Business initiative, and 3) the capitals-related ideas and protocols developed by the Capitals Coalition. Each of these reflect, if not embody aspects of generative fields.

10.3.4 Pedagogy of Evaluation

Learning-related ideas, concepts, and practices have been recurring themes throughout this book, particularly in Chapter 5 (§5.6.2). In Chapter 4, I made a connection between the learning context, performance evaluation, and auditing. That chapter begins with a quotation by one of the key contributors to evaluation science, Michael Quinn Patton, who notes that "... all evaluation approaches constitute a pedagogy of some kind." In 2017, a volume of the American Evaluation Association's journal, *New Directions for Evaluation*, focused on the "Pedagogy of Evaluation."[24] I briefly draw your attention to this work in evaluation science as it has relevance to building generative fields and offers an important leverage point for EHSS/ORM professionals through existing auditing and assessment practices, and structures.

---

24 *New Directions for Evaluation*, no. 155, Fall 2017.

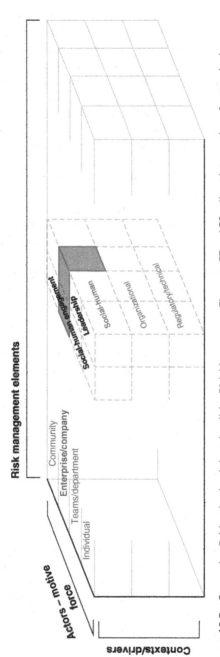

Figure 10.5 Generative field activation (trim tab cells) – Risk Management Elements E7 and E8 cells at intersection of enterprise/company and social–human elements.

Michael Quinn Patton served as the *New Directions* volume editor. He opened the volume with:

> Pedagogy is the study of teaching. *Pedagogy of evaluation* entails examining how and what evaluation teaches. There is no singular or monolithic pedagogy of evaluation. Embedded in different evaluation approaches are varying assumptions, values, premises, priorities, and sense-making processes. Those who participate in an evaluation are experiencing sometimes explicit, more often implicit and tacit, pedagogical principles. Evaluation invites stakeholders involved to see the world in a certain way, to make sense of what is being evaluated through a particular lens, to make judgments based on certain kinds of evidence and values.[25]

He indicates that the volume was "... inspired by and builds on the works of Paulo Freire, especially his classic, *Pedagogy of the Oppressed*."

Aware of Quinn's stature, and the significance of Freire's literacy work, I had a sense of the ground-breaking nature of this issue of *New Directions*. At the time I read it, I was involved with the Robert Wood Johnson Foundation/GRI (RWJF/GRI), Culture of Health for Business (COH4B) initiative, and the GSF events. At one of the COH4B development meetings, I spoke about the role that EHSS auditing plays in organizational learning. It is these observations that evolved into the development of the Risk Matrix.

While the term generative field is not used by Quinn or Freire, characteristics of each of their pursuits reflect what Mette Boell, Diane Walsh and others point to in their GSF work. Quinn states: "Freire reports having learned with colleagues that the people in communities only become deeply interested and engaged when the inquiry related directly to their felt needs. Any deviation... produced silence and indifference."[26] Quinn continues with offering 10 "Freirean Principles" that he generated, "with an eye toward particular influences on and relevance for evaluation." These are:

- Principle 1. Use Evaluative Thinking to Open Up, Develop, and Nurture *Critical Consciousness*
- Principle 2. Consciousness Resides in Communities of People, Not Just Individuals
- Principle 3. Critical Consciousness Pedagogy Must Be Interactive and Dialogical
- Principle 4. Integrate Reflection and Action

---

25 Patton, M.Q. (2017). Pedagogical principles of evaluation: Interpreting Freire. In: *Pedagogy of Evaluation. New Directions for Evaluation 155* (ed. M.Q. Pat- ton), 9.
26 Patton (2017), pp. 52–53.

- Principle 5. Value and Integrate the Objective and Subjective
- Principle 6. Integrate Thinking and Emotion
- Principle 7. Critical Consciousness Pedagogy Is Co-Intentional Education Among Those Involved in Whatever Roles
- Principle 8. Critical Consciousness Is Both Process and Outcome, Both Method and Result, Both Reflection and Action, Both Analytical and Change-Oriented
- Principle 9. All Pedagogy Is Political
- Principle 10. Critical Pedagogy Is Fundamentally and Continuously Evaluative.[27]

Shifting EHSS/ORM assessments (e.g. mainly auditing) into a pedagogical context, keeping in mind Patton's 10 Freirean principles, opens up an opportunity to transcend historic EHSS/ORM paradigms. Embedded in Quinn's interpretation of Freirean principles are interiority, value generation, integrated thinking, stakeholder engagement, and escalating impact.

10.4 Carriers of a Field[28]

This book offers a wide field perspective, with a field sequence identified, progressing from social field to an organizational field to a risk field, introducing a new generative field construct. I suggest, that EHSS/ORM professionals are carriers of many fields. Certainly, there are foundational technical aspects of the E, H, S, and S represented in the Risk Matrix's regulatory/technical element that are embodied 24/7/360 in decision-making and actions. Furthermore, I have asserted that aspects of our practices include the identified characteristics of generative fields.

The central phenomena in field theory are carriers of meaning, and that there "… are fundamental mechanisms that allow us to account of for how ideas move through space and time, who or what is transporting them, and how they may be transformed by their journey."[29] Expanding on this involves considering carriers of a field, particularly in terms of people as the carriers. As suggested, we (EHSS/ORM professionals) do this on the technical front. I invite you to consider this phenomenon, carrier of a field, and what it would look like, and what would be possible to accomplish by expanding it into the Risk Matrix's multidimensional perspectives.

Try considering this in the output, outcome, impact frame: (1) output, distinguishing and defining what this would look like, perhaps as a written training guideline

---

27 Patton (2017), pp. 53–69.
28 The term "carriers of a field" was coined by Dr. JoAnne Kellert.
29 Scott, W.R. (2014). *Institutions and Organizations*, 95. Sage Publications, Inc.

within one of the 14 Risk Management Elements; (2) outcome, realizing the dynamics identified by Mette Boell and Diana Walsh, and seeing greater alignment between intended and actual ORM results; and, (3) impact, realizing results related to ORM outcomes, such as those related to purpose and value generation (from a double materiality perspective), possibly with capitals (human, social, natural).

10.4.1 Portal to Future-Ready[30]

In 2007, I developed a "future-ready" construct to depict EHSS/ORM maturity in an organization. The maturity curve is depicted in Figure 10.6 and reflects the progression from a regulatory compliance orientation to one beyond compliance that is generative and driven by vision and positive impact. This shift in thinking has been present throughout this book and was highlighted in Chapter 5 (§5.6).

At that time, I had not consolidated my thinking on field-related distinctions, but I has on EHSS/ORM professionals as drivers in creating future-ready

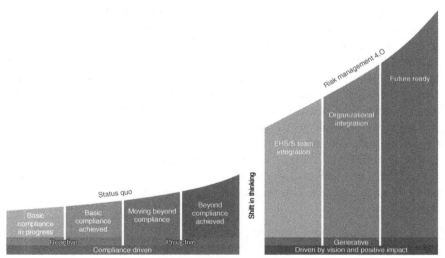

Figure 10.6 Future-ready company construct.

---

30 Future-ready defined as: When a company is future-ready, there is forward-looking, wider awareness about company health and risks, driven by vision and commitment to positive impact rather than compliance. A chief attribute is a culture of health—organizational, human, environmental, community, and global. Reframing risk management in an organizational health context fuels the future-ready company. Engagement and collaboration increase, silos dissolve, and risk perspective widens.

companies, even though they are not always identified as such. I also thought that this organizational function and its professionals were unrecognized "secret weapons" in increasing sustained value, impact, and resilience. The constructs offered in this book reinforce this, especially with increasing attention being paid to value generation, protection, and preservation as reflected in sustainability, ESG and organizational governance (e.g. ISO 37000:2021).

10.4.2 The Table and Its Seats

"Seat at the table" is a common metaphor in EHSS/ORM. A question often posed is "how can we get a seat at the table?" While there is an innate sense within EHSS/ORM professionals of our relevance to the value (generation, protection, control) and purpose, we do not always have a seat at the proverbial table. This is especially surprising given the prominence of these EHSS topics in the literature, and society at large

Two questions here are, "what table" and "who's table?" A goal of writing this book is to offer tools, constructs, ideas, etc. that can be used to (1) define the table, and (2) suggest that **the** table is actually the EHSS/ORM table. While the end result may not have us sitting at the "senior" table in an organization, we could at least be at a table connected to it.

The Risk Matrix's multidimensional nature provides several powerful qualities to impact organizational decision-making and governance. Beyond the 14 Risk Management Elements, the Risk Matrix's other two dimensions (Contexts/Drivers, and Actors/Motive Force) can be applied to any organizational function. The organizational risk field/matrix provides a lens through which organizational values and culture come into focus, and through which they can be transformed. The Risk Matrix offers a tool that can be used to increase organizational energy, vitality, creativity, and resilience. It also offers a chance to increase an organization's ability to anticipate, see, and respond to emerging risks with greater speed and effectiveness.

10.4.3 Trim Tab

> Call me Trim Tab.[31]
>
> Buckminster Fuller

We end where we started with Buckminster Fuller's Trim Tab. Throughout this book, I have sought to provide tools to help you be a Trim Tab. I have introduced

---

31 Buckminster Fuller, *Playboy*, February 1972, p. 59 (v.19, No. 22).

new language and constructs to expand your toolbox and increase your effectiveness.

I have been moved over the years by the deep commitment of EHSS/ORM professionals to improve their organization's performance, and to the health and well-being of people, their communities, and the environment. So, to inspire you to build generative fields, and be carriers of the fields reflected in these commitments, I leave you with the words of George Bernard Shaw:

> This is the true joy in life, the being used for a purpose recognized by yourself as a mighty one; the being a force of nature instead of a feverish selfish little clod of ailments and grievances complaining that the world will not devote itself to making you happy. I am of the opinion that my life belongs to the whole community and as long as I live it is my privilege to do for it whatever I can. I want to be thoroughly used up when I die, for the harder I work the more I live. I rejoice in life for its own sake. Life is no 'brief candle' to me. It is a sort of splendid torch which I have got hold of for the moment, and I want to make it burn as brightly as possible before handing it on to future generations.[32]

I leave you with a question.

Can we call you Trim Tab?

Suggested Reading

Bruch, H. and Vogel, B. (2011). *Fully Charged: How Great Leaders Boost their Organization's Energy and Ignite High Performance.* Boston: Harvard Business Review Press.

Cameron, K.S. and Dutton, J. (2003). *Positive Organizational Scholarship, Foundations of a New Discipline.* San Francisco, California: Berrett-Koehler Publishing, Inc.

Capitals Coalition (2019). *Social and human capital protocol.* https://capitalscoalition.org/capitals-ap-proach/social-human-capital-protocol/ (accessed 14 August 2021).

Capitals Coalition (2021). *Principles of integrated capitals assessments.* https://capitalscoalition.org/wp-content/uploads/2021/01/Principles-of-integrated-capitals-assessments_v362.pdf (accessed 23 February 2023).

---

32 Shaw, Bernard *Man and Superman.* Cambridge, Mass.: The University Press, 1903; Bartleby.com, 1999. www.bartleby.com/157/ (accessed 15 November 2022).

Fritz, R. (1999). *The Path of Least Resistance for Managers – Designing Organizations to Succeed*. Berrett-Koehler Publishers.

Meadows, D. (1999). *Leverage Points, Places to Intervene in a System*. Hartland, VT: Sustainability Institute.

Meadows, D. (2008). *Thinking in Systems, A Primer* (ed. D. Wright). White River Junction, Vermont: Chelsea Green Publishing.

Patton, M.Q. (2017). Pedagogical principles of evaluation: interpreting Freire. In: *Pedagogy of Evaluation. New Directions for Evaluation, 155* (ed. M.Q. Pat-ton), 49–77. Wiley.

Scott, W.R. (2014). *Institutions and Organizations, Ideas, Interests and Identities*, 4e. Thousand Oaks, California: Sage Publications, Inc.

Scharmer, C.O. (2016). *Theory U – Leading from the Future as it Emerges*. Oakland, California: Berrett-Koehler Publishers.

Wooten, M. and Hoffman, A. (2017). Organizational fields: past, present and future. In: *The Sage Handbook of Organizational Institutionalism* (ed. R. Greenwood et al.), 55–74. Thousand Oaks, CA: Sage Publications.

Glossary

A → B Shorthand for getting from point (state) A to point (state) B. "A" represents a current state, "B" a future state, and "→" an intervention. Progression from A to B is referred to as a shift.

Acceptable risk Residual risk that is acceptable to a stakeholder(s).

Actors These "include individuals, associations of individuals, associations of individuals, populations individuals, organizations, associations of organizations, and populations of organizations."[1]

Actors/Motive Force An organizational risk management (ORM) dimension. It is represented in the Risk Matrix's z-axis and comprised of four elements: individual; team/department; enterprise; and community.

Artifacts "An artifact is a discrete material object, consciously produced or transformed by human activity, under the influence of the physical and/or cultural environment."[2]

Becoming aware The process of increasing awareness, both individual and organizational. This is a preliminary step in change, transformation, and creating new clearings, and involves increasing knowledge and perceptions of ORM, and the structures and frameworks within which people operate.

Capitals Coalition (CC) An international entity that evolved from The Economics of Ecosystems and Biodiversity (TEEB) for Business Coalition in 2012, which became the Natural Capital Coalition in 2014. In 2018, the Social and Human Capital Coalition was founded, and in 2020, the two coalitions joined to form the Capitals Coalition.

---

1 Richard, S.W. (2014). *Institutions and Organizations*, 228. Sage Publications, Inc.
2 Suchman (2003). The contract as social artifact. *Law and Society Review* 37: 91–142, p. 98. In Richard, S.W. (2014). *Institutions and Organizations*, 102. Sage Publications, Inc.

Organizational Risk Management: An Integrated Framework for Environmental, Health, Safety, and Sustainability Professionals, and their C-Suites, First Edition. Charles F. Redinger.
© 2025 John Wiley & Sons, Inc. Published 2025 by John Wiley & Sons, Inc.

Carriers of meaning "Are fundamental mechanisms that allow us to account for how ideas move through space and time, who or what is transporting them, and how they may be transformed by their journey."[3]

Clearing A state (conceptual or physical), and context where new ideas, thinking, frameworks, structures, policies, procedures, and culture can arise.

Conformity assessment "The determination of whether a product or process conforms to particular standards or specifications. Activities associated with conformity assessment include testing, certification, and quality assurance system registration."[4]

Consensus standard A nongovernmental generated document that provides recommendations or requirements for an activity.

Contexts/Drivers An organizational risk management (ORM) dimension. It is represented in the Risk Matrix's y-axis and comprised of three elements: regulatory/technology; organizational; and social-human.

Committee of the Sponsoring Organizations of the Treadway Commission (COSO) An organization that develops guidelines for businesses to evaluate internal controls, risk management, and fraud deterrence.

Corporate governance As used here, refers to three governance-related drivers that are central in organizational decision-making in general, and specifically ORM risk decision-making, these are purpose, shareholder versus stakeholder, and value creation/protection.

Corporate Social Responsibility (CSR) A term coined in 1953 by Howard Bowen who is often cited as the "father of CSR."[5] This includes the idea of a "social contract," which was put forth by the Committee for Economic Development in 1971. Core to this was the idea of business obligation to serve the needs of society which is embedded in the notion of "license to operate."

Double materiality The impact an organization's activities have on entities and stakeholders external to itself. Commonly referred to in terms of environmental impact.

Environmental, social, and governance (ESG) A term that evolved from a 2004 United Nations-led initiative to "better integrate environmental, social, and corporate governance issues in asset management, securities brokerage services and associated research functions."[6]

---

3 Richard, S.W. (2014). *Institutions and Organizations*, 95. Sage Publications, Inc.
4 National Research Council (1995), p. 206.
5 Bowen, Howard R. (1953). *Social Responsibilities of the Businessman*. University of Iowa Press.
6 The Global Compact (2004). *Who Care Wins: Connecting Financial Markets to a Changing World*. Recommendations by the financial industry to better integrated environmental, social and governance issues in analysis, assessment management and securities brokerage. Swiss Federal Department of Foreign Affairs (Bern) and the United Nations (New York). https://www.unepfi.org/fileadmin/events/2004/stocks/who_cares_wins_global_compact_2004.pdf

Escalating impact Actions that increase the rate and probability that organizational risk management (ORM) outcomes positively support fulfillment of purpose, generation and preservation of value, and increases in health and well-being for the organization and its stakeholders.

Field "An area of open land, especially one planted with crops or pasture, typically bound by hedges and fences. A space or range within which objects are visible from a particular viewpoint or through a piece of apparatus. The region in which a particular condition prevails, especially one in which a force or influence is effective regardless of the presence or absence of a material medium."[7]

Field leadership The practice of integrating context and motive force that generates engagement toward fulfilling organizational purpose, and value creation.

Foundational Five elements Risk Management Elements that require 24/7/365 attention. These include risk assessment; emergency preparedness and response; management of change; communication, systems and practices; and competency and capabilities.

Fourth generation risk management ORM frameworks that add dimensionality by including actors/motive force and expanded context/drivers to create a foundation for integrated thinking, decision-making, and action.

Framework Structures that provide the "container" in which EHSS/ORM activities and processes happen. These include processes, programs, systems, and the Risk Matrix.

Generative organizational risk field An organizational risk field is generative when there is collective energy and leadership creating and preserving value that includes social, human, and natural capital.

Generative social field "Are where families, schools, businesses, governments, and the larger society in which we live can create relational interactions that support the nurturance of compassion, connection, curiosity and well-being. Within such generative fields, a sense of belonging, meaning, and effective action emerges."[8]

Global Reporting Initiative (GRI) An entity formed in 1997 by the Coalition for Environmentally Responsible Economies (Ceres) and Tellus Institute, with support from the United Nations Environmental Programme (UNEP).

Gravitational pull organizational energy and motive force toward a particular direction.

---

7 Apple OS dictionary. New Oxford American Dictionary.
8 April 26, 2019. "2019 IPNB Pre-Conference Workshop Awareness-Based Systems Leadership: Cultivating Generative Social Fields and Systems Change" Marina Del Rey, CA.

Heuristic A mental shortcut for solving problems in a quick way that delivers a result that is sufficient enough to be useful given time constraints.

Human Capital The knowledge, skills, competencies, and attributes embodied in individuals that facilitate the creation of personal, social, and economic well-being.[9]

Impact The effect that an outcome is having. In an emissions example, the installation and functioning of a control system is an output. Reduced emissions is the outcome. Reduced asthma in the community is the impact.

Impact-dependency pathways A central concept in the Capitals Coalition protocols. "Pathways draw links between (natural, social or human) capital issues identified and the business activities that affect or rely on them. These pathways (also-called logical frameworks, results chains or theories of change) outline the potential and empirically testable relationships between your business activities and social and human capital creation, destruction, or reliance."[10]

Institutional logic(s) "A set of material practices and symbolic constructions which constitutes its organizing principles, which is available to organizations and individuals to elaborate."[11] "Many of the most important tensions and change dynamics observed in contemporary organizations and organization fields can be fruitfully examined by competition and struggle among various categories of actors committed to contrasting institutional logics. An important source of institutional tension and change experienced by both organizations and individuals in everyday life involves jurisdictional disputes among the various institutional logics."[12]

Interiority The phenomena of individual, team, or organizational engagement and generation of motive force that produces outcomes and impacts.

International Organization for Standardization (ISO) A international organization founded in 1947 that develops nongovernmental consensus standards.

International Sustainability Standards Board (ISSB) An entity established within the International Financial Reporting Standards (IFRS) Foundation in 2021.

---

9 Capitals Coalition (2019). Social & Human Capital Protocol. p. 11.
10 Capitals Coalition (2019). Social and Human Capitals Protocol. p. 33.
11 Feidland and Alford (1991). Bringing society back in: symbols, practices, and institutional contradictions. In: *The New Institutionalism in Organizational Analysis* (eds. Powell, W. and DiMaggio, P.), 248. Chicago University Press.
12 Scott (2014), p. 91.

Integrate Combine one thing with another so that they become whole. Latin root (integrare) – to make whole.[13]

Integrated capitals assessment An assessment measuring and valuing all relevant capitals in terms of impacts and dependencies on them, which explicitly considers systems thinking including the interconnections both within and between all of the capitals.[14]

Integrated decision-making Processes that consider the risk matrix's three dimensions and their elements to achieve outcomes and impacts.

Integrative thinking "The ability to face constructively the tension of opposing ideas, and instead of choosing one at the expense of the other, generate a creative resolution of the tension in the form of a new idea that contains elements of the opposing ideas but is superior to each."[15]

Internal state Orientations, perspectives, and mental models of primarily of individuals, but also teams, and an organization itself.

Language as currency Characterizing language and words as currency that help access to new clearings and capacities that increase the ability to generate and preserve value and increase resilience. They impact personal and collective thinking, which in turn impact systems, structures, and outcomes.

Materiality "an accounting principle which states that all items that are reasonably likely to impact investors' decision-making must be recorded or reported in detail in a business's financial statements using Generally Accepted Accounting Principles (GAAP) standards Essentially, materiality is related to the significance of information within a company's financial statements. If a transaction or business decision is significant enough to warrant reporting to investors or other users of the financial statements, that information is 'material' to the business and cannot be omitted."[16]

---

13 New Oxford American Dictionary (in Apple OS 10.15.7).
14 Capitals Coalition (2021). Principles of integrated capitals assessment. p. 26. https://capitalscoalition.org/wp-content/uploads/2021/01/Principles_of_integrated_capitals_assessments_final.pdf (accessed 15 May 2023).
15 Martin, Roger L. (2007). *The Opposable Mind: How Successful Leaders Win Through Integrative Thinking*, 15. Boston, MA: Harvard Business School Press.
16 Harvard Business School: https://online.hbs.edu/blog/post/what-is-materiality (accessed 16 February 2023).

Mental model "The semipermanent tacit 'maps' of the world which people hold in their long-term memory, and the short-term perceptions which people build up as part of their everyday reasoning process."[17]

Motive Force The generation of physical action by individuals, teams, or organizations that cause action.

Natural Capital The stock of renewable and nonrenewable natural resources (e.g. plants, animals, air, water, soils, minerals) that combine to yield a flow of benefits to people.[18]

New clearing A state, or way of being, aspired to be reached. The integrated organizational framework (Risk Matrix) offers such a state designed to increase the ability to generate and preserve value, resilience, and fulfillment for an organization and its stakeholders.

Operational elements Risk Management Elements that are operationally oriented. These include frameworks; auditing/metrics; operation; escalating impact; and future-ready strategy.

Organizational energy "The extent to which an organization, division, or team has mobilized its emotional, cognitive, and behavioral potential to pursue its goals. Simply put, it is the force with which a company (or division or team) works."[19]

Organizational field "Those organizations that, in the aggregate, constitute an area of institutional life: key suppliers, resource and product consumers, regulatory agencies, and other organizations that produce similar services or products."[20] // "A community of organizations that partakes of a common meaning system and whose participants interact more frequently and fatefully with one another than with actors outside of the field."[21]

Organizational risk field The relational and contextualized space that creates the interactions and collective behavior, which in turn produces the organization's risk management governance, strategy, execution, and outcomes.[22]

---

17 Senge, P. et al. (1994). *The Fifth Discipline Field Book*. Doubleday. p. 237. Quote attributed to Art Kleiner.
18 Capitals Coalition (2016). Natural capital protocol. p. 2.
19 Bruch, Hieke and Vogel, B. (2011). *Fully Charged: How Great Leaders Boost their Organization's Energy and Ignite High Performance*, 5. Boston: Harvard Business Review Press.
20 DiMaggio and Powell, 1983: 148.
21 Richard, S.W. (1994), pp. 207–208.
22 Redinger, C. (2019). Organizational Risk as a Generative Social Field – New Promise for Increasing Risk Awareness, Resilience, and Improving Decision-Making." Poster presented at the Society for Risk Analysis Annual Meeting, 9–11 December 2019.

Orient "Align or position (something) relative to the points or a compass or other specified position. Find one's position in relation to new and strange surroundings."[23]

Orientation "The determination of the relative position of something or someone (especially oneself). A person's basic attitude, belief, or feelings in relation to a particular subject or issue."[24]

ORM evolution sequence Organizational risk management is fluid and dynamic. Five evolution sequences are identified in Chapter 2 – these are – 1) EHSS management, 2) frameworks, 3) sustainability, 4) object/foci, and 5) ORM generations.

Outcome The effect or impact of the output. In an emissions example, the installation and functioning of a control system is an output. Reduced emissions is the outcome.

Outcome Space The intersection of system/frameworks and social and human capital where integrated thinking, integrated decision-making and integrated actions take place, which lead to outcomes. The expansion of this outcome space reflects the increasing frequency of integrated thinking, integrated decision-making and integrated action.

Output This is the result of an activity. Examples include the implementation and functioning of a risk management system as verified by an audit.

Pedagogy The method and practice of teaching, especially as an academic subject or theoretical concept. New Oxford American Dictionary (in Apple OS 10.15.7).

Perspective "A particular attitude toward or way of regarding something; a point of view."[25]

Purpose (organization) An "organizations meaningful reason to exist."[26]

Residual risk Risk that remains after an intervention.

Risk *ISO 31000* – "Effect of uncertainty on objectives."[27] *ISO Annex SL* – "Effect of uncertainty," with the following notes, "An effect is a deviation from the expected – positive or negative. Uncertainty is the state, even partial, of deficiency of information related to, understanding or knowledge of, an event, its consequence, or likelihood." And, "Risk is often characterized by reference to potential; *events*' (as defined in ISO Guide 73:2009, 3.5.1.3) and '*consequences*' (as defined in ISO Guide 73:2009, 3.6.1.3), or a combination of

---

23 New Oxford American Dictionary (in Apple OS 10.15.7).
24 New Oxford American Dictionary (in Apple OS 10.15.7).
25 New Oxford American Dictionary (in Apple OS 10.15.7).
26 ISO 37000:2021. *Governance of Organizations – Guidance*. Definition 3.2.10, p. 4. Geneva, Switzerland.
27 ISO 31000 (2018), p. 1.

these."[28] *COSO ERM* – "The possibility that events will occur and affect the achievement of strategy and business objectives. NOTE: 'Risks' (plural) refers to one or more potential events that may affect the achievement of objectives. 'Risk' (singular) refers to all potential events collectively that may affect the achievement of objectives."[29]

Risk analysis "Systematic process to comprehend the nature of risk and to express the risk, with the available knowledge."[30]

Risk assessment "Systematic process to comprehend the nature of risk, express and evaluate risk, with the available knowledge."[31]

Risk-based thinking A concept reflected in ISO Annex SL-based standards used promotes to integrate risk considerations in decision-making.

Risk capacity "The maximum amount of risk that an entity is able to absorb in the pursuit of strategy and business objectives."[32]

Risk communication "Exchange or sharing of risk-related data, information, and knowledge between and among different target groups (such as regulators, stakeholders, consumers, media, and general public)."[33]

Risk decision-making kernel A heuristic that contains five common elements of risk decision-making – source, impact, tolerance, acceptability, and control/transfer.

Risk field The relational and contextualized space that creates the interactions and collective behavior, that in turn produce the organization's risk management governance, strategy, execution, and outcomes.

Risk logics Approaches, principles, ideas, and contexts that establish foundations for risk-related systems and frameworks, decision-making processes, and subsequent actions.

Risk management "Activities to handle risk such as prevention, mitigation, adaptation or sharing. It often includes trade-offs between costs and benefits of risk reduction and choice of a level of tolerable risk."[34]

Risk Management Element An organizational risk management (ORM) dimension. It is represented in the Risk Matrix's *x*-axis and is comprised of fourteen elements: purpose and scope; risk assessment; emergency preparedness and response; management of change; communication: systems and practices; competency and capabilities; social-human engagement; leadership; decision-making; frameworks; auditing/metrics; operation; escalating impact; future-ready strategy.

---

28 Annex SL (2013), p. 138.
29 COSO (2017b), p. 110.
30 SRA (2018). Society for risk analysis glossary. Society for Risk Analysis. p. 8.
31 SRA (2018). Society for risk analysis glossary. Society for Risk Analysis. p. 8.
32 COSO, p. 110.
33 SRA (2018). Society for risk analysis glossary. Society for Risk Analysis. p. 8.
34 SRA (2018). Society for risk analysis glossary. Society for Risk Analysis. p. 8.

Risk Matrix A multidimensional organizational risk management (ORM) framework that reflects the operationalization of an organizational risk field construct. Its three dimensions are Contexts/Drivers (y-axis); Actors/Motive Force (z-axis); and Risk Management Elements (x-axis). It is designed for use in upgrading existing processes, programs, and systems; upgrading measurement activities such as audit programs and metrics; and strengthening external reporting/disclosure needs. It serves as a foundation for creating generative organizational fields and escalating organizational impact.

Risk object refers to a specific risk, such as debt, a new acquisition, a regulation, or a chemical exposure.

Risk profile An organization's risk inventory. COSO defines this as "a composite view of the risk assumed at a particular level of the entity, or aspect of the business that positions management to consider the types, severity, and interdependencies of risks, and how they may affect performance relative to the strategy and business objectives."[35]

Risk Society "a society increasingly preoccupied with the future (and also with safety), which generates the notion of risk."[36]

Risk tolerance "An attitude expressing that the risk is judged tolerable."[37]

Risk transfer "Sharing with another party the benefit of gain, or burden of loss, from the risk. Passing a risk to another party."[38]

Social capital Networks together with shared norms, values, and understanding facilitate cooperation within and among groups.[39]

Social–human Refers to people individually and collectively, both inside and outside an organization's fence line, including those in supply chains, communities, consumers, and the population in general.

Stakeholder A person or organization that can affect, be affected by, or perceive itself to be affected by a decision or activity. ISO 37000:2021, p. 5.

Stocks and Flows Stocks are the number, types and sizes of current risks at a particular point in time. Flows are the risks (and information about them) that travel through impact-dependency pathways and how the number, types and sizes of risks increase or decrease over time as a result of an organization's intervention.

---

35 COSO (2017). *Enterprise Risk Management – Integrating with Strategy and Performance*, 110. Durham, NC: Association of International Certified Professional Accountants.
36 Giddens, A. and Pierson, C. (1998). *Conversations with Anthony Giddens, Making Sense of Modernity*. Stanford University Press, p. 209.
37 SRA glossary, p. 9.
38 SRA, p. 9.
39 Capitals Coalition (2019). Social & Human Capital Protocol. p. 11.

Systems-thinking "A way of seeing and talking about reality that helps us better understand and work with systems to influence the quality of our lives. In this sense, systems thinking can be seen as a perspective. It also involves a unique vocabulary for describing systemic behavior, and so can be thought of as a language as well. And, because it offers a range of techniques and devices for visually capturing and communicating about systems, it is a set of tools."[40]

Tailoring Activities related to an organization adapting or modifying standards (governmental or nongovernmental to their needs.

Templating The act of using knowledge, practices, structures, processes, procedures, and policies developed in one Risk Matrix cell or area, to develop as less evolved Risk Matrix cell or area.

Topographies/ecosystems Construct used to characterize organizational contexts, structures, and cultures.

Transformational learning Individual, team, or organizational learning that results from, "an exploration of how deep-seated values, beliefs, and assumptions shape the ways in which [they] frame and react to situations."[41]

Trim Tab elements Risk Management Elements that provide leverage for A → B shifts. These include purpose and scope; social-human engagement; leadership; and decision-making.

Uncertainty *COSO ERM* – "The state of not knowing how or if potential events may manifest."[42] *ISO 45001* – "Uncertainty is the state, even partial, of deficiency of information related to, understanding or knowledge of, an event, its consequence, or likelihood."[43]

---

40 Kim, D. H. (1999). *Introduction to Systems Thinking*, 2. Waltham, MA: Pegasus Communications.
41 Kofman F. Transformational learning: a blueprint for organizational change. The Systems Thinker. https://thesystemsthinker.com/transformational-learning-a-blueprint-for-organizational-change/ (accessed 18 November 2022).
42 COSO ERM (2017), p. 110.
43 *45001 (note 2) to risk definition, 3.20, p. 5.*

Index

a
absenteeism rate   123
Accenture   197
acceptable risk   225–226
Actors/Motive Force dimension
   240–243, 299–302
agency theory   48
American National Standards Institute
   (ANSI)   81
artifacts   168, 180
auditing   139
   conformity assessment
      guidelines   144
      history   143–144
      hybrid approaches   145–146
      types   144–146
   operational elements, of Risk
      Management Elements
      (x-axis)
      COSO   273–275
      ISO 31000:2018   272
      ISO 37000:2021   272
      ISO MSS-14001 and 45001   272
      planned intervals   272
automatic decisions   204
Awareness-Based Risk Management
   (ABRM)   57, 152–154

awareness, in risk management   152
   being and doing   162
   current iteration risk field   155–156
   decision-making prequel   161
   defined, 156
      paying attention   159
      standards and
         frameworks   157–159
   early years
      integrated model   153
      second-order change   153–155
      stakeholder domains   155
      360 degree perspective   155
   fourth-generation risk
      management   152
   genesis   152
   language as currency   166
      carriers of meaning   168
      future-based language   167–168
   leverage and risk awareness elements
      awareness   163
      decision-making processes   165
      generative field   165
      internal state   163
      risk and purpose   164
      value creation and
         preservation   164

Organizational Risk Management: An Integrated Framework for Environmental, Health, Safety, and Sustainability Professionals, and their C-Suites, First Edition. Charles F. Redinger.
© 2025 John Wiley & Sons, Inc. Published 2025 by John Wiley & Sons, Inc.

338 | Index

awareness, in risk management (cont'd)
  shifting mindset and paradigms 169–170
    A→B causal model 170–171
    anatomy and physiology of shifts 176
    Capitals Coalition four shifts model 175–176
    learning 171–175
  stakeholder domains 156

b

Bache, Christopher 70
Beck, Ulrick 56
Boell, Mette 3, 68, 310, 312, 319, 321, 323
BlackRock 126
blind spots 155
Board for Global EHS Credentialing (BGC) 140
Board of Certified Safety Professionals (BCSP) 140
British Standards Institute (BSI) 81
business model 217

c

capitals
  coalition 190, 191
  defined 188–189
  IIRC 189–190
  value 190–191
capitals assessment spectrum 19
Capitals Coalition (CapCo) 18, 131–133, 175–176
capitals thinking 122
carrier of a field 322–323, 325
carriers of meaning 72, 223
center of gravity 287
Certified Industrial Hygienists (CIH) 140
Certified Safety Professionals (CSP) 139

change management 247
Chief Risk Officer (CRO) 38
Climate Disclosure Standards Board (CDSB) 124
Committee of Sponsoring Organizations of the Treadway Commission (COSO) 78, 82, 113
  enterprise risk management 222
  ERM framework 293
    evolution 113–114
    principles 115, 116
    2017 version 114–117
    2004 vs. 2017 114
Committee on Conformity Assessment (CASCO) 140
communication channels 249
communication-systems and practices
  COSO ERM 249–250
  ISO 31000:2018 250–251
  ISO 37000:2021 248–249
  ISO MSS-ISO 45001:2018 251–252
company-centered model 48
compliance 54
conformity assessment 137, 139
  auditing
    guidelines 144
    history 143–144
    hybrid approaches 145–146
    types 144–146
  defined 139
  frameworks and guidelines
    decision-making currency 140–141
    inference guidelines 140–141
    National Research Council 139–140
  measurement 141–143
    qualitative terms 141
    quantitative terms 141
conservative Bayesian 205
Contexts/Drivers dimension slice 297–299

Index | 339

contexts, drivers, orientations
  corporate governance 47–51
  culture of health 53–54
  defined 43
  ESG 51–52
  evolutions
    EHSS management (compliance, performance, impact) 44–45
    frameworks (process, program, system, field) 45–46
    generations 47
    object/foci 46
    organizational risk management 47
    sustainability 46
  social and human capital 52–53
corporate social responsibility (CSR) 4, 10, 46
CRA *see* cumulative risk assessment (CRA)
Creating value 186
  capitals 188–191
  COSO's ERM Framework 186–187
  ISO 31000:2018 186–187
  ISO 37000:2021 187–188
  value generation 188
CSR *see* corporate social responsibility (CSR)
Culture of Health (COH) 24, 130
Culture of Health for Business (COH4B) initiative 12, 24, 130–131, 183, 321
cumulative risk assessment (CRA) 85

d
decision-making currency 17, 224–225
decision-making precursors 159
deGeus, Arie 16
diversity, equity, and inclusion (DE&I) 2, 43
double-loop learning 138, 172–173

double materiality 121, 122

e
ecosystem 288
electromagnetic fields 70
Electronics Data Systems (EDS) 173
emergency preparedness and response (EPR) 245, 246
Enabling Purpose Initiative (EPI) 50
enterprise risk management (ERM) 42, 113, 293
entity theory 48
environmental, health, and safety management systems (EHSMS) 257
environmental, health, safety, and sustainability (EHSS) 2, 38, 304
Environmental Protection Agency (EPA) 83, 84
environmental, social, and governance (ESG) 2, 4
  contexts, drivers, orientations 51–52
  frameworks
    CSR 118
    double/impact materiality 120–121
    GRI 124–125
    human capital 121–123
    ISSB 125–126
    materiality 118–121
    overview 117–118
    reporting and performance criteria 123–124
    sustainability 117
    Value Reporting Foundation 126–127
escalating impact 309
  A→B causal model 306
  carriers of a field 322–323
    portal to future-ready 323–324
    table and its seats 324
    trim tab 324–325

escalating impact (cont'd)
  creating generative fields
    interiority 314–316
    pedagogy of evaluation 319–322
    social-human engagement and leadership 316–319
    value generation and health 313
  generative fields 308–312
  transcending paradigms
    evolutions 307
    new clearings 307–308
    shifts 308
ethical leadership 192
European Sustainability Reporting Standards (ESRS) 126
external auditors 145

f

field activation 316
field approach 69
field leaders 180
field leadership 7
  defined 180
  field actors
    accountability 185
    interiority 185
  leadership and participation, in frameworks
    COSO ERM framework 193–194
    ISO 14001:2015 195–196
    ISO 31000:2018 194–195
    ISO 37000:2021 191–193
    ISO 45001:2018 195–196
  model
    responsible leadership 197–198
    system leadership 196–197
  motive force
    culture of health 183–184
    organizational energy 181–184
  value creation 186
    capitals 188–191

COSO's ERM Framework 186–187
ISO 31000:2018 186–187
ISO 37000:2021 187–188
value generation 188
Fifth Discipline 171
financial advisories 52
financial capital 189
financial materiality 118
first-party audits 144–145
Food and Drug Administration (FDA) 84
formal management systems 44
Forrester, Jay 15
Foundational Five 27, 243, 245–253, 286, 311
frameworks 78
  awareness 79
  COSO ERM (2017)
    principles 134–135
  ESG
    CSR 118
    double/impact materiality 120–121
    GRI 124–125
    human capital 121–123
    ISSB 125–126
    materiality 118–121
    overview 117–118
    reporting and performance criteria 123–124
    sustainability 117
    Value Reporting Foundation 126–127
  ISO
    history 86
    ISO 31000 87–88
    ISO 37000 89
    management system standards 90–91
    quality management standards 90
    risk management evolution 86–87
    3100:2018 principles 133–134

learning 80
NAS and EPA 83–85
next generation frameworks 80
structure power 79
transcending paradigms 128
  Capitals Coalition 131–133
  COH4B 130–131
  NIOSH Total Worker Health 129–130
types of 80
  consensus standards 81
  organizational and professional practices 82
  regulatory 81
  tailoring 82–83
Freirean principles 321–322
Freire, Paulo 319
Fuller, Buckminster 14
future-based language 167–168
future-ready company 323–324

g

generative field 6, 12, 15, 31, 33, 40, 68-69, 150, 155, 163, 165, 167–168, 172, 184-186, 197, 206, 218, 226, 265, 279, 280, 289, 302, 304, 306, 308-314, 319-320
generative field engine 316–319
generative organizational risk field 70
generative social fields (GSF) 68, 309, 310
Giddens, Anthony 56
Global Reporting Initiative (GRI) 12, 78, 124–126, 183–184
Goss, Tracy 38, 162, 304, 307
governance of organizations 49
gravitational pulls 240
gravity 287
GSF *see* generative social fields (GSF)

h

hot-cold decision triangle 209
human capital (HC) 40, 52, 189
human capital management 11
Human Health Risk Assessment (HHRA) 85
human role, in RDM
  brain function 209–211
  perception 208–209
  systems 1 and 2, properties 207
  two-system brain 207–208

i

impact-dependency pathways 165
inference guidelines 140–141, 225
influence 27
institutional theory 69, 280, 308
integrated reporting 22
integrated risk management system (RMS) 151
integrated risk matrix 9
integrated thinking 289, 290
intellectual capital 189
interiority 6, 33, 165, 185, 314, 316, 319, 322
internal audit 110, 144–145
internal auditors 145
International Financial Reporting Standards (IFRS) Foundation 119, 125, 292
International Integrated Reporting Council (IIRC) 22, 40, 122, 124, 189–190, 217, 292
International Labor Organization (ILO) 81
International Occupational Hygiene Association (IOHA) 57
International Organization for Standardization (ISO) 78
  frameworks
    history 86
    ISO 31000 87–88
    ISO 37000 89
    management system standards 90–91

342 | Index

International Organization for Standardization (ISO) (*cont'd*)
   quality management standards   90
   risk management evolution   86–87
   ISO 14001:2015   247
   ISO 45001:2018   247–248
   ISO 3100:2018 principles   133–134
   management system standards   2, 44, 53, 90, 91
      high-level structure   91, 92
      occupational health and safety   92–112
International Sustainability Standards Board (ISSB)   4, 119, 123–126
interval variables   142
ISO technical committees (TC)   140

j

Joint Technical Coordination Group (JTCG)   86

k

Kahneman, Daniel   17, 18, 161, 166, 204–205, 207, 216, 223, 228, 309
Kim, Daniel   65, 66, 68
Kurt Lewin's field theory   3

l

Leading Learning Communities (LLC) program   173
legacy or historic field   70
leverage
   current and future strategies   14–15
   Levine's lever   12–14
   shifts   15–19
      communication language   17–18
      frameworks and structures   19–20
      integrated capitals   18–19
      metrics and indicators   20
      perspective and awareness   16–17
   thinking in systems   15

Levine, Steven   12
Levine's lever   12–14, 160
Lewin, Kurt   70

m

management of change (MOC)   106, 245
management system standard (MSS)   138
manufactured capital   189
materiality   118–120, 122, 126
matrix
   integrate and integration
      integrated thinking   291–293
      integrating   289–290
      templating   290–291
   matrix dynamics
      Actors/Motive Forces (Z-axis)   284–285
      complexity and tight coupling   283–284
      Contexts/Drivers (Y-axis)   285–286
      gravitational pulls   287
      Risk Management Elements (X-axis)   286–287
      topographies/ecosystems   288
   Risk Matrix applications   282–283
   scorecards and dashboards
      decision-making dashboard/slice   295–297
      enterprise/company element scorecard/slice   299–302
      social-human element dashboard/slice   297–299
Meadows, Donnella   15, 128, 150, 165, 167–169, 171, 307, 316
member stakeholder   51, 258–259
mental model   65, 67, 160
MIT, system dynamics   15, 65
model uncertainty   41
modernity   56
motive force   151

field leadership
  culture of health   183–184
  organizational energy   181–184
  leadership training   26
MSS *see* management system standard (MSS)
multi-capital approach   22

n

National Academies of Sciences, Engineering, and Medicine (NASEM)   84
National Academy of Sciences (NAS)   78
National Environmental Policy Act   52
National Institute for Occupational Safety and Health (NIOSH)   25, 129–130
National Research Council (NRC)   63, 78
natural capital   18, 41, 190
Natural Capital Protocol (NCP)   132
NFRM *see* nonfinancial risk management (NFRM)
nominal variables   142
nonfinancial risk management (NFRM)   150
nongovernmental organizations (NGO)   52, 78

o

occupational exposure limits (OEL)   141
occupational health and safety (OHS)   13, 43
occupational health and safety management system (OHSMS)
  frameworks   13, 43, 83, 191
  improvement
    continual   112
    incident, nonconformity, and corrective action   111–112
  leadership and worker participation
    leadership and commitment   97–98
    non-managerial workers   99
    OH&S policy   99–100
    organizational roles, responsibilities, and authorities   100
    worker consultation and participation   98–99
  operation   105
    emergency preparedness and response   108–109
    MOC   106–107
    procurement   107–108
    reduction of, hazards and OH&S risks   106
  organizational context   96–97
  PDCA and ISO 45001, relationship between   95
  performance evaluation
    compliance evaluation   109–110
    internal audit   110
    management review   111
    monitoring, measurement, analysis and   109
  planning   100–101
    action   102
    addressing risks and opportunities   101–102
    legal and other requirements   102
    objectives   102–103
  scope   94–95
  support   103
    awareness   104
    communication   104–105
    competence   104
    documented information   105
    resources   103
  terminology   96

Occupational Safety and Health Act   52
operational elements, of Risk Management Elements dimension (*x*-axis)
  auditing and metrics
    COSO   273–275
    ISO 31000:2018   272
    ISO 37000:2021   272
    ISO MSS-14001 and 45001   272
    planned intervals   272
  frameworks
    COSO   271
    ISO 31000:2018   271
    ISO 37000:2021   269–270
    ISO MSS   268–269
  operation   275–276
ordinal variables   142
organizational element   240
organizational energy   232–233
organizational entity   191
organizational field   69–70
organizational governance   2, 30, 38, 155, 187, 254
organizational health   3, 8, 12, 25, 48
organizational learning   138, 171–172
organizational risk   55, 113
organizational risk field   2, 7, 29, 69, 70, 73, 287–288, 303, 307
organizational risk management (ORM)   2, 38, 39, 53, 150, 232, 282, 304
  see also risk decision-making (RDM)
  awareness   5
  culture of health   24–25
  finding leverage
    complexities   27–28
    developing procedures   28
    project creation   28–29
  goals   7–8
  integration   20, 21
    defined   21
    integrated risk management   22–23
    integrated thinking   22
  leadership training   26, 27
    motive force   26
    organizational energy   26
  leverage
    current and future strategies   14–15
    Levine's lever   12–14
    shifts   15–19
    thinking in systems   15
  new generation
    diversity, equity, and inclusion   11
    environmental, social, and governance (ESG)   10
    health-organizational and human   11–12
    social-human dimension, of risk   9–10
    value and purpose   8–9
  postulates and design principles   5–7
organizational risk management system (ORMS) standard   87
outcome space   22, 39

p

paradox   54, 57, 170, 203, 204
parameter uncertainty   41
pedagogy   138, 317, 319–320
pedagogy of evaluation   319–322
plan-do-check-act (PDCA)   94
playbook, developing   28
produced capital   190
profit and loss (P&L) statements   217
Public Law 102-245   139

q

qualitative and quantitative data   214, 215

r

ratio variables   142
RDM *see* risk decision-making (RDM)

Index

Red Book 84
reference stakeholder 51, 259
regulatory compliance 45, 311
regulatory/technical element 240
relationship capital 189
residual risk 225–226
responsible leadership 197–198
return on investment (ROI) 273
risk
  analysis 58
  assessment 58
  communication 58
  conformity assessment 63–64
  defined, 54–56
  fourth-generation risk management 57–58
  management 58, 59
  overview 54–55
  owner 59–60
  risk-based thinking 62–63
  risk profile 59
  risk-reward and opportunity 56–57
  risk transfer 62
  system-perspectives
    deeper levels 68
    system dynamics iceberg 66–67
    systems thinking 65–66
  tolerance 60–61
risk aware decision making 117
risk capacity 61
risk decision-making (RDM) 28, 38, 44, 202, 232
  awareness 203–204
  background
    decision science 205–206
    role of, human 206–211
  carriers of meaning 223
  context, framing, and narrative 223–224
  data 214, 215
  decision-making currency
    acceptable risk 225–226
    inference guidelines 225
    materiality 226
    organizational purpose 226
    residual risk 225–226
    value generation and preservation 226
  decision-making sequence 212
  decisions types 204
  decision theory 205
  delays and buffers 227
  expanding platform 204
  frameworks
    COSO enterprise risk management 222
    ISO 31000:2018 (risk management-guidelines) 220–221
    ISO 37000:2021 (governance of organizations) 219–220
    ISO management system standards 222
  measurement levels 214
  normative *vs.* observed decision process 206
  organizational learning 204
  rates and cycles 226
  risk 54–58, 224
  risk decision-making kernel 227–228
  systems perspective
    feedback 216
    impact-dependency pathways 218–219
    inputs and process 214–215
    output, outcome, and impact 215–216
    stocks and flows 216–218
    systems 101 212–214
risk field 68–74, 290
  background 70
  field elements 71
  framework 3

risk field (cont'd)
  institutional logic 72
  operationalization 74
  organizational field, characterization 71–73
risk governance 187
risk logics
  contexts, drivers, orientations 42
    defined 43
    evolutions 44–47
  core logic 39–41
  defined 39
  uncertainty 41–42
Risk Management Elements (x-axis) 243–245
  foundational elements (Foundational Five)
    communication-systems and practices 248–252
    competency and capabilities 252–253
    EPR 246–247
    management of change 247–248
    risk assessment 245–246
  operational elements 268–276
    auditing and metrics 271–275
    escalating impact 276
    frameworks 268–271
    future-ready strategy 276
    operation 275–276
  trim tabs 254–268
    decision-making 266–268
    leadership 266
    purpose and scope 254–257
    social-human engagement 257–266
Risk Management-Guidelines, ISO 31000 133–134
risk management system (RMS) 142
Risk Matrix 57, 78
  Actors/Motive Force (z-axis) 240, 242, 243
  Contexts/Drivers (y-axis) 239–241
  dimensionality 233
  language 233
  organizational energy 232–233
  risk field 234–236
  Risk Management Elements (x-axis) 243–245
  structure
    classification 236–238
    dimensions and elements 236
    matrix cells, rows and columns 238
  tailoring 233
risk society 56
risk stock 217
risk tolerance 61
Robert Wood Johnson Foundation (RWJF) 12, 128, 183–184
robust communication system 104
Rotman School of Management 291–292
"rule-based" approaches 40
rules-based risk-management system 40

s

Scharmer, Otto 3, 68, 150, 160, 184
Schein, Edgar 280
scorecards and dashboards 293–295
Scott, W. Richard 68, 69, 72, 280
SDG *see* sustainable development goals (SDG)
seat at the table 25–27, 324
second-and third-party external audits 145
Security and Exchange Commission (SEC) 54
seeing systems 165
Senge, Peter 3, 68, 171, 196, 304, 310, 314, 319
Silver Book 84
Singapore Standards Council (SSC) 81

Index

single materiality   120
Slovic, Paul   54, 62, 67, 161
Social and Human Capital Protocol (SHCP)   132, 217, 218
social capital   40, 52, 189
social contract   118
social determinants of health (SDH)   10
social fields   315
social field theory   185
social-human element   240
social regulations   10, 52
Society for Organizational Learning   150, 152, 196
Society for Risk Analysis (SRA)   58
spring model   22
stakeholder   50–51, 257, 264, 265
stakeholder capitalism   52, 118, 120
subconscious spinal reflexes   210
supply chain   43
sustainability   52
Sustainability Accounting Standards Board (SASB)   124, 126, 292
sustainable development goals (SDG)   117
system leadership   196–197
system 1/system 2 (S1/S2)   204, 207, 208

t

technical committee (TC)   89
templating   175, 238, 247, 284, 290
The Economics of Ecosystems and Biodiversity (TEEB)   122
Theory U model   68, 315
Thinking Fast and Slow   166, 204, 207, 208
third-party rating agencies   52
three-dimensional risk matrix   3, 232–233

tight coupling   4–5, 8, 33, 283
time horizon   226
topography   288
total recordable incidence rate (TRIR)   123
Total Worker Health®   129–130
transcending paradigms   15–16, 30, 306–307
transformational learning   173–175
transformative change model   175
trim tab elements, of the Risk Management Element dimension (x-axis)   254–268
  decision-making   266–268
    COSO ERM framework   266
    ISO 37000:2021   265–266
    ISO 31000:2018   266
    ISO example (14001 and 45001)   267
  purpose and scope   254–257
    COSO   255
    ISO 31000:2018   255–256
    ISO 37000:2021   255
    ISO MSS-45001:2018   256–257
  leadership   265
  social-human engagement   257–265
    COSO ERM framework   263–264
    ISO 31000:2018   264–266
    ISO 37000:2021   258–260
    ISO-45001:2018   260–263
triple bottom line (TBL)   117

u

U.S. Securities and Exchange Commission (SEC)   126

v

value of statistical life (VSL)   225

Value Reporting Foundation (VRF) 78, 292
volatile, uncertain, complex, and ambiguous (VUCA) 3

w
water cooler conversations 24

World Health Organization (WHO) 10, 11–12

z
zero risk 54